DESIGN OF JOINTS IN STEEL AND COMPOSITE STRUCTURES

ECCS Eurocode Design Manuals

ECCS Editorial Board
Luís Simões da Silva (ECCS)
António Lamas (Portugal)
Jean-Pierre Jaspart (Belgium)
Reidar Bjorhovde (USA)
Ulrike Kuhlmann (Germany)

Design of Steel Structures – 2ND Edition
Luís Simões da Silva, Rui Simões and Helena Gervásio

Fire Design of Steel Structures – 2ND Edition
Jean-Marc Franssen and Paulo Vila Real

Design of Plated Structures
Darko Beg, Ulrike Kuhlmann, Laurence Davaine and Benjamin Braun

Fatigue Design of Steel and Composite Structures
Alain Nussbaumer, Luís Borges and Laurence Davaine

Design of Cold-formed Steel Structures
Dan Dubina, Viorel Ungureanu and Rafaelle Landolfo

Design of Joints in Steel and Composite Structures
Jean-Pierre Jaspart and Klaus Weynand

Design of Steel Structures for Buildings in Seismic Areas
Raffaele Landolfo, Federico Mazzolani, Dan Dubina, Luís Simões da Silva and Mario d'Aniello

Available Soon

Design of Composite Structures
Markus Feldman and Benno Hoffmeister

ECCS – SCI Eurocode Design Manuals

Design of Steel Structures, U. K. Edition
Luís Simões da Silva, Rui Simões, Helena Gervásio and Graham Couchman

Information and Ordering Details

For price, availability, and ordering visit our website **www.steelconstruct.com**.
For more information about books and journals visit **www.ernst-und-sohn.de**.

DESIGN OF JOINTS IN STEEL AND COMPOSITE STRUCTURES

Eurocode 3: Design of steel structures
Part 1-8 – Design of Joints
Eurocode 4: Design of composite steel and concrete structures
Part 1-1 – General rules and rules for buildings

Jean-Pierre Jaspart
Klaus Weynand

Design of Joints in Steel and Composite Structures

2016

Published by:
ECCS – European Convention for Constructional Steelwork
publications@steelconstruct.com
www.steelconstruct.com

Sales:
Wilhelm Ernst & Sohn Verlag für Architektur und technische Wissenschaften
GmbH & Co. KG, Berlin

All rights reserved. No parts of this publication may be reproduced, stored in a retrieval system, or transmitted in any form or by any means, electronic, mechanical, photocopying, recording or otherwise, without the prior permission of the copyright owner.

ECCS assumes no liability with respect to the use for any application of the material and information contained in this publication.

Copyright © 2016 ECCS – European Convention for Constructional Steelwork

ISBN (ECCS): 978-92-9147-132-4
ISBN (Ernst & Sohn): 978-3-433-02985-5
Legal dep.: 405451/16   Printed in Multicomp Lda, Mem Martins, Portugal
Photo cover credits: Klaus Weynand

TABLE OF CONTENTS

FOREWORD	XIII
PREFACE	XV
LIST OF SYMBOLS AND ABBREVIATIONS	XIX

Chapter 1

INTRODUCTION			1
1.1	General		1
	1.1.1	Aims of the book	1
	1.1.2	Brief description of the contents of the book	10
	1.1.3	Types of structural systems and joints covered	11
	1.1.4	Basis of design	12
1.2	Definitions		12
	1.2.1	Joint properties	14
	1.2.2	Sources of joint deformability	15
	1.2.3	Beam splices and column splices	20
	1.2.4	Beam-to-beam joints	21
	1.2.5	Column bases	22
	1.2.6	Composite joints	23
	1.2.7	Hollow section joints	24
1.3	Material choice		26
1.4	Fabrication and erection		28
1.5	Costs		29
1.6	Design approaches		29
	1.6.1	Application of the "static approach"	29
	1.6.2	Component approach	31
	1.6.3	Hybrid connection aspects	38
1.7	Design tools		39
	1.7.1	Types of design tools	39

TABLE OF CONTENTS

		1.7.2	Examples of design tools	40
	1.8		Worked examples	44

Chapter 2

STRUCTURAL ANALYSIS AND DESIGN — 47

	2.1		Introduction	47
		2.1.1	Elastic or plastic analysis and verification process	48
		2.1.2	First order or second order analysis	49
		2.1.3	Integration of joint response into the frame analysis and design process	51
	2.2		Joint modelling	51
		2.2.1	General	51
		2.2.2	Modelling and sources of joint deformability	54
		2.2.3	Simplified modelling according to Eurocode 3	54
		2.2.4	Concentration of the joint deformability	55
	2.3		Joint idealisation	60
		2.3.1	Elastic idealisation for an elastic analysis	61
		2.3.2	Rigid-plastic idealisation for a rigid-plastic analysis	62
		2.3.3	Non-linear idealisation for an elastic-plastic analysis	63
	2.4		Joint classification	63
		2.4.1	General	63
		2.4.2	Classification based on mechanical joint properties	63
	2.5		Ductility classes	66
		2.5.1	General concept	66
		2.5.2	Requirements for classes of joints	69

Chapter 3

CONNECTIONS WITH MECHANICAL FASTENERS — 71

	3.1		Mechanical fasteners	71
	3.2		Categories of connections	73
		3.2.1	Shear connections	73
		3.2.2	Tension connections	75

3.3	Positioning of bolt holes		76
3.4	Design of the basic components		78
	3.4.1	Bolts in shear	78
	3.4.2	Bolts in tension	80
	3.4.3	Bolts in shear and tension	80
	3.4.4	Preloaded bolts	81
	3.4.5	Plates in bearing	89
	3.4.6	Block tearing	90
	3.4.7	Injection bolts	91
	3.4.8	Pins	92
	3.4.9	Blind bolting	95
	3.4.10	Nails	97
	3.4.11	Eccentricity of angles	98
3.5	Design of connections		100
	3.5.1	Bolted lap joints	100
	3.5.2	Bolted T-stubs	105
	3.5.3	Gusset plates	117
	3.5.4	Long joints	121

Chapter 4
WELDED CONNECTIONS — 123

4.1	Type of welds		123
	4.1.1	Butt welds	123
	4.1.2	Fillet welds	124
	4.1.3	Fillet welds all round	126
	4.1.4	Plug welds	126
4.2	Constructive constraints		127
	4.2.1	Mechanical properties of materials	127
	4.2.2	Welding processes, preparation of welds and weld quality	128
	4.2.3	Geometry and dimensions of welds	132
4.3	Design of welds		135
	4.3.1	Generalities	135
	4.3.2	Fillet welds	136

TABLE OF CONTENTS

		4.3.3	Fillet welds all round	140
		4.3.4	Butt welds	140
		4.3.5	Plug welds	142
		4.3.6	Concept of full strength fillet weld	142
	4.4	Distribution of forces in a welded joint		145
		4.4.1	Generalities	145
		4.4.2	Particular situations	147

Chapter 5

SIMPLE JOINTS — 153

	5.1	Introduction		153
	5.2	Steel joints		155
		5.2.1	Introduction	155
		5.2.2	Scope and field of application	156
		5.2.3	Joint modelling for frame analysis and design requirements	159
		5.2.4	Practical ways to satisfy the ductility and rotation requirements	162
		5.2.5	Design rules for joint characterisation	174
	5.3	Composite joints		187
		5.3.1	Composite joints for simple framing	187
	5.4	Column bases		189
		5.4.1	Introduction	189
		5.4.2	Basis for the evaluation of the design resistance	190
		5.4.3	Resistance to axial forces	191
		5.4.4	Resistance to shear forces	200

Chapter 6

MOMENT RESISTANT JOINTS — 205

	6.1	Introduction		205
	6.2	Component characterisation		206
		6.2.1	Column web panel in shear in steel or composite joints	206

	6.2.2	Column web in transverse compression in steel or composite joints	208
	6.2.3	Column web in transverse tension	212
	6.2.4	Column flange in transverse bending	213
	6.2.5	End-plate in bending	218
	6.2.6	Flange cleat in bending	221
	6.2.7	Beam or column flange and web in compression	223
	6.2.8	Beam web in tension	225
	6.2.9	Plate in tension or compression	226
	6.2.10	Bolts in tension	227
	6.2.11	Bolts in shear	228
	6.2.12	Bolts in bearing (on beam flange, column flange, end-plate or cleat)	229
	6.2.13	Concrete in compression including grout	230
	6.2.14	Base plate in bending under compression	230
	6.2.15	Base plate in bending under tension	230
	6.2.16	Anchor bolts in tension	231
	6.2.17	Anchor bolts in shear	232
	6.2.18	Anchor bolts in bearing	232
	6.2.19	Welds	232
	6.2.20	Haunched beam	232
	6.2.21	Longitudinal steel reinforcement in tension	233
	6.2.22	Steel contact plate in compression	234
6.3		Assembly for resistance	235
	6.3.1	Joints under bending moments	235
	6.3.2	Joints under axial forces	243
	6.3.3	Joints under bending moments and axial forces	244
	6.3.4	*M-N-V*	251
	6.3.5	Design of welds	252
6.4		Assembly for rotational stiffness	257
	6.4.1	Joints under bending moments	257
	6.4.2	Joints under bending moments and axial forces	266
6.5		Assembly for ductility	268
	6.5.1	Steel bolted joints	269
	6.5.2	Steel welded joints	271

6.6	Application to steel beam-to-column joint configurations		272
	6.6.1	Extended scope	272
	6.6.2	Possible design simplifications for endplate connections	275
	6.6.3	Worked example	277
6.7	Application to steel column splices		300
	6.7.1	Common splice configurations	300
	6.7.2	Design considerations	302
6.8	Application to column bases		303
	6.8.1	Common column basis configurations	303
	6.8.2	Design considerations	306
6.9	Application to composite joints		314
	6.9.1	Generalities	314
	6.9.2	Design properties	318
	6.9.3	Assembly procedure under M and N	320

Chapter 7

LATTICE GIRDER JOINTS — 329

7.1	General		329
7.2	Scope and field of application		330
7.3	Design models		333
	7.3.1	General	333
	7.3.2	Failure modes	334
	7.3.3	Models for CHS chords	335
	7.3.4	Model for RHS chords	336
	7.3.5	Punching shear failure	338
	7.3.6	Model for brace failure	339
	7.3.7	M-N interaction	339

Chapter 8

JOINTS UNDER VARIOUS LOADING SITUATIONS — 341

8.1	Introduction	341
8.2	Composite joints under sagging moment	342

8.3	Joints in fire	343
8.4	Joints under cyclic loading	344
8.5	Joints under exceptional events	346

Chapter 9
DESIGN STRATEGIES 349

9.1	Design opportunities for optimisation of joints and frames	349
	9.1.1 Introduction	349
	9.1.2 Traditional design approach	352
	9.1.3 Consistent design approach	355
	9.1.4 Intermediate design approaches	357
	9.1.5 Economic considerations	358
9.2	Application procedures	364
	9.2.1 Guidelines for design methodology	364
	9.2.2 Use of a *good guess* for joint stiffness	365
	9.2.3 Required joint stiffness	366
	9.2.4 Use of the fixity factor concept (traditional design approach)	369
	9.2.5 Design of non-sway frames with rigid-plastic global frame analysis	370

BIBLIOGRAPHIC REFERENCES	375
Annex A Practical values for required rotation capacity	385
Annex B Values for lateral torsional buckling strength of a fin plate	387

Foreword

With this ECCS book "Joints in Steel and Composite Structures" the authors succeeded in placing the joints on the rightful place they deserve in the structural behaviour of steel and composite steel-concrete structures. The many times used word "details" for the joints in structures by far underestimates the importance of joints in the structural behaviour of buildings and civil engineering structures. In their chapter "Aim of the book" the authors clearly explain how the design and safety verification of structures runs in an integral manner where all structural components, including the joints, play balanced roles leading to economic structures.

This book can be seen as a background document for Eurocode 3 "Design of Steel Structures" and for Eurocode 4 "Design of Composite Steel and Concrete Structures" as far as it concerns structural joints. The central theme in describing the behaviour of joints is using the component method and this is leading all over in this book. The book contain many aspects such as design, fabrication, erection and costs.

In this book attention is paid on joint modelling and idealisation, joint classification for strength and stiffness and deformation capacity. This all for connections with mechanical fasteners and for welded connections, for simple joints and moment resistant joints. Also lattice girder joints are described.

The book provides the designer with design strategies to arrive at economic structures.

The authors based themselves on many bibliographic references covering a time span of about 65 years. Many of these references present research of the authors themselves and of the other members of the ECCS-Technical Committee TC10 "Structural Connections".

It was really a privilege to have been the chairperson of this committee from 1998 till the end of 2012 and I thank the authors Prof. Dr. Ir. Jean-Pierre Jaspart and Dr.-Ing. Klaus Weynand for their large effort in writing this book.

Prof. ir. Frans Bijlaard

PREFACE

Steel constructions and composite steel-concrete constructions are generally erected on site by the assembly of prefabricated structural parts prepared at workshop. These parts may themselves be the result of an assembly of individual elements. An example is the assembly by bolting on site of built-up sections welded in the workshop.

In these construction types, joints and connections play a key role and recommendations and guidelines are required for engineers and constructors faced to the conception and design, the fabrication and the erection of such structures. In the Structural Eurocodes, all these aspects are mainly covered in the execution standard EN 1090-2 and in the design standards EN 1993-1-8 (Eurocode 3 for steel structures) and EN 1994-1-1 (Eurocode 4 for composite structures).

In the present book which is part of the series of ECCS Eurocode Design Manuals, the main focus is given to design aspects, but references are also made to EN 1090-2 when necessary.

In comparison to some other fields, the design procedures for joints and connections have significantly evolved in the last decades as a result of the progressive awareness by practitioners of the significant contribution of joints and connections to the global cost of structures. Design for low fabrication and erection costs and high resistance is therefore the targeted objective of modern design codes, the achievement of which has justified the development of new calculation approaches presently integrated into the two afore-mentioned Eurocodes. This situation justifies the writing of the present manual with the main goal to demystify the design by explaining the new concepts to design the joints and to integrate their mechanical response into the structural frame analysis and design process, by providing "keys" for a proper application in practice and finally by providing well documented worked examples.

To refer to "modern" or "new" design approaches and philosophies does not mean that traditional ways are old-fashioned or no more valid. It

Preface

should be understood that the design methods recommended in the Eurocodes are a collection of European practices including the results of intensive research efforts carried out in the last decades and so give many options and alternatives to the engineers to elaborate safe and economic solutions.

Chapter 1 introduces generalities about joint properties, aspects of materials, fabrication, erection and costs, design approach - and especially the so-called component method - and design tools available to practitioners for easier code application. The integration of the response of the joints into the structural analysis and design process is addressed in chapter 2. In chapter 3, the attention is paid to the design of common connections with mechanical fasteners. Preloaded bolts and non-preloaded bolts are mainly considered but the use of some less classical connectors is also briefly described. Welded connections are covered in chapter 5.

The three next chapters relate to three specific types of joints, respectively simple joints, moment resisting joints and lattice girder joints. For these ones, substantial novelties are brought in the Eurocodes in comparison to traditional national codes; and more especially for simple and moment resisting joints. A significant number of pages is therefore devoted to these topics in this manual.

The design of joints under static loading, as it is addressed in the seven first chapters, is essential in all cases but further checks or different conceptual design of the joints are often required in case of load reversal, fire, earthquake or even exceptional events like impact or explosion. Chapter 8 summarises present knowledge in this field.

Traditionally joints were designed as rigid or pinned, what enabled – and still enables – a sort of dichotomy between the design of the frame, on the one hand, and the design of the joints, on the other hand. The clear economical advantage associated in many situations to the use of semi-rigid and/or partial-strength joints leads however to "structure-joints" interactions that have to be mastered by the engineer so as to fully profit from the beneficial generated cost effects. The Eurocodes do not at all cover this aspect which is not falling within the normalisation domain but within the application by engineers and constructors in daily practice. From this point of view, chapter 9 may be considered as "a première" even if the content had already been somewhat described years ago in an ECSC publication.

Before letting the reader discover the contents of this book, we would like to express acknowledgment. We are very grateful to Prof. Frans Bijlaard for all the comments, suggestions and corrections he made through the review process of the present manual. Warm thanks are also addressed to José Fuchs and Sönke Müller who helped us in preparing the drawings. Last but not least we would like to thank our wives for their patience when we worked "on our project" during innumerable evenings and week-ends.

Jean-Pierre Jaspart
Klaus Weynand

LIST OF SYMBOLS AND ABBREVIATIONS

SYMBOLS

b_{eff}	effective width
d_n	nominal diameter of the bolt shank
e_o	magnitude of initial out-of-straightness
g	gap (in a lattice girder gap joint)
f_u	material ultimate tensile strength
f_y	material yield strength
h_b	depth of beam cross section
h_c	depth of column cross section
h_r	the distance from bolt-row r to the centre of compression
h_t	distance between the centroids of the beam flanges
k_{eq}	equivalent stiffness coefficient
k_i	stiffness coefficient of component i
l_{eff}	effective length (of a T stub flange)
$m_{pl,Rd}$	design plastic moment of a plate per unit length
r_c	fillet radius of the structural shape used as column
$t_{f,b}$	thickness of the beam flange
$t_{f,c}$	thickness of the column flange
t_p	thickness of the end-plate
$t_{w,c}$	thickness of the column web
t_0	thickness of the chord
t_i	thickness of the braces $i = 1,2$
z	lever arm of the resultant tensile and compressive forces in the connection
z_{eq}	equivalent lever arm
xx	longitudinal axis of a member
yy	major axis of a cross section
zz	minor axis of a cross section
A_o	original cross sectional area
A_s	shear area of the bolt shank
$A_{v,c}$	shear area of the column web

List of Symbols and Abbreviations

E	Young modulus for steel material
F_b	tensile and compressive forces in the connection, statically equivalent to the beam end moment
$F_{b,Ed}$	design bearing force (bolt hole)
$F_{c,Ed}$	design compressive force
$F_{c,fb,Rd}$	design compression resistance of a beam flange and the adjacent compression zone of the beam web
$F_{c,wc,Rd}$	design resistance of a column web subject to transverse compression
$F_{t,Rd}$	design tension resistance per bolt
$F_{t,r,Rd}$	design tension resistance per bolt r
$F_{T,Rd}$	design tension resistance of a T-stub flange
$F_{t,wc,Rd}$	design resistance of a column web subject to transverse tension
$F_{t,wb,Rd}$	design tension resistance of the beam web
$F_{t,Ed}$	design tensile force
$F_{v,Ed}$	design shear force (bolts)
$F_{v,Rd}$	the design shear resistance per bolt
$F_{wp,Rd}$	plastic shear resistance of a column web panel
H	horizontal load
I	second moment of area
I_b	second moment of area of the beam section (major axis bending)
I_c	second moment of area in the column section (major axis bending)
L	member length
L_b	beam span (system length)
L_c	column height (system length measured between two consecutive storeys)
M	bending moment
M_b	bending moment at the beam end (at the location of the joint)
M_c	bending moment in the column (at the location of the joint)
$M_{j,Rd}$	design moment resistance of a joint
$M_{j,Ed}$	design bending moment experienced by the joint
$M_{j,u}$	ultimate bending moment resistance of the joint
$M_{pl,Rd}$	design plastic moment resistant of a cross section
M_u	ultimate bending moment
M_{Ed}	design bending moment

LIST OF SYMBOLS AND ABBREVIATIONS

N	axial force
N_b	axial force in the beam (at the location of the joint)
N_c	axial force in the column (at the location of the joint)
$N_{j,Rd}$	axial design resistance of the joint
Q	prying force
P	axial compressive load
R_k	characteristic value of resistance
S	rotational joint stiffness
S_j	nominal rotational joint stiffness
$S_{j,app}$	approximate rotational joint stiffness (estimate of the initial one)
$S_{j,ini}$	initial rotational joint stiffness
$S_{j,post-limit}$	post-limit rotational joint stiffness
V	shear force or gravity load
V_b	shear force at the beam end
V_c	shear force in the column
V_{cr}	critical value of the resultant gravity load
V_{Ed}	design shear force in the connection
V_n	shear force experienced by the column web panel
$V_{wp,Rd}$	design plastic shear resistance of a web panel
W	gravity load
β	transformation parameter
ε_y	material yield strain
ε_u	material ultimate strain
δ	magnitude of the member deflection or local second-order effect
ϕ	the rotation of a joint (relative rotation between the axis of the connected members or sum of the rotations at the beam ends)
ϕ_{cd}	design rotation capacity of a joint
ϕ_b	rotation of the beam end
ϕ_t	rotation of the (beam + joint) end
γ	shear deformation
γ_F	partial safety factor for the loads (actions)
γ_M	partial safety factor for the resistance (strength function)
η	stiffness reduction factor $\left(S_j/S_{j,ini}\right)$
λ	slenderness
λ_L	load parameter
λ_{cr}	elastic critical load parameter (gravity loads)

LIST OF SYMBOLS AND ABBREVIATIONS

λ_p	plastic load parameter
$\bar{\lambda}$	reduced slenderness
θ_b	absolute rotation of the beam end
θ_c	absolute rotation of the column axis
σ_u	average ultimate stress
$\sigma_{com,Ed}$	maximum longitudinal compressive stress in a column due to axial force and bending
ψ	load combination factor
γ_{M0}	partial safety factor for plastic resistance of members or sections
γ_{M1}	partial safety factor for resistance to instability
γ_{M2}	partial safety factor for resistance of cross sections in tension to fracture or bolts or welds
Δ	sway displacement or global second-order effect

ABBREVIATIONS

SLS	Service limit state(s)
ULS	Ultimate limit state(s)
RHS	rectangular hollow section
CHS	circular hollow section

Chapter 1

INTRODUCTION

1.1 GENERAL

1.1.1 Aims of the book

The aim of the present book is threefold:

- To provide designers with practical guidance and tools for the design of steel and composite joints;
- To point out the importance of structural joints on the response of steel and composite structures and to show how the actual behaviour of joints may be incorporated into the structural design and analysis process;
- To illustrate the possibilities of producing more economical structures using the new approaches offered in Eurocode 3 and Eurocode 4 as far as structural joints are concerned.

The organisation of the book reflects the belief that, in addition to the sizing of the members (beams and columns), consideration should also be given to the joint characteristics throughout the design process. This approach, despite the novelty it may present to many designers, is shown to be relatively easy to integrate into everyday practice using present day design tools.

Hence the present book addresses design methodology, structural analysis, joint behaviour and design checks, at different levels:

- Presentation and discussion of concepts;
- Practical guidance and design tools.

1. Introduction

1.1.1.1 The traditional common way in which joints are modelled for the design of a frame

Generally speaking, the process of designing building structures has been up to now made up of the following successive steps:
- Frame modelling including the choice of rigid or pinned joints;
- Initial sizing of beams and columns;
- Evaluation of internal forces and moments (load effects) for each ultimate limit state (ULS) and serviceability limit state (SLS) load combination;
- Design checks of ULS and SLS criteria for the structure and the constitutive beams and columns;
- Iteration on member sizes until all design checks are satisfactory;
- Design of joints to resist the relevant members end forces and moments (either those calculated or the maximum ones able to be transmitted by the actual members); the design is carried out in accordance with the prior assumptions (frame modelling) on joint stiffness.

This approach was possible since designers were accustomed to considering the joints to be either pinned or rigid only. In this way, the design of the joints became a separate task from the design of the members. Indeed, joint design was often performed at a later stage, either by other personnel or by another company.

Recognising that most joints have an actual behaviour which is intermediate between that of pinned and rigid joints, Eurocode 3 and Eurocode 4 offer the possibility to account for this behaviour by opening up the way to what is presently known as the semi-continuous approach. This approach offers the potential for achieving better and more economical structures.

1.1.1.2 The semi-continuous approach

The rotational behaviour of actual joints is well recognised as being often intermediate between the two extreme situations, i.e. rigid or pinned.

In sub-chapter 1.2, the difference between joints and connections will be introduced. For the time being, examples of joints between one beam and one column only will be used.

Let us now consider the bending moments and the related rotations at a joint (Fig. 1.1):

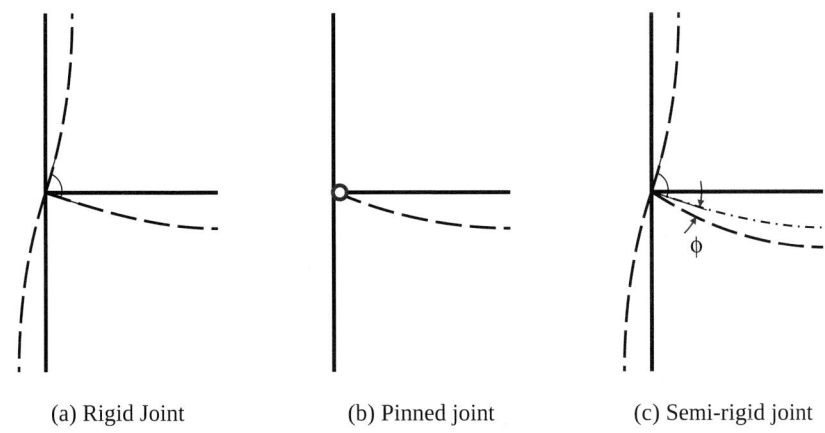

(a) Rigid Joint   (b) Pinned joint   (c) Semi-rigid joint

Figure 1.1 – Classification of joints according to stiffness

When all the different parts in the joint are sufficiently stiff (i.e. ideally infinitely stiff), the joint is rigid, and there is no difference between the respective rotations at the ends of the members connected at this joint (Fig. 1.1a). The joint experiences a single global rigid-body rotation which is the nodal rotation in the commonly used analysis methods for framed structures.

Should the joint be without any stiffness, then the beam will behave just as a simply supported beam, whatever the behaviour of the other connected member(s) (Fig. 1.1b). This is a pinned joint.

For intermediate cases (non-zero and non-infinite stiffness), the transmitted moment will result in a difference ϕ between the absolute rotations of the two connected members (Fig. 1.1c). The joint is semi-rigid in these cases.

The simplest way for representing this concept is a rotational (spiral) spring between the ends of the two connected members. The rotational stiffness S_j of this spring is the parameter that links the transmitted moment M_j to the relative rotation ϕ, which is the difference between the absolute rotations of the two connected members.

1. INTRODUCTION

When this rotational stiffness S_j is zero, or when it is relatively small, the joint falls back into the pinned joint class. In contrast, when the rotational stiffness S_j is infinite, or when it is relatively high, the joint falls into the rigid joint class. In all the intermediate cases, the joint belongs to the semi-rigid joint class.

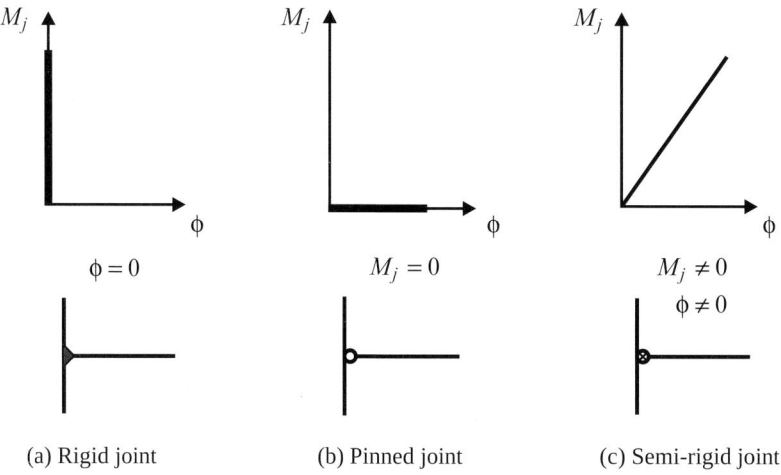

Figure 1.2 – Modelling of joints (case of elastic global analysis)

For semi-rigid joints the loads will result in both a bending moment M_j and a relative rotation ϕ between the connected members. The moment and the relative rotation are related through a constitutive law depending on the joint properties. This is illustrated in Figure 1.2 where, for the sake of simplicity, an elastic response of the joint is assumed in view of the structural analysis to be performed (how to deal with non-linear behaviour situations will be addressed later on, especially in chapter 2).

It shall be understood that the effect, at the global analysis stage, of having semi-rigid joints instead of rigid or pinned joints is to modify not only the displacements, but also the distribution and magnitude of the internal forces throughout the structure.

As an example, the bending moment diagrams in a fixed-base simple portal frame subjected to a uniformly distributed load are given in Figure 1.3 for two situations, where the beam-to-column joints are respectively either pinned or semi-rigid. The same kind of consideration holds for deflections.

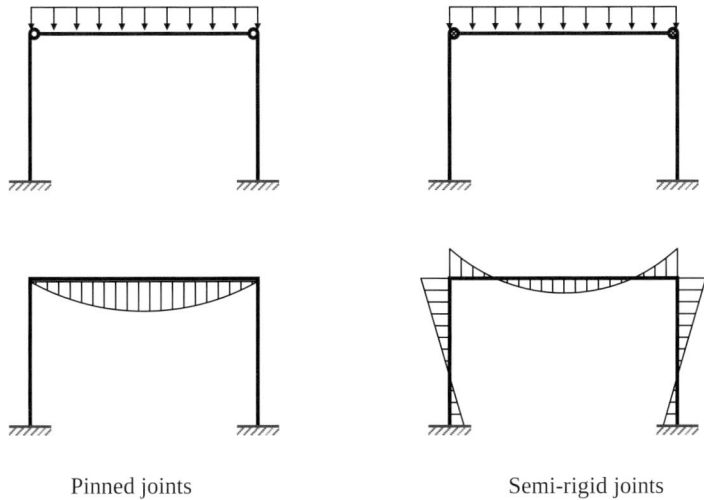

Figure 1.3 – Elastic distribution of bending moments in a simple portal frame

1.1.1.3 The merits of the semi-continuous approach

Both the Eurocode requirements and the desire to model the behaviour of the structure in a more realistic way leads to the consideration of the semi-rigid behaviour when necessary.

Many designers would stop at that basic interpretation of Eurocode 3 and Eurocode 4 and hence would be reluctant to confront the implied additional computational effort involved. Obviously a crude way to deal with this new burden will be for them to design joints that will actually continue to be classified as being either pinned or fully rigid. However such properties will have to be proven at the end of the design process; in addition, such joints will certainly be found to be uneconomical in a number of situations.

It shall be noted that the concept of rigid and pinned joints still exists in Eurocode 3 and Eurocode 4. It is accepted that a joint which is almost rigid or, on the contrary, almost pinned, may still be considered as being truly rigid or truly pinned in the design process. How to judge whether a joint can be considered as rigid, semi-rigid or pinned depends on the comparison between the joint stiffness and the beam stiffness, which latter depends on the second moment of area and length of the beam.

The designer is strongly encouraged to go beyond this "all or nothing" attitude. Actually it is possible, and therefore of interest, to consider the

1. INTRODUCTION

benefits to be gained from the semi-rigid behaviour of joints. Those benefits can be brought in two ways:

1. The designer decides to continue with the practice of assuming - sometimes erroneously - that joints are either pinned or fully rigid. However, Eurocode 3 and Eurocode 4 require that proper consideration be given to the influence that the actual behaviour of the joints has on the global behaviour of the structure, i.e. on the precision with which the distribution of forces and moments and the displacements have been determined. This may not prove to be easy when the joints are designed at a late stage in the design process since some iteration between global analysis and design checking may be required. Nevertheless, the following situations can be foreseen:

   - So that a joint can be assumed to be rigid, it is common practice to introduce, for instance, column web stiffeners in a beam-to-column joint. Eurocode 3 and Eurocode 4 now provide the means to check whether such stiffeners are really necessary for the joint to be both rigid and have sufficient resistance. There are practical cases where they are not needed, thus permitting the adoption of a more economical joint design.
   - When joints assumed to be pinned are later found to have fairly significant stiffness (i.e. to be semi-rigid), the designer may be in a position to reduce beam sizes. This is simply because the moments carried out by the joints reduce the span moments and deflections in the beams.

2. The designer decides to give consideration, at the preliminary design stage, not only to the properties of the members but also to those of the joints. It will be shown that this new approach is not at all incompatible with the sometimes customary separation of the design tasks between those who have the responsibility for conceiving the structure and carrying out the global analysis and those who have the responsibility for designing the joints. Indeed, both tasks are very often performed by different people, indeed, or by different companies, depending on national or local industrial habits. Adopting this novel manner towards design requires a good

understanding of the balance between, on the one hand, the costs and the complexity of joints and, on the other hand, the optimisation of the structural behaviour and performance through the more accurate consideration of joint behaviour for the design as a whole. Two examples are given to illustrate this:

- It was mentioned previously that it is possible in some situations to eliminate stiffeners, and therefore to reduce costs. Despite the reduction in its stiffness and, possibly, in its strength, the joint can still be considered to be rigid and be found to have sufficient strength. This is shown to be possible for industrial portal frames with rafter-to-column haunch joints, in particular, but other cases can be envisaged.
- In a more general way, it is worthwhile to consider the effect of adjusting the joint stiffness so as to strike the best balance between the cost of the joints and the costs of the beams and the columns. For instance, for braced frames, the use of semi-rigid joints, which are probably more costly than the pinned joints, leads to reducing the beam sizes. For unbraced frames, the use of less costly semi-rigid joints, instead of the rigid joints, leads to increased column sizes and possibly beam sizes.

Of course the task may seem a difficult one, and this is why the present book, in chapter 9, is aimed at providing the designer with a set of useful design strategies. The whole philosophy could be summarised as "*As you must do it, so better make the best of it*".

Thus Eurocode 3 and Eurocode 4 now offer to the designer the choice between a traditionalist attitude, where however something may often be gained, and an innovative attitude, where the economic benefits may best be sought.

It is important to stress the high level of similarity that exists between response of member cross sections and the one of structural joints. This topic is addressed in the next section

1.1.1.4 A parallel between member cross sections and joints

Member cross section behaviour may be considered through a $M-\phi$ curve for a simply supported beam loaded at mid span (M is the bending-moment at mid-span; ϕ is the sum of rotations at the span ends). Joint

1. INTRODUCTION

behaviour will be considered through a similar relationship, but with $M = M_j$ being the bending moment transmitted by the joint and ϕ being the relative rotation between the connected members. Those relationships have a similar shape as illustrated in Figure 1.4.

According to Eurocode 3 and Eurocode 4, member cross sections are divided into four classes, especially according to their varying ability to resist local instability, when partially or totally subject to compression, and the consequences this may have on the possibility for plastic redistribution. Therefore their resistance ranges from the full plastic resistance (class 1 and 2) to the elastic resistance (class 3) or the sub-elastic resistance (class 4).

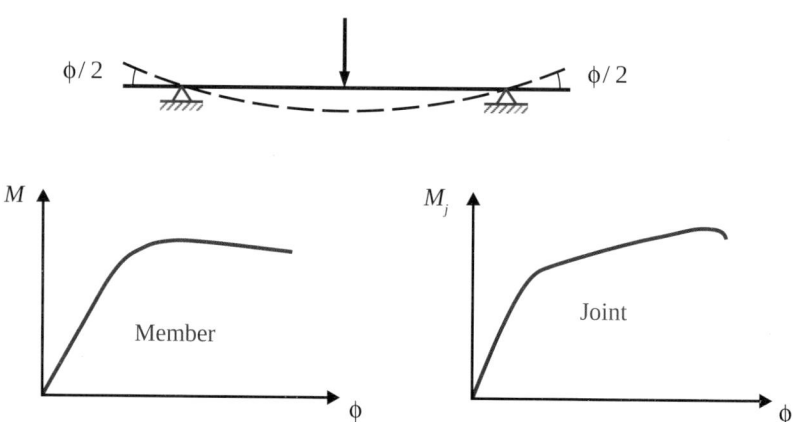

Figure 1.4 – $M - \phi$ characteristics for member cross section and joint

The specific class to which a cross section belongs governs the assumptions on:

- The behaviour to be idealised for global analysis (i.e. class 1 will allow the formation of a plastic hinge and permit the redistribution of internal forces in the frame as loads are increased up to or beyond the design loads);
- The behaviour to be taken into account for local design checks (i.e. class 4 will imply that the resistance of the cross section is based on the properties of a relevant effective cross section rather than of the gross cross section).

The classification of a cross section is based on the width-to-thickness ratio of the steel component walls of the section. Ductility is directly related to the amount of rotation during which the design bending resistance will be sustained. This results in a so-called rotation capacity concept.

In a similar manner, joints are classified in terms of ductility or rotation capacity. This classification is a measure of their ability to resist premature local instability and, even more likely, premature brittle failure (especially due to bolt or weld failure) with due consequences on the type of global analysis allowed.

The practical interest of such a classification for joints is to check whether an elasto-plastic global analysis may be conducted up to the formation of a plastic collapse mechanism in the structure, which implies such hinges in at least some of the joints.

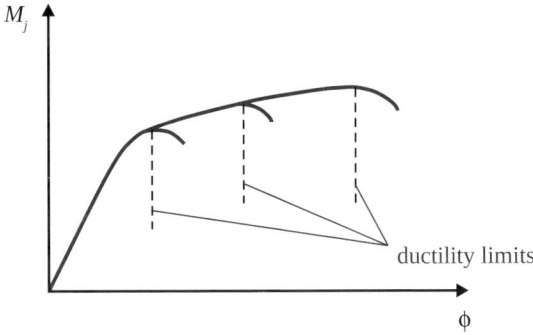

Figure 1.5 – Ductility or rotation capacity in joints

As will be shown, this classification of joints by ductility, while not explicitly stated in the Eurocodes, may be defined from the geometric and mechanical properties of the joint components (bolts, welds, plate thickness, etc.).

Joints may therefore be classified according to both their stiffness and their ductility. Moreover, joints may be classified according to their strength.

In terms of strength, joints are classified as full-strength or partial-strength according to their resistance compared to the resistance of the connected members. For elastic design, the use of partial-strength joints is well understood. When plastic design is used, the main use of this classification is to foresee the possible need to allow a plastic hinge to form in the joint during the global analysis. In order to permit the increase of loads, a partial-strength joint may be required to act as a plastic hinge from the moment when its plastic bending resistance is reached. In that case, the joint must also have sufficient ductility (plastic rotation capacity).

1. INTRODUCTION

The final parallel to be stressed between joints and members is that the same kind of link exists between the global analysis stage and the ultimate limit states design checking stage. The latter has to extend to all aspects that were not implicitly or explicitly taken into account at the global analysis stage. Generally speaking, one can state that the more sophisticated the global analysis is, the simpler are the ultimate limit states design checks required.

The choice of global analysis will thus depend not only of what is required by Eurocodes 3 or 4 but also on personal choices, depending on specific situations, available software's, etc. A particular choice means striking a balance between the amount of effort devoted to global analysis and the amount of effort required for the check of remaining ULS (Figure 1.6).

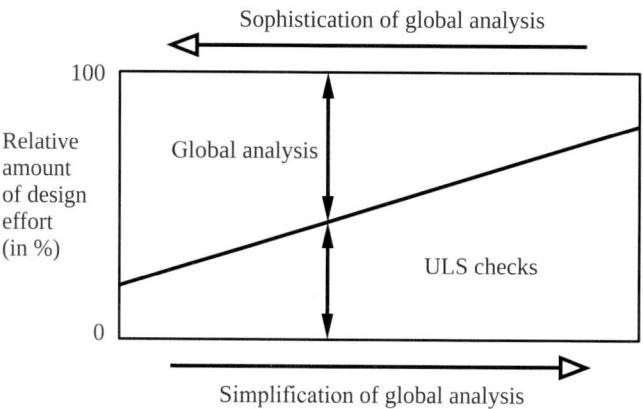

Figure 1.6 – Schematic of the proportion of effort for global analysis and for ULS checks

1.1.2 Brief description of the contents of the book

This book is divided into three main parts:

- The first part covered by chapter 2 is a recall of the available methods of structural analysis and design, but with a special focus on the integration of the joint response into the whole design process.
- The second part includes chapter 3 to chapter 8. It concerns the characterisation of the joint properties in terms of resistance, stiffness, ductility. Chapter 3 to chapter 7 concentrate on the response of joints under static loading while chapter 8 provides

information about the extension of the previously described joint characterisation methods to fire, seismic and fatigue loading.
- In the third part covered by chapter 9, design opportunities to optimise the structural frames and the constitutive joints are presented and guidelines are provided on how to implement them into daily practice.

Practical design recommendations and tools are presented throughout the whole book. This is intended to help the reader to "materialise" the possibly new concepts and the way to "operate" them in their design activities.

In order to achieve a full consistency with the European norms, the notations adopted in the book are basically those suggested in Eurocode 3 and Eurocode 4. But for sure, all the design rules implemented in the codes are not repeated in the present book which is aimed at commenting the joint related aspects of Eurocodes 3 and 4, but certainly not at "replacing" the normative documents.

Finally, it has to be mentioned that only one load case is generally considered in the book, when referring to the design of a specific joint or frame, while, in actual projects, several load combination cases are obviously to be dealt with.

1.1.3 Types of structural systems and joints covered

No particular type of structure is excluded but the book is focusing more particularly on steel and composite building frames.

As far as joints are concerned, the configuration and the loading conditions addressed in this book are, strictly speaking, those directly covered by Eurocode 3 Part 1-8 (CEN, 2005c) and Eurocode 4 Part 1-1 (CEN, 2004b); however, as the Eurocode design principles are also valid for many other types of joints and loading conditions, the scope the present book extends to the following field:
- Member cross sections:

    Open H or I rolled and welded sections as well as tubular sections.

- Connections with mechanical fasteners.
- Welded connections.

- Beam-to-column joints, beam-to-beam joints and column or beam splices:

    Simple and moment resisting joints with welded or bolted steel connections (with or without haunch) and using various connecting elements (endplates, angles, fin plates, splices, ...) and stiffening systems.

- Column bases:

    Bolted end-plate connections.

- Lattice girder joints.
- Joints under static and non-static loading, including fire, seismic and fatigue loading.

1.1.4 Basis of design

EN 1993-1-8 (CEN, 2005c) on which the present book focusses is one of the twelve constitutive parts of Eurocode 3. Accordingly, its basis of design is fully in line with the one of Eurocode 3, in terms of basic concepts, reliability management, basic variables, ultimate limit states, serviceability limit states, durability or sustainability. Reference will therefore be done here to the introductory chapter of the ECCS Design Manual "Design of steel structures" (Simões da Silva *et al*, 2010) where the here-above mentioned aspects are extensively discussed.

1.2 DEFINITIONS

One of the advantages of the steel construction lies in the prefabrication of steel pieces. These pieces, according to the needs or the transportation requirements, will have to be assembled either on site or directly in the workshop. The connections resulting from this fabrication and erection process will sometimes be there just for constructional reasons; but in many cases, they will have to transfer forces between the connected pieces and so play an important structural role. One speaks than about "structural connections". These ones are more particularly covered in this book.

In the past, forging has been used as a convenient way to assemble steel pieces together. This was achieved through the melting of the steel at the interface between the connected pieces. Nowadays, in modern constructions, use is made of welds and mechanical fasteners, including various bolt types (TC bolts, injection bolts, flow drill bolting ...), pins or even nails. Beside the connectors, a great variety of configurations is also required in daily practice, sometimes as a result of the architectural demand or for economic reasons. This leads to a quite infinite amount of situations to which the practitioner is likely to be faced. Moreover various loading situations (static, dynamic, seismic, fire, ...) and structural systems (building, bridge, tower ...) are also to be considered.

In this book, an exhaustive consideration of all the existing connection and joint configurations is obviously not foreseen; therefore the authors have tried to deliver the most update and complete information in the domain by structuring chapter 3 to chapter 8 as follows:

- Chapter 3: Connections with Mechanical Fasteners
- Chapter 4: Welded Connections
- Chapter 5: Simple Joints
- Chapter 6: Moment Resistant Joints
- Chapter 7: Lattice Girder Joints
- Chapter 8: Joints under Various Loading Situations

Chapter 3 and chapter 4 refer to basic joints where mainly tension and/or shear forces are to be transferred between the connected pieces. The most important design property is their level of resistance.

Chapter 5 is devoted to connections aimed at transferring shear forces, possibly together with axial forces, between two elements, but additionally should allow a free rotation between these two ones. The word "simple" refers to this specific ability.

Chapter 6 concentrates on moment resisting joints where a combination of axial/shear forces and moments are to be carried over by the joint. Their rotational stiffness plays an important role as far as the global structural response is considered. In case of plastic frame analysis, their rotational capacity will also have to be assessed.

In chapter 7, lattice girder joints are discussed while, in chapter 8, the extension of the static design rules to fire, fatigue or earthquake situations will be discussed.

1. INTRODUCTION

The design of the basic connections (covered by chapter 3 and chapter 4) and of lattice girder joints according to the Eurocodes is quite similar to what practitioners have been used to apply in the last decades at the national level while news concepts are now available for the design of simple and moment resisting joints. The next pages will allow the designer to familiarise with important definitions used all along the book pages.

1.2.1 Joint properties

The specific case of building frames is addressed below, just for sake of illustration.

Building frames consist of beams and columns, usually made of H, I or tubular shapes that are assembled together by means of connections. These connections are between two beams, two columns, a beam and a column or a column and the foundation (Figure 1.7).

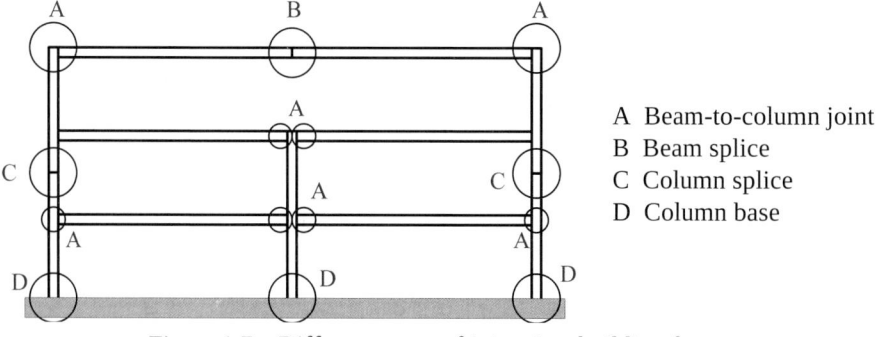

A Beam-to-column joint
B Beam splice
C Column splice
D Column base

Figure 1.7 – Different types of joints in a building frame

A *connection* is defined as the set of the physical components which mechanically fasten the connected elements. One considers the connection to be concentrated at the location where the fastening action occurs, for instance at the beam end/column interface in a beam-to-column joint. When the connection as well as the corresponding zone of interaction between the connected members are considered together, the wording *joint* is used (Figure 1.8a).

Depending on the number of in-plane elements connected together, single-sided and double-sided joint configurations are defined (Figure 1.8). In a double-sided configuration (Figure 1.8b), two joints - left and right - have to be considered.

The definitions illustrated in Figure 1.8 are valid for other joint configurations and connection types.

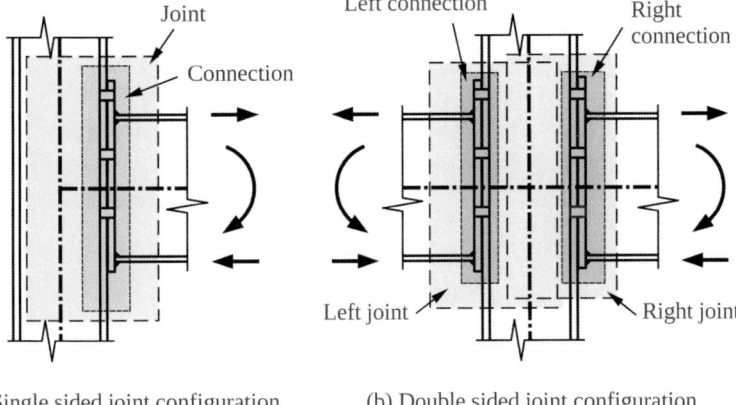

(a) Single sided joint configuration (b) Double sided joint configuration

Figure 1.8 – Joints and connections

As explained previously, the joints which are traditionally considered as rigid or pinned and are designed accordingly, possess, in reality, their own degree of flexibility resulting from the deformability of all the constitutive components. Section 1.2.2 is aimed at describing the main *sources of joint deformability*. Sub-chapter 2.2 provides information on how to *model* the joints in view of the frame analysis. This modelling depends on the level of joint flexibility. In sub-chapter 2.3, the way in which the shape of the non-linear joint deformability curves may be *idealised* is given. Section 1.6.2 refers to the *component method* as a general tool for the prediction of the main joint mechanical properties in bending. The concept of *joint classification* is introduced in sub-chapter 2.4. Finally, it is commented on the *ductility classes* of joints sub-chapter 2.5. This aspect, expressed in terms of rotational capacity will be of particular importance in cases where a plastic structural analysis will be performed.

1.2.2 Sources of joint deformability

As said in sub-chapter 1.1, the rotational behaviour of the joints may affect the local and/or global structural response of the frames. In this sub-chapter, the sources of rotational deformability are identified for beam-to-column joints, splices and column bases.

1. Introduction

It is worthwhile mentioning that the rotational stiffness, the joint resistance and the rotation capacity are likely to be affected by the shear force and/or the axial force acting in the joint.

These shear and axial forces may obviously have contributions to the shear and axial deformability within the connections. However it is known that these contributions do not affect significantly the frame response. Therefore, the shear and axial responses of the connection, in terms of rotational deformability, are usually neglected in applications as buildings.

1.2.2.1 Beam-to-column joints

1.2.2.1.1 Major axis joints

In a major axis beam-to-column joint, different sources of deformability can be identified. For the particular case of a single-sided joint (Figure 1.9a and Figure 1.10a), these are:

- *The deformation of the connection*. It includes the deformation of the connection elements: column flange, bolts, end-plate or angles,... and the load-introduction deformation of the column web resulting from the transverse shortening and elongation of the column web under the compressive and tensile forces F_b acting on the column web. The couple of F_b forces are statically equivalent to the moment M_b at the beam end. These deformations result in a relative rotation ϕ_c between the beam and column axes; this rotation, which is equal to $\theta_b - \theta_c$ (see Figure 1.9a), results in a flexural deformability curve $M_b - \phi_c$.
- *The shear deformation of the column web panel* associated to the shear force V_{wp} acting in this panel. It leads to a relative rotation γ between the beam and column axes; this rotation makes it possible to establish a shear deformability curve $V_{wp} - \gamma$.

The deformability curve of a connection may obviously be influenced by the axial and shear forces possibly acting in the connected beam.

Similar definitions apply to double-sided joint configurations (Figure 1.9b and Figure 1.10b). For such configurations, two connections and a sheared web panel, forming two joints, must be considered.

1.2 DEFINITIONS

In short, the main sources of deformability which must be contemplated in a beam-to-column major axis joint are:

Single-sided joint configuration:

- the connection deformability $M_b - \phi_c$ characteristic;
- the column web panel shear deformability $V_{wp} - \gamma$ characteristic.

Double-sided joint configuration:

- the left hand side connection deformability $M_{b1} - \phi_{c1}$ characteristic;
- the right hand side connection deformability $M_{b2} - \phi_{c2}$ characteristic;
- the column web panel shear deformability $V_{wp} - \gamma$ characteristic.

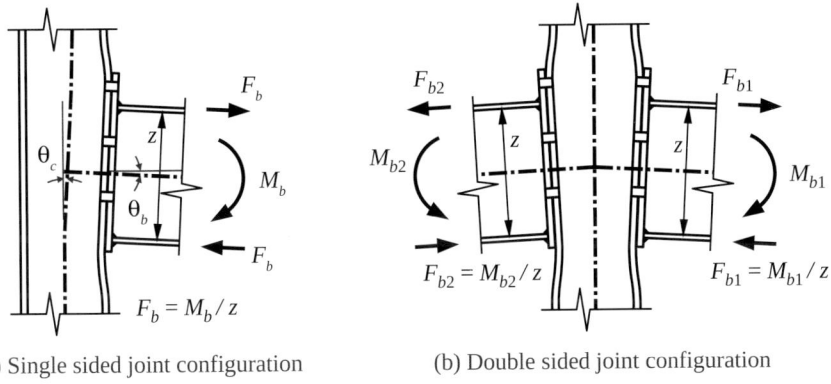

(a) Single sided joint configuration (b) Double sided joint configuration

Figure 1.9 – Sources of joint deformability

The deformability of the connection (connection elements + load-introduction) is only due to the couple of forces transferred by the flanges of the beam (equivalent to the beam end moment M_b). The shear deformability of the column web panel results from the combined action of these equal but opposite forces and of the shear forces in the column at the level of the beam flanges. Equilibrium equations of the web panel provide the shear force V_{wp} (for the sign convention, see Figure 1.10, where the directions indicated by the arrows are positive):

$$V_{wp} = \frac{M_{b1} - M_{b2}}{z} - \frac{V_{c1} - V_{c2}}{2} \quad (1.1)$$

1. INTRODUCTION

Another formula to which it is sometimes referred, i.e.:

$$V_{wp} = \frac{M_{b1} - M_{b2}}{z} \qquad (1.2)$$

is only a rough and conservative approximation of Eq. (1.1).

In both formulae, z is the lever arm of the resultant tensile and compressive forces in the connection(s). How to derive the value of z is explained in chapter 6.

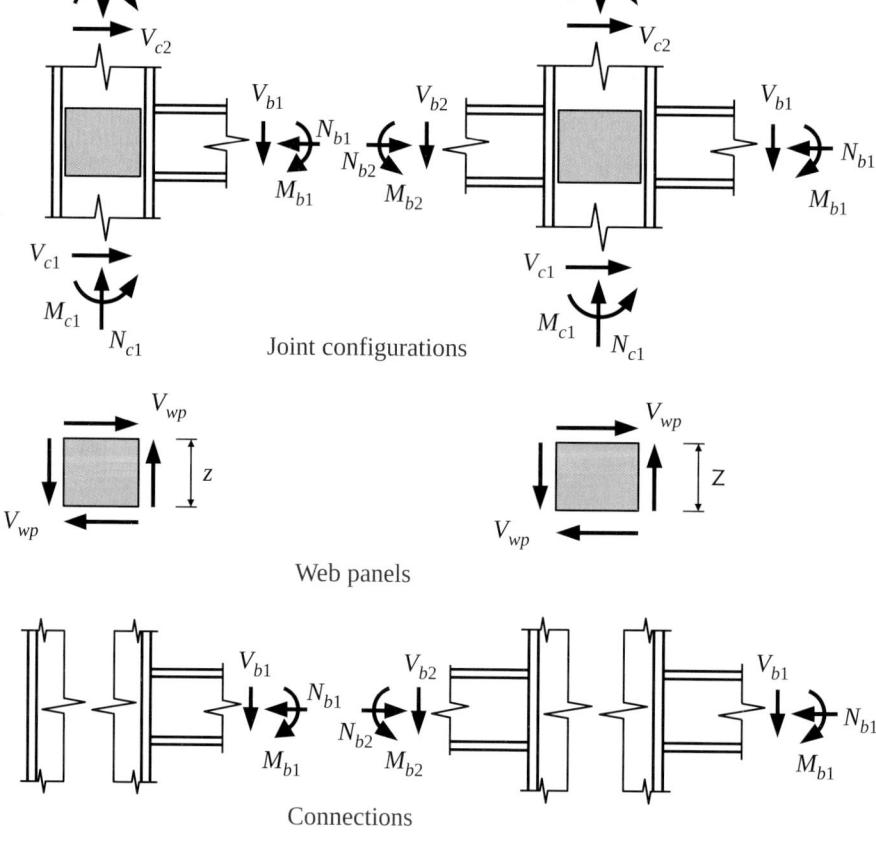

(a) Single sided joint configuration   (b) Double sided joint configuration

Figure 1.10 – Loading of the web panel and the connections

1.2.2.1.2 Minor axis joints

A similar distinction between *web panel* and *connection* shall also be made for a minor axis joint (Figure 1.11). The column web exhibits a so-called out-of-plane deformability while the connection deforms in bending as it does in a major axis joint. However no load-introduction deformability is involved.

In the double-sided joint configuration, the out-of-plane deformation of the column web depends on the bending moments experienced by the right and left connections (see Figure 1.12):

$$\Delta M_b = M_{b1} - M_{b2} \tag{1.3}$$

For a single-sided joint configuration (Figure 1.11), the value of ΔM_b equals that of M_b.

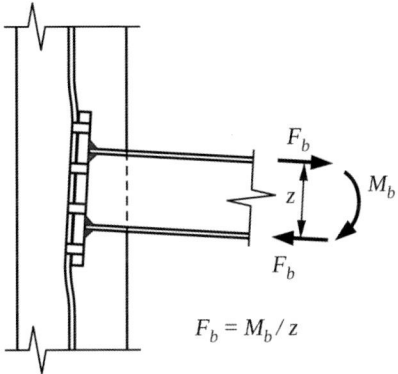

Figure 1.11: Deformability of a minor axis joint

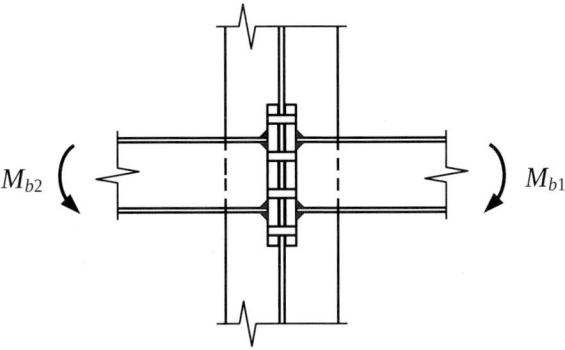

Figure 1.12 – Loading of a double-sided minor axis joint

1. INTRODUCTION

1.2.2.1.3 Joints with beams on both major and minor column axes

A 3-D joint is (Figure 1.13) characterised by the presence of beams connected to both the column flange(s) and web. In such joints, a shear deformation (see 1.2.2.1.1) and an out-of-plane deformation (see 1.2.2.1.2) of the column web develop coincidently.

The loading of the web panel appears therefore as the superimposition of the shear loading given by formulae (1.1) or (1.2) and the out-of-plane loading given by Eq. (1.3).

The joint configuration of Figure 1.13 involves two beams only; configurations with three or four beams can also be met.

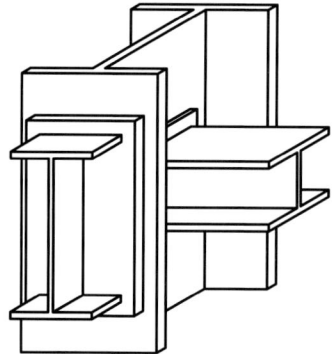

Figure 1.13 – Example of a 3-D joint

1.2.3 Beam splices and column splices

The sources of deformability in a beam splice (Figure 1.14) or in a column splice (Figure 1.15) are less than in a beam-to-column joint; indeed they are concerned with connections only. The deformability is depicted by the sole $M_b - \phi$ curve.

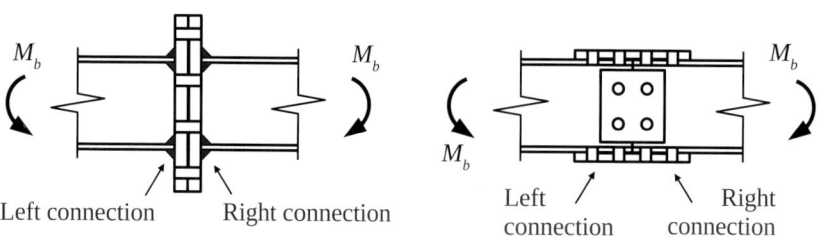

Figure 1.14 – Deformation of a beam splice

This curve corresponds to the deformability of the whole joint, i.e. the two constituent connections (left connection and right one in a beam splice, upper connection and lower one in a column splice).

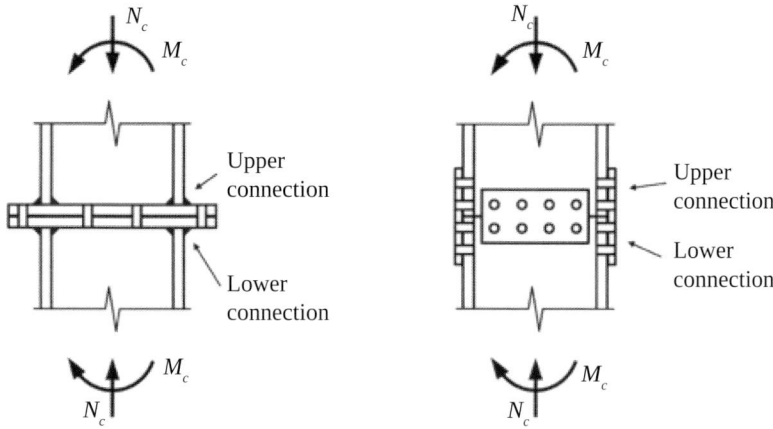

Figure 1.15 – Deformation of a column splice

In a column splice where the compressive force is predominant, the axial force affects in a significant way the mechanical properties of the joint, i.e. its rotational stiffness, its strength and its rotation capacity. The influence, on the global frame response, of the axial deformability of splices is however limited; therefore it may be neglected provided that specific requirements given in sub-chapter 6.7 are fulfilled.

1.2.4 Beam-to-beam joints

The deformability of a beam-to-beam joint (Figure 1.16) is quite similar to the one of a minor axis beam-to-column joint; the loadings and the sources of deformability are similar to those expressed in 1.2.2.1.2 and can therefore be identified.

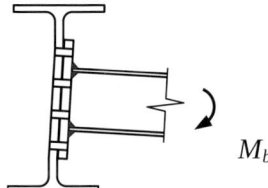

Figure 1.16 – Deformation of a beam-to-beam joint

1. INTRODUCTION

1.2.5 Column bases

In a column base, two connection deformabilities need to be distinguished (Figure 1.17):

- the deformability of the connection between the column and the concrete foundation (*column-to-concrete connection*);
- the deformability of the connection between the concrete foundation and the soil (*concrete-to-soil connection*).

For the column-to-concrete connection, the bending behaviour is represented by a $M_c - \phi$ curve, the shape of which is influenced by the ratio of the bending moment to the axial load at the bottom of the column.

For the connection between the concrete foundation and the soil, two basic deformability curves are identified:

- a $N_c - u$ curve which mainly corresponds to the soil settlement due to the axial compressive force in the column; in contrast with the other types of joint, this deformability curve may have a significant effect on the frame behaviour;
- a $M_c - \phi$ curve characterising the rotation of the concrete block in the soil.

As for all the other joints described above, the deformability due to shear forces in the column may be neglected in the case of column bases.

The column-to-concrete connection and concrete-to-soil connection $M_c - \phi$ characteristics are combined in order to derive the rotational stiffness at the bottom of the column and conduct the frame analysis and design accordingly.

Similar deformability sources exist in column bases subjected to biaxial bending and axial force. The connection $M_c - \phi$ characteristics are then defined respectively for both the major and the minor axes.

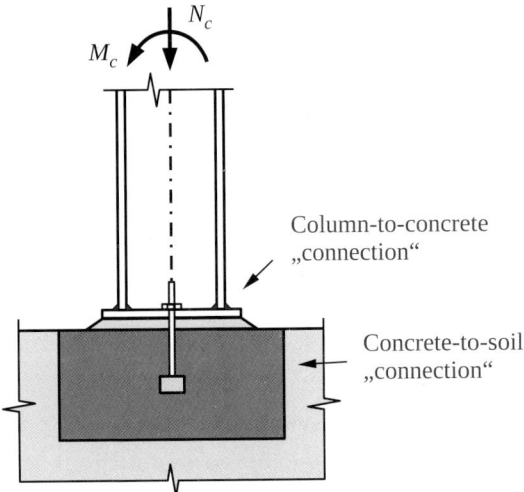

Figure 1.17 – The two connections in a column base

1.2.6 Composite joints

In terms of general behaviour and sources of deformability, composite joints present generally a complete similarity with steel joints (beam-to-column, beam-to-beam, splice, column base). Reference may so simply be made to the previous sections by replacing steel members by composite ones (column and/or beams).

Some specific joint configurations may however exhibit different joint responses. An example of such a situation is illustrated in Figure 1.18 where no bending moments are transferred to the column through the concrete slab. In this case, the column is just seen as a local vertical support of the continuous slab and therefore the steel connections (lower flange cleats here) only transfer the slab reaction to the column.

The designer has so to adapt the joint detailing to the required local joint response.

1. INTRODUCTION

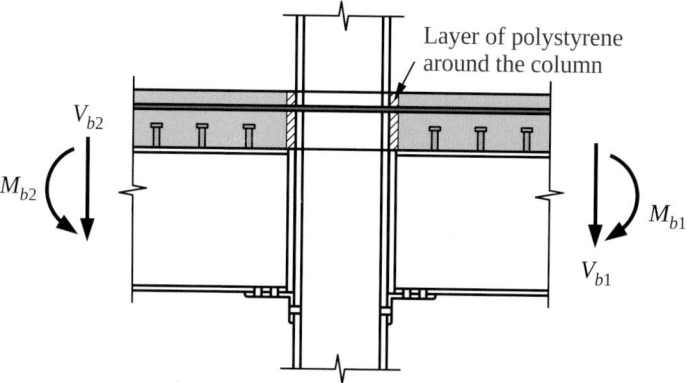

Figure 1.18 – Particular response of a composite joint configuration

1.2.7 Hollow section joints

Design rules in Part 1-8 of EN 1993 for hollow section joints were originally developed and published as design recommendations by CIDECT (International Committee for Research and Technical Support for Hollow Section Structures). Chapter 7 of EN 1993-1-8 gives application rules to determine the static design resistances of joints in lattice structures.

Design formulae for hollow section joints are based on semi-empirical investigations in which analytical models were fitted with test results. The complex geometry of the joints, local influences of the corners of rectangular sections and residual stresses, for instance due to welding, lead to non-uniform stress distributions. Strain hardening and membrane effects are also influencing the local structural behaviour. Simplified analytical models which consider the most relevant parameters were developed. In contrast to the design models for open section joints, where the so-called component method (see section 1.6.2) is used, the design of hollow section joints is based on the study of failure modes. Note that the terminology is not fully consistent here. In the previous sections, the terms of *joint configuration*, *joint* and *connection* respectively are defined. For the design of hollow section joints, different types of joints are covered by Eurocode 3, see Table 1.1. However, here the word "joint" means *joint* and *joint configuration* at the same time. Hence, the resistance of a joint is also the resistance of the whole node, i.e. joint configuration, typically expressed by the resistances of the braces, see Figure 1.19.

1.2 DEFINITIONS

Table 1.1 – Types of hollow section joints in lattice structures

K joint	KT joint	N joint
T joint	X joint	Y joint
DK joint	KK joint	
X joint	TT joint	
DY joint	XX joint	

1. INTRODUCTION

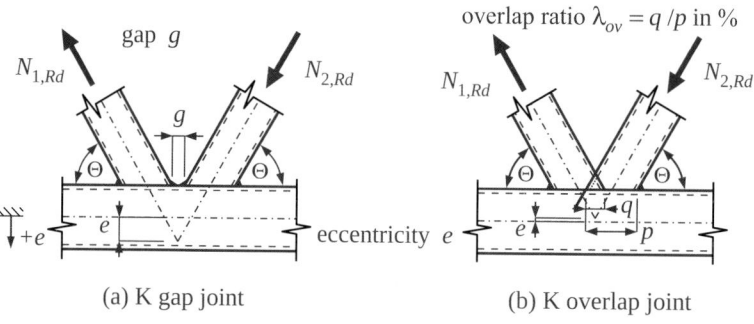

Figure 1.19 – Definition of design resistance, gap and overlap of a K joint

1.3 MATERIAL CHOICE

The design provisions given in the Eurocodes are only valid if material or products comply with the reference standards given in the appropriate parts of the Eurocodes. Both, in EN 1993-1-1 (CEN, 2005a) and in EN 1993-1-8 (CEN, 2005c), material and product standards are listed in sections 1.2. EN 1993-1-1, as a master document for the other parts of EN 1993, covers the design of steel structures fabricated from steel material conforming to the steel grades listed in Table 3.1 of EN 1993-1-1. According to this table, also for Part 1-8 of Eurocode 3, only steel grades from S 235 to S 460 are covered. For hot rolled structural steel (I or H sections and welded build-up sections), reference is made to EN 10025. For structural hollow sections, reference is made to EN 10210 for hot finished sections and to EN 10219 for cold formed sections. Especially with regard to the design of joints where, in many cases, a local plastic redistribution of stresses will be assumed to determine the resistance of a joint, sufficient ductility of the material is required. Steel conforming with steel grades listed in Table 3.1 of EN 1993-1-1 can be assumed to have sufficient ductility.

Due to the fabrication process, the nominal values of the yield strength f_y and also of the ultimate strength f_u depend on the thickness of the elements. It should be noted that different values for the relation between the material thickness and the material strength are given in the delivery condition specified in the product standards (for example EN 10025) and in Eurocode 3. Eurocode 3 allows to define the strength values either by adopting the values directly from the product standards or by using simplified values given in Table 3.1 of Eurocode 3 Part 1-1. The user should check the National Annex where a choice could have been made.

To avoid brittle fracture of tension elements, the material should have sufficient facture toughness. EN 1993-1-10 (CEN, 2005e) provides rules to select the appropriate steel grade. Several aspects should be taken into account for the choice of material:

- Steel material properties:

    The yield strength depending on the material thickness $f_y(t)$ and the toughness quality which is expressed in terms of T_{27J} or T_{40J}.

- Member characteristics:

    The member shape, detail and element thickness (t), stress concentrations according to the details in EN 1993-1-9 (CEN, 2005d) as well as fabrication flaws (for example through-thickness)

- Design situations:

    Design value at lowest member temperature as well as maximum stresses for permanent and imposed actions derived from the design residual stress should be considered. If relevant, assumptions for crack growth for fatigue loading during an inspection interval, the strain rate $\dot{\varepsilon}$ for accidental actions and the degree of cold forming ε_{cf}.

In steel assemblies where the plate is loaded in tension perpendicular to its plane, i.e. in the through-thickness direction, the material should have adequate through-thickness properties to prevent lamellar tearing, see Figure 1.20. EN 10164 (CEN, 2004e) defined quality classes in terms of Z-values.

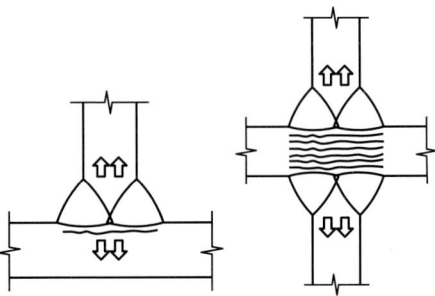

Figure 1.20 – Lamellar tearing

1. INTRODUCTION

EN 1993-1-10 (CEN, 2005e) gives guidance on the quality class to be used where steel with improved through-thickness properties is necessary. This it for example the case for welded beam-to-column joints (where the column flange is loaded in its through-thickness direction) or in beam-to-column joints with bolted end-plates (where the end-plate is loaded in its through-thickness direction).

Finally, for steel grades ranging from S460 to S690 (S700) reference will have to be made to EN 1993-1-12 (CEN, 2007a).

1.4 FABRICATION AND ERECTION

The design process, on one side, and the fabrication and erection of a structure, on the other side, cannot be seen as independent. In many, many cases, in particular for the joints, the design procedures are directly dependent on the way how a structure is built and vice versa. Eurocodes, strictly speaking, are design codes, or even verification codes, i.e. they provide the designer with rules to verify the resistance, stability, deformations and ductility requirements. But no guideline related to fabrication and erection is given. For these specific aspects, the engineer must refer to the so-called execution standard EN 1090-2 (CEN, 2011). This standard gives quite a number of detailed requirements to be fulfilled.

For bolted and welded joints, the link between design rules and fabrication/execution rules are particularly strong. For instance, the design of a bolt in bearing depends on the diameter of the bolt, but also on the diameter of the bolt hole. In practice, the hole clearance, i.e. the difference of diameter between the hole and the bolt diameter, is not specified EN 1993-1-8, but the design rules provided there is only valid if the hole is made in accordance with EN 1090-2 provisions. If it is not the case, then the design rule proposed in EN 1993-1-8 cannot be applied.

Even if it is not explicitly specified in EN 1993-1-8, most of the design rules are only valid if the structural elements comply with the regulations specified in the execution/fabrication/erection standards EN 1090-2. Such requirements concern for example imperfections, geometrical tolerances, e.g. tolerances on hole diameter for bolts and pins, methods of tightening of preloaded bolts, preparation of contact surfaces in slip resistant connections, welding plan and preparation and execution of welding etc.

In this book, not specific focus is given to the fabrication and erection aspects. To know more about that, reference should be made to commentaries and background documentation of the execution standard EN 1090-2. But for sure all the design rules provided in this book may only be applied under the condition that the requirements of EN 1090-2 are satisfied. Similarly, all connected or connecting elements (bolts, welding materials, steel for endplates, etc.) are assumed to respect the EN product norms listed in the first pages of EN 1993-1-8.

1.5 COSTS

The price of a steel structure depends significantly on the fabrication, transportation and erection costs in countries where labour costs are the dominant factor. Hence, in order to design steel structures in an economical manner, special attention should be made to the joints which are known to be expensive. However, the joint detailing must also be chosen in such a way that the actual behaviour of the joint, i.e. its structural properties (strength, stiffness and ductility), is in line with the assumptions made when the joints have been modelled for the global frame analysis. In chapter 9, some strategies are presented which allow to minimise the effects of the joints on the global cost of the steel structures.

1.6 DESIGN APPROACHES

1.6.1 Application of the "static approach"

The design of joints may, as for any other cross section, be performed on an elastic or a plastic basis.

In a pure elastic approach, the joints should be designed in such a way that the generalised Von Mises stress nowhere exceeds the elastic strength of the constitutive materials. To achieve it, and because of the geometrical complexity of the connection elements, a refined stress analysis will be required, often requiring sophisticated numerical approaches like FEM ones. In this process, the presence of residual stresses or of any other set of self-equilibrated stresses

1. INTRODUCTION

which could, for instance, result from lack-of-fit, is usually neglected, but would have normally to be integrated. As a result, the elastic resistance of the joint may be evaluated but generally appears to be relatively low.

Steel generally exhibits significant ductility. The designer may therefore profit from this ability to deform steel plastically and so to develop "plastic" design approaches in which local stress plastic redistributions in the joint elements are allowed.

The use of plastic design approaches in steel and composite construction is explicitly allowed in the Eurocodes. For joints and connections, reference has to be made to clause 2.5(1) of EN 1993-1-8. Obviously, limitations to the use of such plastic approaches exist. They all relate to the possible lack of ductility of the steel material, on the one hand, and of some connections elements like bolts, welds, reinforced concrete slab in tension or concrete in compression, on the other hand. As far as steel material itself is concerned, the use of normalized steels according to Euronorms (for instance EN 10025 (CEN, 2004d)) guarantees a sufficient material ductility. For the here above listed possibly non ductile joint elements, EN 1993-1-8 introduces specific requirements which will be detailed later on when describing the design methods in dedicating chapters.

Clause 2.5(1) of EN 1993-1-8 is simply paraphrasing the so-called "static" theorem of the limit analysis. The application of this theorem to cross sections, for instance in bending, is well known as it leads to the concept of bi-rectangular stress patterns and to the notion of plastic moment resistance used for Class 1/2 cross sections (as no limit of ductility linked to plate buckling phenomena prevents these sections from developing a full plastic resistance). For connections and joints, the implementation of the static theorem is probably less direct, but this does not at all prevent designers to use it.

The static theorem requires first the determination of an internal statically admissible distribution of stresses (for cross sections) or forces (for connections and joints), i.e. of a set of stresses or forces in equilibrium with the external forces acting on the cross section or on the joint/connection and resulting from the global structural analysis. The second requirement is to be plastically admissible, which means that nowhere in the cross section or joint/connection, the plastic resistance and ductility criteria should be violated.

A large number of statically and plastically admissible distributions may exist. In most of the cases, these ones will not respect the "kinematically admissible" criterion; in fact, only the actual distribution has the ability to satisfy the three requirements: equilibrium, plasticity and compatibility of displacements. But in fact, this is not a problem as long as sufficient local deformation capacity (ductility) is available at the places where plasticity develops in the cross section or in the joint/connection. As long as this last condition is fulfilled, the static theorem ensures that the resistance of the cross section or joint/connection associated to any statically and plastically is lower than the actual one (and therefore on the safe side). The closer the assumed distribution will be compared to the actual one, the closer the estimated resistance will be compared to the actual resistance.

1.6.2 Component approach

1.6.2.1 General

The characterisation of the response of the joints in terms of stiffness, resistance and ductility is a key aspect for design purposes. From this point of view, three main approaches may be followed:

- experimental;
- numerical;
- analytical.

The only practical one for the designer is usually the analytical approach. Analytical procedures enable a prediction of the joint response based on the knowledge of the mechanical and geometrical properties of the so-called "joint components".

In this section a general analytical procedure, termed *component method*, is introduced. It applies to any type of steel or composite joints, whatever the geometrical configuration, the type of loading (axial force and/or bending moment, ...) and the type of member cross sections.

The method is nowadays widely recognised, and particularly in the Eurocodes, as a general and convenient procedure to evaluate the mechanical properties of joints subjected to various loading situations, including static and dynamic loading conditions, fire, earthquake....

1. Introduction

1.6.2.2 Introduction to the component method

A joint is generally considered as a whole and studied accordingly; the originality of the component method is to consider any joint as a set of individual basic components. For the particular joint shown in Figure 1.9a (joint with an extended end-plate connection mainly subject to bending), the relevant components (i.e. zones of transfer of internal forces) are the following:

- column web in compression;
- beam flange and web in compression;
- column web in tension;
- column flange in bending;
- bolts in tension;
- end-plate in bending;
- beam web in tension;
- column web panel in shear.

Each of these basic components possesses its own strength and stiffness either in tension, compression or shear. The column web is subject to coincident compression, tension and shear. This coexistence of several components within the same joint element can obviously lead to stress interactions that are likely to decrease the resistance of the individual basic components.

The application of the component method requires the following steps:

1. *identification* of the active components in the joint being considered;
2. *evaluation of the stiffness and/or resistance characteristics* for each individual basic component (specific characteristics - initial stiffness, design resistance, ... - or whole deformability curve);
3. *assembly* of all the constituent components and evaluation of the stiffness and/or resistance characteristics of the whole joint (specific characteristics - initial stiffness, design resistance, ... - or whole deformability curve).

The assembly procedure is the step where the mechanical properties of the whole joint are derived from those of all the individual constituent components. That requires, according to the static theorem introduced in section 1.6.1, to define how the external forces acting on the joint distribute into internal forces acting on the components in a way that satisfies equilibrium and respects the behaviour of the components.

1.6 DESIGN APPROACHES

In EN 1993-1-8, guidelines on how to apply the component method for the evaluation of the initial stiffness and the design moment resistance of the joints are provided; the aspects of ductility are also addressed.

The application of the component method requires a sufficient knowledge of the behaviour of the basic components. Those covered for static loading by EN 1993-1-1 and EN 1994-1-1 are listed respectively in Table 1.2 and in Table 1.3. The combination of these components allows one to cover a wide range of joint configurations and should be largely sufficient to satisfy the needs of practitioners. Examples of such joints are given in Figure 1.21 and Figure 1.22, for steel and composite joints respectively.

Table 1.2 – List of components covered by Eurocode 3 Part 1-8

No	Component	
1	Column web panel in shear	
2	Column web in transverse compression	
3	Column web in transverse tension	
4	Column flange in bending	

1. Introduction

Table 1.2 – List of components covered by Eurocode 3 Part 1-8, (continuation)

No	Component	
5	End-plate in bending	$F_{t,Ed}$
6	Flange cleat in bending	$F_{t,Ed}$
7	Beam or column flange and web in compression	$F_{c,Ed}$
8	Beam web in tension	$F_{t,Ed}$
9	Plate in tension or compression	$F_{t,Ed}$ $F_{t,Ed}$ $F_{c,Ed}$ $F_{c,Ed}$
10	Bolts in tension	$F_{t,Ed}$ $F_{t,Ed}$
11	Bolts in shear	$F_{v,Ed}$

Table 1.2 – List of components covered by Eurocode 3 Part 1-8, (continuation)

No	Component	
12	Bolts in bearing (on beam flange, column flange, end-plate or cleat)	$F_{b,Ed}$
13	Concrete in compression including grout	$F_{c,Ed}$
14	Base plate in bending under compression	
15	Base plate in bending under tension	$F_{t,Ed}$
16	Anchor bolts in tension	$F_{t,Ed}$
17	Anchor bolts in shear	$F_{v,Ed}$
18	Anchor bolts in bearing	$F_{b,Ed}$

1. Introduction

Table 1.2 – List of components covered by Eurocode 3 Part 1-8, (continuation)

No	Component	
19	Welds	
20	Haunched beam	$F_{c,Ed}$

Table 1.3 – List of components covered by Eurocode 4 Part 1-1

No	Component	
1	Longitudinal steel reinforcement in tension	$F_{t,Ed}$
2	Steel contact plate in compression	contact plate $F_{c,Ed}$
3	Column web panel in shear (encased column)	V_{Ed} V_{Ed}
4	Column web in transvers compression (encased column)	$F_{c,Ed}$

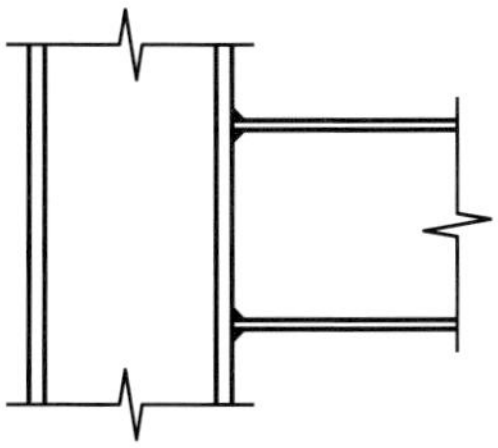

1.6 DESIGN APPROACHES

(a) Welded joint

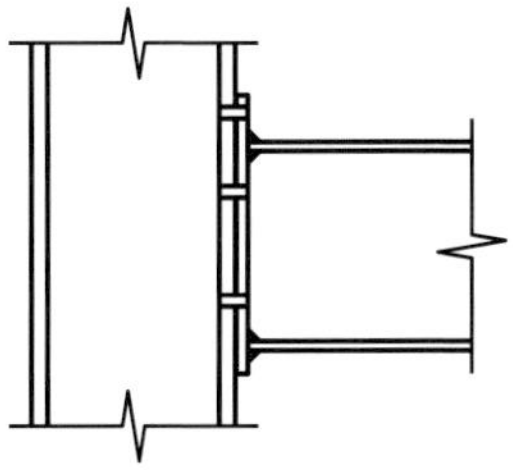

(b) Bolted joint with extended end plate

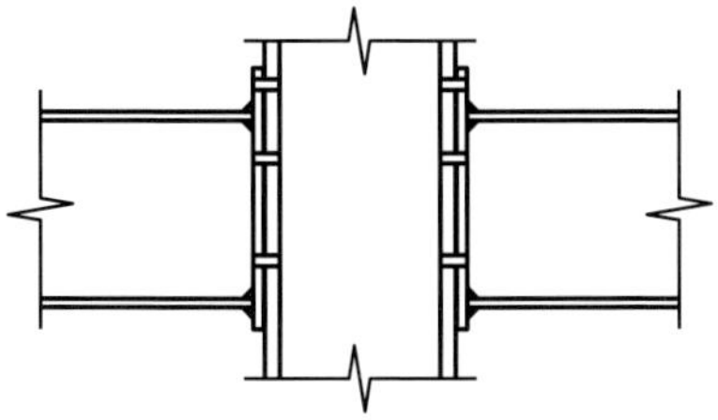

(c) Two joints with extended end plates (double-sided configuration)

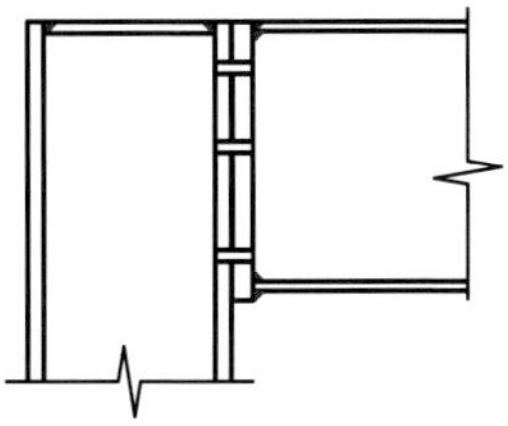

(d) Joint with flush end plate

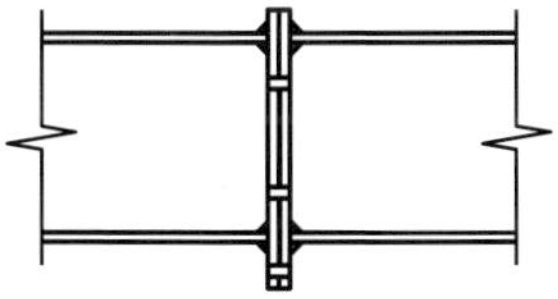

(e) End-plate type beam splice

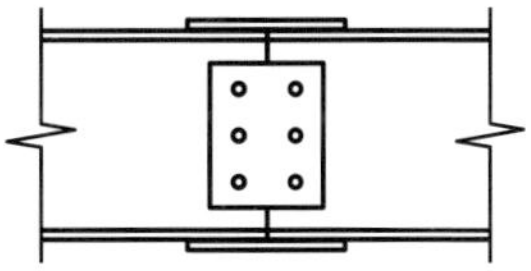

(f) Cover-joint type beam splice

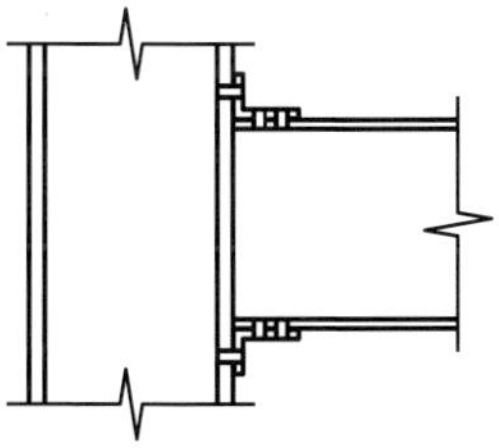

(g) Bolted joint with angle flange cleats

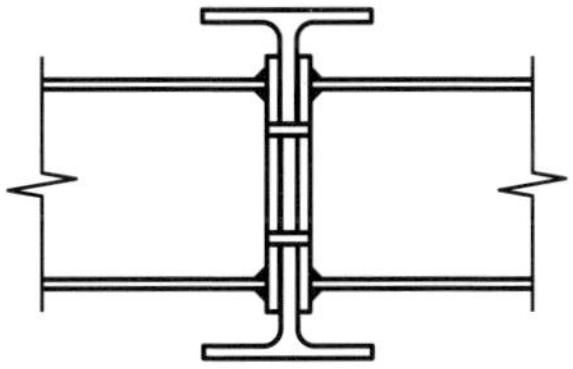

(h) Two beam-to-beam joints (double-sided configuration)

Figure 1.21 – Examples of steel beam-to-column joints, beam-to-beam joints and beam splices joints covered by Eurocode 3 Part 1-8

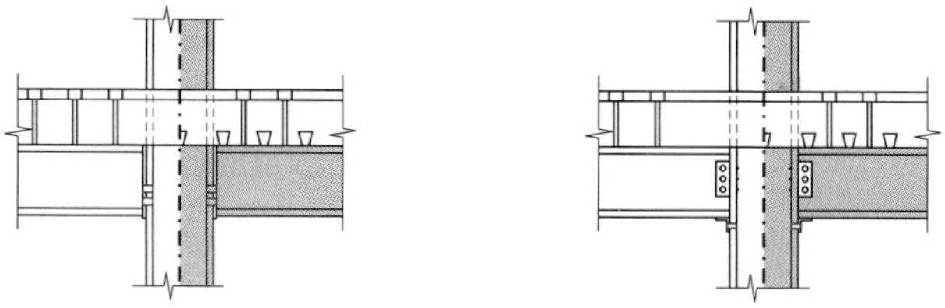

1. INTRODUCTION

(a) Composite beams, partial depth end plates

(b) Composite beams, angle cleats + contact plates

(c) Flush end plate (moment resisting)

(d) Fin and contact plates (moment resisting)

Figure 1.22 – Examples of composite joints covered by Eurocode 4 Part 1-1

1.6.3 Hybrid connection aspects

In the present book, the design of welded and bolted connections is mainly considered. All the principles ruling the design of such connections are globally similar for hybrid ones in which the transfer of forces between two connected elements is commonly achieved by two different connectors, for instance welds and bolts.

Hybrid connections are however not widely used in practice and many experts in the field of connections are not recommending them, because of the different level of stiffness which often characterises their respective behaviours and which generally prevents the user to "add up" the contributions of the two connection systems to the resistance and so to profit from the higher expected connection resistance.

For these reasons, hybrid connections will not be explicitly addressed here.

In Eurocode 3 Part 1-8, few recommendations are only given: "where fasteners with different stiffnesses are used to carry a shear load, the fasteners with the highest stiffness should be designed to carry the design load. As an exception, preloaded class 8.8 and 10.9 bolts in connections designed as slip-resistant at the ultimate limit state may be assumed to share load with welds, provided that the final tightening of the bolts is carried out after the welding is complete."

1.7 DESIGN TOOLS

The Eurocodes give new and advanced options to design efficient and economic steel structures. The design of joints plays a major role in that process. Thus the detailing of joints and the methods of considering the joints properties in the frame analysis will significantly influence the costs of a steel structure. This has been demonstrated by various investigations.

However, the exploitation of the advanced possibilities is rather time consuming for the designer if no appropriate tools for a quick and easy design are available. Different opinions have been discussed in Europe concerning the development of the Eurocodes. On one side it is expected that the Eurocodes provide design methods which will allow safe, robust and economic solutions. Of course this requires more sophisticated approaches for the design rules. On the other side the users of the Eurocodes are requesting simple codes for practice. But this is in conflict with the major request to make steel structures more economic. It would be unfortunate to make standards too simple as there is the loss of many possibilities to take profit of the new and advanced options mentioned above.

The message is quite clear: there was and there still is a need for sophisticated standards which form an accepted basis to design steel structures. Based on the methods given in these standards simple design tools need to be developed and provided to practitioners. This is an optimal way to bring more economic solutions on the market with an acceptable effort needed by the designers. As often said: *"Not simple rules sell steel, but simple tools sell steel."*

1.7.1 Types of design tools

Beside the need for background information the engineer requires simple design tools to be able to design joint in an efficient way. Three different types of design aids can be provided. The most appropriate type depends on various aspects.

- Design tables

1. INTRODUCTION

Design tables are ready-to-use tables containing standardised joint layouts including dimension details and all relevant mechanical properties like resistance, stiffness and ductility. The use of tables is certainly the quickest way to design a joint. However, any change in the layout will require further calculations and tables are no more helpful. Here design sheets may be used.

– Design sheets

Design sheets are set of simple design formulae. The aim is to allow a simple and rather quick hand calculation. Due to simplifications, the results could be more conservative or the range of validity more limited. Both design tables and design sheets can be published in handbooks.

– Software

The most flexible way is the use of software. Of course it takes a few minutes to enter all joint details, but there will be only few limitations in the range of validity and any re-calculation, for example due to a change in the layout, is a matter of a few seconds.

1.7.2 Examples of design tools

In the recent years, efforts have been achieved in various countries so as to develop specific design tools for joints. To be exhaustive in the referencing of these tools appears definitively as quite impossible and is certainly not targeted here. Therefore only few rather characteristic examples (drafted in English) will be briefly described hereunder so as to illustrate the variety of the available information sources.

One of the difficulties, sometimes, for the designer will be often to know about the existence of such tools, but also to select the one which would be the most appropriate for his daily practice. This will request some investment in time, but which will be quickly counterbalance by an efficient and economical design of the joints, a quick fabrication in workshop or an easy erection on site.

– Design tables

As an example, so so-called DSTV/DASt Ringbuch (Weynand, Oerder, 2013) can be mentioned here. The book is a publication of Stahlbau Verlags- und Service GmbH in Germany but is published in two languages (English and German). It covers a very wide range of moment resisting and simple (pinned) joints in various joint configurations: beam-to-column, beam splices and beam-to-beam. Bolted connections with flush and extended endplates equipped two or four bolts per row as well as header plates and web cleats are considered, as well as different bolt and steel grades.

The book consists of design tables, which can be used in a straightforward manner to select a joint or, for instance, to check the resistance of another one. They are established for a selection of standard combinations of connected member sections and, in more details, provide the designer with the following data:

- the detailing of the joint (geometrical and mechanical properties of the sections and connection elements);
- indications about requirements for stiffeners;
- for simple joints: the design shear resistance of the joint;
- for moment resistant joints: the design moment resistance under positive and negative bending moments and the shear resistance of the joint as well as the initial stiffness;
- an identification of the joint component which is governing the resistance moment;

Similar books are published in UK as a joints publication of SCI, BCSA and Tata Steel, one for moment resisting joints (Brown *et al*, 2013) and one for simple joints (Moreno *et al*, 2011).

– Design sheets

One of the first publication providing design sheets for a simple design of joints in accordance with Eurocode 3, is published by the European Commission (CRIF *et al*; Maquoi, Chabrolin, 1998). These sheets bear the name of an EC-funded project, called SPRINT project, in which they have been developed. Each individual set of sheet is devoted a particular joint configuration and connection type, see Figure 1.23.

1. INTRODUCTION

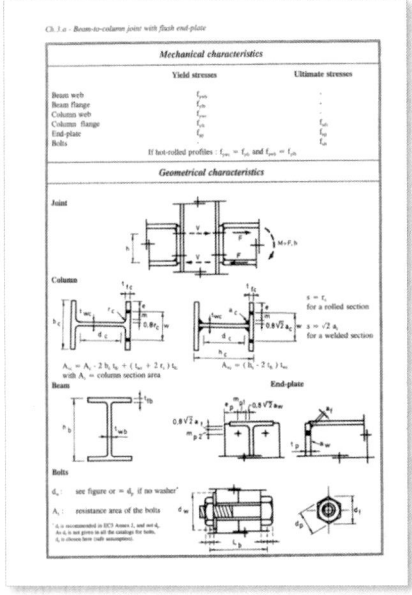
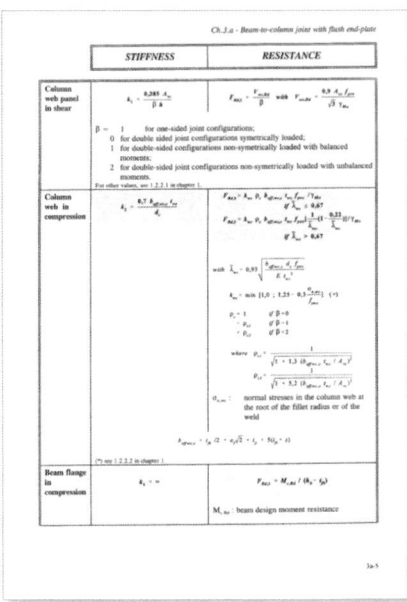

Figure 1.23 – Design sheets (CRIF *et al*)

The joint design procedure included in this design tool is aimed at assisting the designer who wishes to take account of the full potential of semi-rigidity and/or partial-strength joints, without having to go through the more general but often complex approach provided by Eurocode 3 Part 1-8. In reality, to derive these sheets, profit has been taken of all possible "shortcuts" allowed by the standards so as to safely but more easily perform the calculations.

In each set, the first design sheet summarises all the data requested concerning the joint configuration and the connection type. In the remaining sheets, the calculation procedure first provides all the expressions for the evaluation both stiffness and resistance for each of the joint components in a logical order and finally shows how to derive the mechanical properties of the whole joint, i.e. the initial stiffness and/or the design moment resistance. The failure mode corresponds to the component whose resistance determines the design moment resistance of the joints. It gives an indication on the level of rotational capacity of the joint.

1.7 DESIGN TOOLS

The shear resistance of the joint is an important value. For the sake of clarity, it is not dealt with in the design sheets but relevant information is provided just after the sheets.

In (Maquoi *et al*, 1998), sheets are provided for beam splices and beam-to-column steel joints with endplates or flange cleats. In (Anderson *et al*, 1999), similar sheets are available for composite beam-to-column joints.

Such sheets may be prepared for any joint configuration, connections types or loading situations. The designer may easily program the sheet so as to develop its own set of sheets or even to establish his own design tables (i.e. corresponding to the particular need of his company).

- Software

Of course, quite a number of software tools are offered on the market. As an example, the program COP (Weynand *et al*, 2014), could be mentioned here. COP is an innovative computer program for the design of joints in steel and composite structures. The calculations are made in full accordance with the new Eurocode 3 (EN 1993-1-8) using the so-called component method. This new calculation method not only leads to a more economic structural design, it also gives a better insight into the behaviour of the joint. The joint is central here, as it is an important factor in determining the costs of a structure.

In close cooperation and with the support of the steel industry, a user-friendly calculation program has been developed. Using modern software technics the software fits perfectly with the needs of engineers and draughtsmen. All joint details will be defined in clearly arranged data input masks. During input, all data will be visualised in scaled 2D or 3D views and a data check module monitors if the requirements of Eurocode 3 like bolt end distances, required weld sizes, etc. are fulfilled and, if needed, valid values are proposed.

Four editions of COP are available, respectively for:

- Composite joints
- Hollow section joints

1. INTRODUCTION

- Simple joints
- Moment resisting joints and simple joints

The two first ones are freely available at http://cop.fw-ing.com.

COP provides a full calculation note (PDF) and it is available in English, German and French (output language independent of user interface language)

1.8 WORKED EXAMPLES

Worked examples represent another significant support for designers applying new design approaches. However, in the field of joints, the variety of the joint configurations, connection types, connected elements and loading situations is such that infinity of worked examples would be required to cover the whole range of European practical applications and design practices.

One could summarise the needs in terms of worked examples as follows if reference is made to the contents of the present book:

- bolted connections;
- welded connections;
- simple joints in buildings;
  - beam splice, beam-to-beam and beam-to-column joints;
  - column splices;
  - column bases;
- moment resisting joints in buildings;
  - beam splice, beam-to-beam and beam-to-column joints;
  - column splices;
  - column bases;
- joints in lattice girders;

knowing that, in each category, several sub-categories should be considered according to the connected and the connecting elements and the loading situation.

In the present book, the choice has been made to limit the worked examples to two well defined situations and to refer to some existing material in the literature for all the other cases. The two selected cases correspond to rather complex cases for which the use of the design provisions is traditionally leading to questions and requests for information from the designers:

- beam-to-column joint with an endplate connection including several bolt-rows and subjected to bending and shear (section 6.6.3);
- beam-to-column joint with an endplate connection including several bolt-rows and subjected to bending and axial force (section 6.9.3).

The first case illustrates, in the case of a steel joint, the way on how all the components are characterised and the way on how the assembly of the latter is achieved to derive stiffness and the resistance of the whole joint. The second worked example goes one step further by explaining how the assembly of the components has to be achieved under bending moments and axial forces, as it could also be the case in a column base.

For the other situations which can be met, and as said before, reference may be made to various publications available at national or European levels. Hereunder, few ones are selected (non-exhaustive list) according to the two following criteria: easily available to designers and drafted in English.

- bolted connections: (Veljkovic, 2015)
- welded connections: (Veljkovic, 2015)
- simple joints in buildings - beam splice, beam-to-beam and beam-to-column joints: (Anderson *et al*, 1999; Moreno *et al*, 2011; Veljkovic, 2015)
- simple joints in buildings - column splices: (Brettle, 2009; Moreno *et al*, 2011)
- simple joints in buildings - column bases: (Brettle, 2009; Moreno *et al*, 2011; Veljkovic, 2015)
- moment resisting joints in buildings - beam splice, beam-to-beam and beam-to-column joints: (Anderson *et al*, 1999; Brettle, 2009; Brown *et al*, 2013; Veljkovic, 2015)
- moment resisting joints in buildings - column splices: (Brettle, 2009; Brown *et al*, 2013)

- moment resisting joints in buildings - column bases: (Brettle, 2009; Brown *et al*, 2013; Veljkovic, 2015)
- joints in lattice girders: (Wardenier *et al*, 1991; Packer *et al*, 1992; Kurobane *et al*, 2004; Wardenier *et al*, 2008; Brettle, 2008; Packer *et al*, 2009)

Chapter 2

STRUCTURAL ANALYSIS AND DESIGN

2.1 INTRODUCTION

The design of a frame and of its components consists of a two-step procedure involving a global frame analysis followed by individual cross section and member design checks.

Global frame analysis is aimed at deriving the values of the internal forces and moments and of the displacements in the considered structure when it is subjected to a given set of loads. It is based on assumptions regarding the component behaviour (elastic or plastic) and the geometrical response (first-order or second-order theory) of the frame. The ECCS Eurocode Design Manual (Simões da Silva *et al*, 2010) devoted to the Design of Steel Structures provides detailed information about the different available analysis procedures. Once the analysis is complete, i.e. all relevant internal forces and moments and displacements are determined in the whole structure, then the design checks of all the frame components can be performed. These ones consist in verifying whether the structure satisfies all the required design criteria under service loads (serviceability limit states – SLS) and under factored loads (ultimate limit states – ULS).

Globally speaking, four main analysis approaches may be contemplated according to Eurocode 3. They are reported in Table 2.1:

- linear elastic first order analysis;
- plastic (or elasto-plastic) first order analysis;
- linear elastic second order analysis;
- plastic (or elasto-plastic) second order analysis;

2. STRUCTURAL ANALYSIS AND DESIGN

Table 2.1 – Four main approaches for frame analysis

		Assumed material law (linear or non-linear)	
		Elastic	Plastic
Assumption in terms of geometrical effects (linearity or non-linearity)	1st order	Linear elastic first order analysis	Plastic (or elasto-plastic) first order analysis
	2nd order	Linear elastic second order analysis	Plastic (or elasto-plastic) second order analysis

The selection of the most appropriate one by the user is a quite important step which will strongly influence the number and the nature of the checks to be achieved further on, as already stated in section 1.1.1. This choice is however not free. Next paragraphs tend to summarise the recommendations of Eurocode 3 in this domain.

2.1.1 Elastic or plastic analysis and verification process

The selection of an elastic or plastic analysis and design process depends on the class of the member cross sections (Class 1, 2, 3 or 4).

A class expresses the way on how the possible local plate buckling of cross section walls subjected to compression may or not affect the resistance or the ductility of the cross sections.

An elastic analysis is required for frames made of members with Class 2, Class 3 or Class 4 cross sections. For frames constituted of Class 1 member cross sections (at least at the locations of the plastic hinges), a plastic (elasto-plastic) analysis may be also performed.

The cross section and member design checks are also depending on the cross section class. Table 2.2 summarises the various possibilities offered by Eurocode 3.

Table 2.2 – Selection between elastic and plastic (elasto-plastic) analysis and design process

Cross section class	Frame analysis	Cross section/member verifications	Designation of the global approach (E = elastic; P = plastic)
Class 1	Plastic	Plastic	P-P
	Elastic	Plastic	E-P
	Elastic	Elastic	E-E
Class 2	Elastic	Plastic	E-P
	Elastic	Elastic	E-E
Class 3 – Class 4	Elastic	Elastic	E-E

2.1.2 First order or second order analysis

Geometrical non-linearities may affect the response of members or the response of the full frame and lead respectively to local (flexural buckling, torsional buckling, lateral torsional buckling ...) or global (global sway instability) instability phenomena.

A structural analysis may be defined as the expression of the equilibrium between the external forces acting on the structure and the resulting internal forces and members in the member cross sections. According to the geometry of the structure taken as a reference to express this equilibrium, one will speak about first order or second order analysis:

- in a first order analysis, reference is made to the initial undeformed shape of the structure;
- in a second order analysis, reference is made to the actual deformed shape of the structure.

Two types of second order behaviour are to be considered, respectively linked to so-called $P-\Delta$ effects and $P-\delta$ effects. $P-\Delta$ effects determine possible sway instability phenomena whereas $P-\delta$ effects govern the possible development of member instabilities. A full second order analysis will be characterised by an explicit consideration of all geometrical second order effects (sway and member). In this case, no member check has to be performed further to the structural analysis. However, in most cases $P-\delta$ effects are integrated in the analysis to cover buckling phenomena but not the

2. STRUCTURAL ANALYSIS AND DESIGN

lateral torsional buckling of the members. So, an additional check for this potential instability mode may still be needed.

Usually the designer prefers not to integrate the so-called $P-\delta$ effects (second order effects related to the member buckling) in the frame analysis and so to check member buckling phenomena further to the analysis, when verifying ULS. It just then remains to decide whether the $P-\Delta$ effects are to be or not considered. This is achieved through the verification of a specific criterion defining the "sway" or "non sway" character of the structure ("α_{cr}" criterion in EN 1993-1-1).

The description "non-sway frame" applies to a frame when its response to in-plane horizontal forces is so stiff that it is acceptable to neglect any additional forces or moments arising from horizontal displacements of its storeys. This means that the global second-order effects may be neglected. When the second order effects are not negligible, the frame is said to be a "sway frame".

Figure 2.1 summarises the choices offered to the designer, not only in terms of frame analysis, but also of type of cross section and/or member verification at ULS further to the frame analysis.

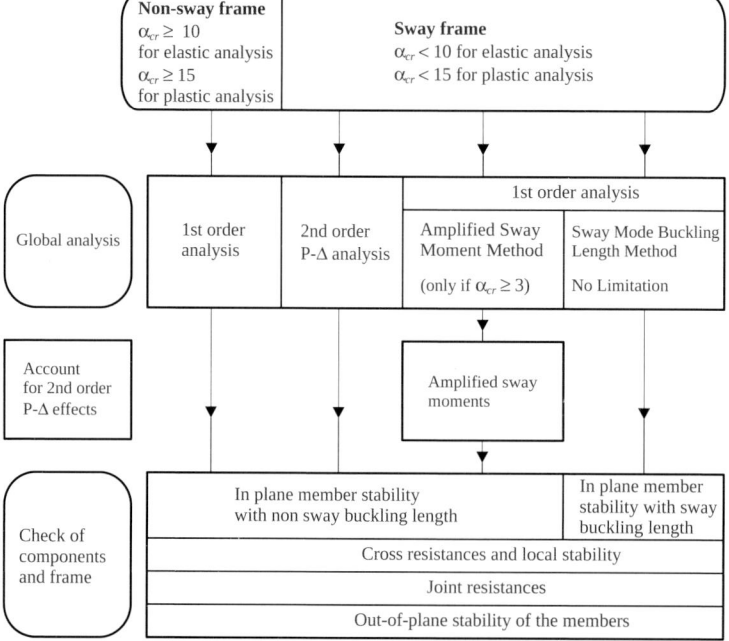

Figure 2.1 – Various ways for the global analysis and design process

2.1.3 Integration of joint response into the frame analysis and design process

Traditionally frame analysis is performed by assuming joints as perfectly pinned or perfectly rigid. Further to the analysis, joints have therefore to be designed to behave accordingly. In chapter 1 it has been pointed out that significant global economy may often result from what has been presented as a semi-continuous approach for structural joints.

A cross section being nothing else than a particular cross section, characterised by its own stiffness, resistance and ductility, the integration of the actual joint properties into the frame analysis and design process appears as natural. And it is the case. For instance, all what has been said in the previous sections remains unchanged at the condition that the meaning of the word "cross section" is extended to "member or joint cross section". From an operational point of view, the attention has however to be drawn to some specific concepts respectively named "joint modelling", "joint classification" and joint "idealisation". These ones are presented in the next sections.

2.2 JOINT MODELLING

2.2.1 General

Joint behaviour affects the structural frame response and shall therefore be modelled, just like beams and columns are, for the frame analysis and design. Traditionally, the following types of *joint modelling* are considered:

- *For rotational stiffness:*
  - rigid
  - pinned
- *For resistance:*
  - full-strength
  - partial-strength
  - pinned

When the joint rotational stiffness is of concern, the wording *rigid* means that no relative rotation occurs between the connected members

2. STRUCTURAL ANALYSIS AND DESIGN

whatever is the applied moment. The wording *pinned* postulates the existence of a perfect (i.e. frictionless) hinge between the members. In fact these definitions may be relaxed, as explained in sub-chapter 2.4 devoted to the joint classification. Indeed rather flexible but not fully pinned joints and rather stiff but not fully rigid joints may be considered as fairly pinned and fairly rigid respectively. The stiffness boundaries allowing one to classify joints as rigid or pinned are examined in sub-chapter 2.4.

For what regards the joint resistance, a *full-strength joint* is stronger than the weaker of the connected members, what is in contrast with a *partial-strength joint*. In the daily practice, partial-strength joints are used whenever the joints are designed to transfer the internal forces and not to resist the full capacity of the connected members. A *pinned joint* transfers no moment. Related classification criteria are conceptually discussed in sub-chapter 2.4.

Consideration of rotational stiffness and resistance joint properties leads to three significant joint modellings:

- rigid/full-strength;
- rigid/partial-strength;
- pinned.

However, as far as the joint rotational stiffness is considered, joints designed for economy may be neither rigid nor pinned but semi-rigid. There are thus new possibilities for joint modelling:

- semi-rigid/full-strength;
- semi-rigid/partial-strength.

With a view to simplification, EN 1993-1-8 accounts for these possibilities by introducing three joint models (Table 2.3):

- *continuous*:

    covering the rigid/full-strength case only;

- *semi-continuous*:

    covering the rigid/partial-strength, the semi-rigid/full-strength and the semi-rigid/partial-strength cases;

- *simple*:

    covering the pinned case only.

Table 2.3 – Types of joint modelling

Stiffness	Resistance		
	Full-strength	Partial-strength	Pinned
Rigid	Continuous	Semi-continuous	*
Semi-rigid	Semi-continuous	Semi-continuous	*
Pinned	*	*	Simple
	* : Without meaning		

The following meanings are given to these terms:

– *continuous:*
   the joint ensures a full rotational continuity between the connected members;
– *semi-continuous:*
   the joint ensures only a partial rotational continuity between the connected members;
– *simple:*
   the joint prevents from any rotational continuity between the connected members.

The interpretation to be given to these wordings depends on the type of frame analysis to be performed. In the case of an elastic global frame analysis, only the stiffness properties of the joint are relevant for the joint modelling. In the case of a rigid-plastic analysis, the main joint feature is the resistance. In all the other cases, both the stiffness and resistance properties govern the manner the joints shall be modelled. These possibilities are illustrated in Table 2.4.

Table 2.4 – Joint modelling and frame analysis

Modelling	Type of frame analysis		
	Elastic analysis	Rigid-plastic analysis	Elastic-perfectly plastic and elasto-plastic analysis
Continuous	Rigid	Full-strength	Rigid/full-strength
Semi-continuous	Semi-rigid	Partial-strength	Rigid/partial-strength Semi-rigid/full-strength Semi-rigid/partial-strength
Simple	Pinned	Pinned	Pinned

2. STRUCTURAL ANALYSIS AND DESIGN

2.2.2 Modelling and sources of joint deformability

As an example, the difference between the loading of the connection and that of the column web in a beam-to-column joint (see section 1.2.2) requires, from a theoretical point of view, that account be taken separately of both deformability sources when designing a building frame.

However doing so is only feasible when the frame is analysed by means of a sophisticated computer program which enables a separate modelling of both deformability sources. For most available software, the modelling of the joints has to be simplified by concentrating the sources of deformability into a single rotational spring located at the intersection of the axes of the connected members.

2.2.3 Simplified modelling according to Eurocode 3

For most applications, the separate modelling of the connection and of the web panel behaviour is neither useful nor feasible; therefore only the simplified modelling of the joint behaviour (see section 2.2.2) will be considered in the present document. This idea is the one followed by EN 1993-1-8. Table 2.5, excerpted from this EC3 Part 1-8, shows how to relate the simplified modelling of typical joints to the basic wordings used for the joint modelling: simple, semi-continuous and continuous.

Table 2.5 – Simplified modelling for joints according to EC3

Joint modelling	Beam-to-column joints major axis bending	Beam splices	Column bases
Simple			
Semi-continuous			

Table 2.5 – Simplified modelling for joints according to EC3 (continuation)

Joint modelling	Beam-to-column joints major axis bending	Beam splices	Column bases
Continuous			

2.2.4 Concentration of the joint deformability

For the daily practice a separate account of both the flexural behaviour of the connection and the shear (major axis beam-to-column joint) or out-of-plane behaviour of the column web panel (minor axis beam-to-column joint configurations or beam-to-beam configurations) is not feasible. This section is aimed at explaining how to concentrate the two deformabilities into a single flexural spring located at the intersection of the axes of the connected members.

2.2.4.1 Major axis beam-to-column joint configurations

In a single-sided configuration, only one joint is concerned. The characteristic deformability curve of the column web panel in shear (see Figure 1.9 and Figure 2.2b) is first transformed into a $M_b - \gamma$ curve through the use of the *transformation parameter* β. This parameter, defined in Figure 2.3a, relates the web panel shear force to the (load-introduction) compressive and tensile connection forces, see Eq. (1.1) and Eq. (1.2).

The $M_b - \phi$ spring characteristic which represents the joint behaviour is shown in Figure 2.2c; it is obtained by summing the contributions of rotation, from the connection (ϕ_c) and from the shear panel (γ). The $M_j - \phi$ characteristic of the joint rotational spring located at the beam-to-column interaction is assumed to identify itself to the $M_b - \phi$ characteristic obtained as indicated in Figure 2.2c.

2. Structural Analysis and Design

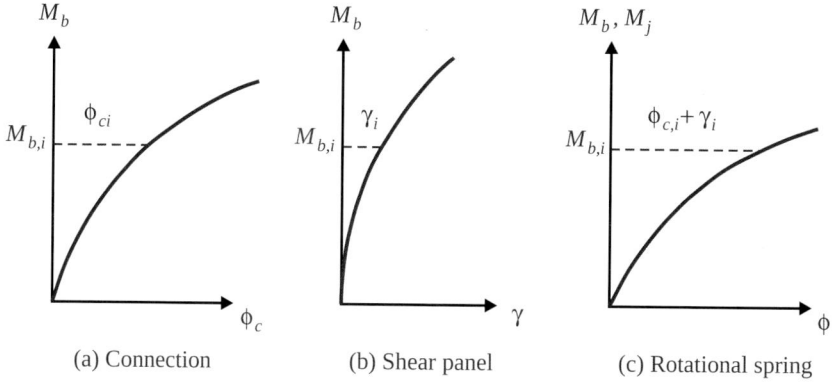

(a) Connection (b) Shear panel (c) Rotational spring

Figure 2.2 – Flexural characteristic of the rotational spring

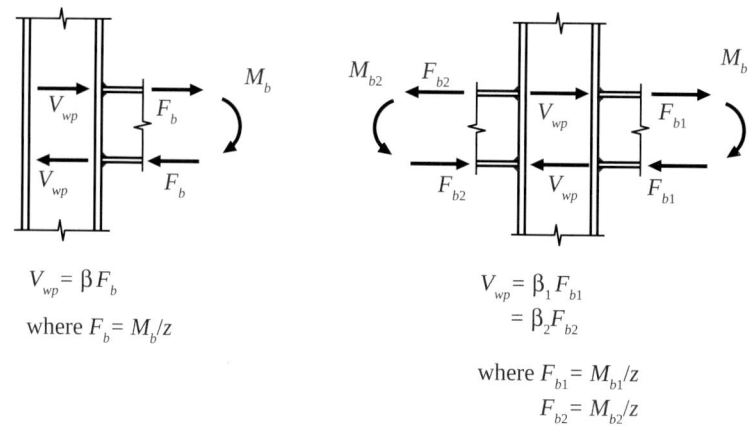

(a) Single-sided joint configuration (b) Double-sided joint configuration

Figure 2.3 – Definition of the transformation parameter β

In a double-sided configuration, two joints - the left one and the right one - are concerned. The derivation of their corresponding deformability curves is conducted similarly as in a single-sided configuration by using transformation parameters β_1 and β_2 (Figure 2.3b).

As already said in section 1.2.2, significant shear forces are likely to develop in the column web panels. They result from all the forces acting in the adjacent beams and columns, i.e. in a double-sided joint configuration (see Figure 2.4) the shear force in the web panel $V_{wp,Ed}$ is determined as:

$$V_{wp,Ed} = \left(M_{b1,Ed} - M_{b2,Ed}\right)/z - \left(V_{c1,Ed} - V_{c2,Ed}\right)/2 \qquad (2.1)$$

where z designates the level arm of the internal forces (see Figure 2.5 for few examples).

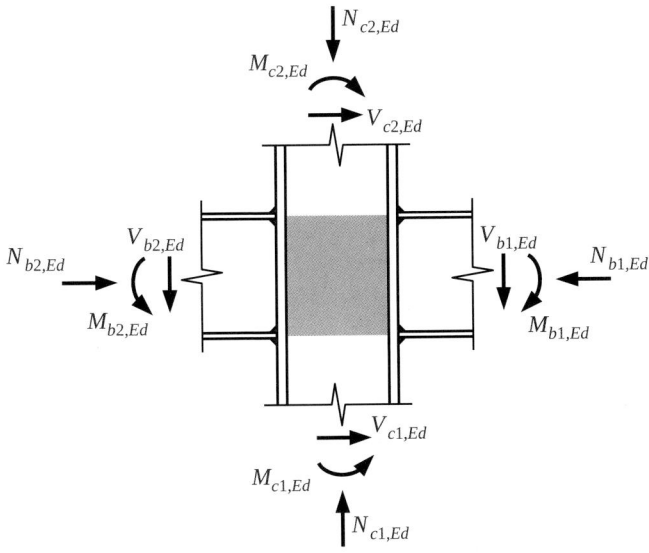

Figure 2.4 – Forces applied at the periphery of a web panel

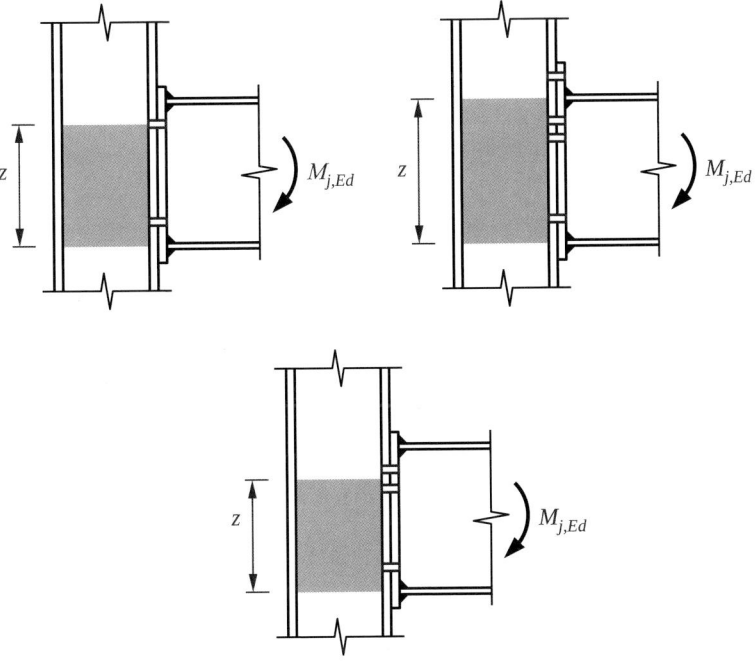

Figure 2.5 – Definition of the level arm z

2. Structural Analysis and Design

Figure 2.6 illustrates the determination of the shear force in a column web panel for few particular cases.

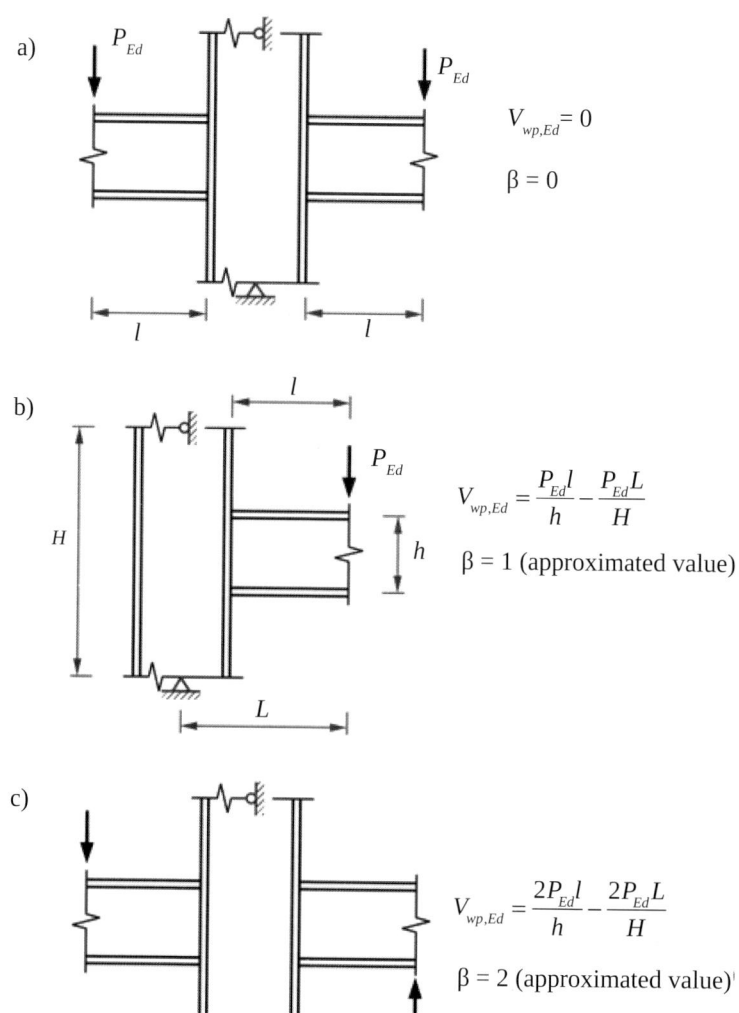

Figure 2.6 – Shear force in a column web panel

The β coefficients are defined as (see Figure 2.3b):

$$\beta_1 = V_{wp,Ed} z / M_{b1,Ed} \quad \text{for the right joint} \tag{2.2}$$

$$\beta_2 = V_{wp,Ed} z / M_{b2,Ed} \quad \text{for the left joint} \tag{2.3}$$

2.2 JOINT MODELLING

Because the values of the β parameters can only be determined once the internal forces are known, their accurate determination requires an iterative process in the global analysis. For practical applications, such an iterative process may be avoided provided safe β values be available. These values may be used a priori to model the joints and, on the basis of such joint modelling, the frame analysis may be performed safely in a non-iterative way.

The recommendations for the approximate values of β are given in Table 5.4 of EN 1993-1-8, see Table 2.6. They vary from $\beta = 0$ (double-sided joint configuration with balanced moments in the beams) to $\beta = 2$ (double-sided joint configuration with equal but unbalanced moments in the beams). These two extreme cases are illustrated in Figure 2.7.

Table 2.6 – Approximate values for the transformation parameter β

Type of joint configuration	Action	Value
(single-sided joints with $M_{b1,Ed}$)	$M_{b1,Ed}$	$\beta \approx 1$
(double-sided joints with $M_{b1,Ed}$ and $M_{b2,Ed}$)	$M_{b1,Ed} = M_{b2,Ed}$	$\beta = 0$ *)
	$M_{b1,Ed}/M_{b2,Ed} > 0$	$\beta \approx 1$
	$M_{b1,Ed}/M_{b2,Ed} < 0$	$\beta \approx 2$
	$M_{b1,Ed} + M_{b2,Ed} = 0$	$\beta \approx 2$

*) In this case the value of β is the exact value rather than an approximation

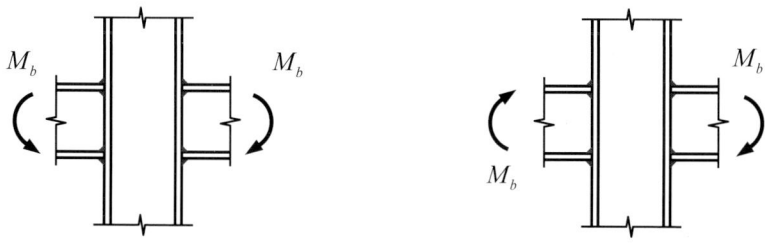

(a) Balanced beam moments   (b) Equal but unbalanced beam moments

Figure 2.7 – Extreme cases for β values

2. STRUCTURAL ANALYSIS AND DESIGN

The values given in Table 2.6 are derived from the following simplified version of Eqs. (2.2) and (2.3) (see Figure 2.8):

$$\beta_1 = \left|1 - M_{j,b2,Ed}/M_{j,b1,Ed}\right| \leq 2 \quad \text{for the right joint} \quad (2.4)$$

$$\beta_2 = \left|1 - M_{j,b1,Ed}/M_{j,b2,Ed}\right| \leq 2 \quad \text{for the left joint} \quad (2.5)$$

In these ones, the contribution of the shear forces acting in the column are is conservatively disregarded (Jaspart, 1991).

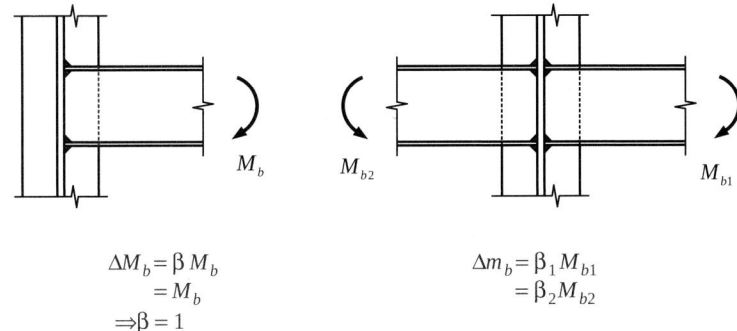

(a) Single-sided joint configuration   (b) Double-sided joint configuration

Figure 2.8 – Definition of the transformation parameter β

2.2.4.2 Minor axis beam-to-column joint configurations and beam-to-beam configurations

Similar concepts as those developed in section 2.2.4.1 are referred to for minor axis beam-to-column joint configurations and beam-to-beam configurations. The definition of the transformation parameter is somewhat different (see Figure 2.8).

2.3 JOINT IDEALISATION

The non-linear behaviour of the isolated flexural spring which characterises the actual joint response does not lend itself towards everyday design practice. However the moment-rotation characteristic curve may be idealised without significant loss of accuracy. One of the most simple idealisations possible is the elastic-perfectly plastic one (Figure 2.9a). This modelling has the advantage of being quite similar to that used traditionally for the modelling of member cross sections subject to bending (Figure 2.9b).

The moment $M_{j,Rd}$ that corresponds to the yield plateau is termed design moment resistance in Eurocode 3. It may be understood as the *pseudo-plastic moment resistance* of the joint. Strain-hardening effects and possible membrane effects are henceforth neglected; that explains the difference in Figure 2.9 between the actual $M-\phi$ characteristic and the *yield plateau* of the idealised one.

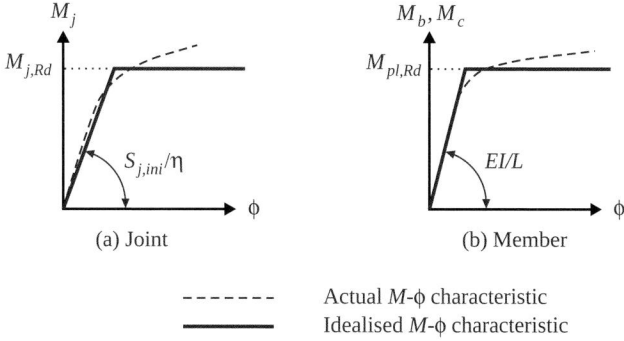

Figure 2.9 – Bi-linearisation of moment-rotation curves

The value of the constant stiffness is discussed below.

In fact there are different possible ways to idealise a joint $M-\phi$ characteristic. The choice of one of them is subordinated to the type of frame analysis which is contemplated, as explained below.

2.3.1 Elastic idealisation for an elastic analysis

The main joint characteristic is the constant rotational stiffness.

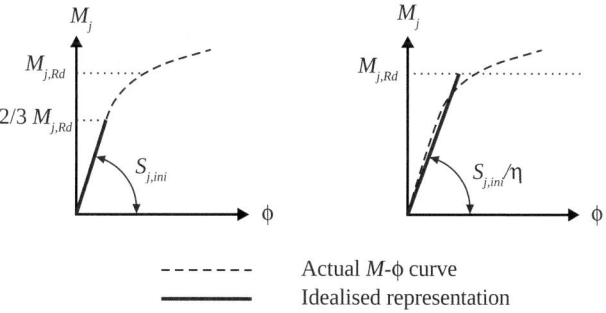

Figure 2.10 – Linear representation of a $M-\phi$ curve

2. STRUCTURAL ANALYSIS AND DESIGN

Two possibilities are offered in Eurocode 3 Part 1-8, see Figure 2.10:

- *Elastic verification of the joint resistance* (Figure 2.10a): the constant stiffness is taken equal to the initial stiffness $S_{j,ini}$; at the end of the frame analysis, it shall be checked that the design moment M_{Ed} experienced by the joint is less than the maximum elastic joint moment resistance defined as $2/3 M_{j,Rd}$;
- Plastic verification of the joint resistance (Figure 2.10b): the constant stiffness is taken equal to a fictitious stiffness, the value of which is intermediate between the initial stiffness and the secant stiffness relative to $M_{j,Rd}$; it is defined as $S_{j,ini}/\eta$. This idealisation is aimed at "replacing" the actual non-linear response of the joint by an "equivalent" constant one; it is valid for M_{Ed} values less than or equal to $M_{j,Rd}$. EN 1993-1-8 recommends values of η reported in in Table 2.7.

Table 2.7 – Stiffness modification factor

Type of connection	Beam-to-column joints	Other types of joints (beam-to-beam joints, beam splices, column base joints)
Welded	2	3
Bolted end-plates	2	3
Bolted flange cleats	2	3.5
Base plates	-	3

2.3.2 Rigid-plastic idealisation for a rigid-plastic analysis

Only the design resistance $M_{j,Rd}$ is needed. In order to allow the possible plastic hinges to form and rotate in the joint locations, it shall be checked that the joint has a sufficient rotation capacity, see Figure 2.11.

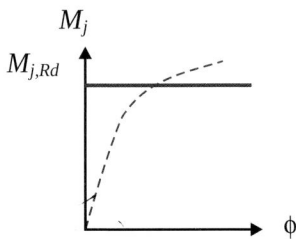

Figure 2.11 – Rigid-plastic representation of a $M-\phi$ curve

2.3.3 Non-linear idealisation for an elastic-plastic analysis

The stiffness and resistance properties are of equal importance in this case. The possible idealisations range from bi-linear, tri-linear representations or a fully non-linear curve, see Figure 2.12. Again rotation capacity is required in joints where plastic hinges are likely to form and rotate.

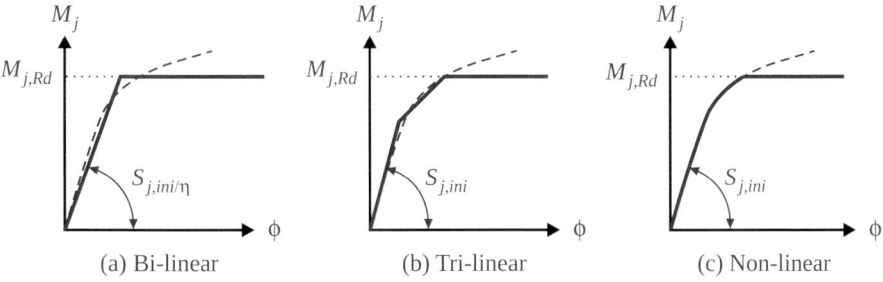

(a) Bi-linear　　　　(b) Tri-linear　　　　(c) Non-linear

Figure 2.12 – Non-linear representations of a $M-\phi$ curve

2.4 JOINT CLASSIFICATION

2.4.1 General

In sub-chapter 2.2, it is shown that the joints need to be modelled for the global frame analysis and that three different types of joint modelling are introduced: simple, semi-continuous and continuous.

It has also been explained that the type of joint modelling to which it shall be referred is dependent both on the type of frame analysis and on the class of the joint in terms of stiffness and/or strength (Table 2.4).

Classification criteria are used to define the stiffness class and the strength class to which the joint belongs and also to determine the type of joint modelling which shall be adopted for analysis. They are described here below.

2.4.2 Classification based on mechanical joint properties

The stiffness classification is performed by comparing simply the design joint stiffness to two stiffness boundaries (Figure 2.13). For the sake of simplicity, the stiffness boundaries have been derived so as to allow a direct

Table 2.8 – Joint classification boundaries (continuation)

Classification by resistance	
Full strength joint:	Joint which possesses a higher resistance than the weakest of the connected members
Pinned joint:	Joint with design resistance lower than 25 % of the full-strength resistance.
Partial-strength joint:	Joint with design resistance lower than the full-strength resistance and which has not been classified as pinned.

2.5 DUCTILITY CLASSES

2.5.1 General concept

Experience and proper detailing result in so-called *pinned* joints which exhibit a sufficient rotation capacity to sustain the rotations imposed on them. This topic is addressed in chapter 5.

For moment resistant joints the concept of ductility classes is introduced to deal with the question of rotation capacity.

For most of these structural joints, the shape of the $M-\phi$ characteristic is rather bi-linear (Figure 2.15a). The initial slope $S_{j,ini}$ corresponds to the elastic deformation of the joint. It is followed by a progressive yielding of the joint (of one or some of the constituent components) until the design moment resistance $M_{j,Rd}$ is reached. Then a post-limit behaviour ($S_{j,post-lim}$) develops which corresponds to the onset of strain-hardening and possibly of membrane effects. The latter are especially important in components where rather thin plates are subject to transverse tensile forces as, for instance, in minor axis joints and in joints with columns made of rectangular hollow sections.

In many experimental tests (Figure 2.15a) the collapse of the joints at a peak moment $M_{j,u}$ has practically never been reached because of high local deformations in the joints involving extremely high relative rotations. In the others (Figure 2.15b) the collapse has involved an excessive yielding (rupture of the material) or, more often, the instability of one of the constituent components (ex: column web panel in compression or buckling

2.5 DUCTILITY CLASSES

of the beam flange and web in compression) or the brittle failure in the welds or in the bolts.

In some joints, the premature collapse of one of the components prevents the development of a high moment resistance and high rotation. The post-limit range is rather limited and the bi-linear character of the $M_j - \phi$ response is less obvious to detect (Figure 2.15c).

As explained in sub-chapter 2.3, the actual $M_j - \phi$ curves are idealised before performing the global analysis. As for beam and column cross sections, the usual concept of plastic hinge can be referred to for plastic global analysis.

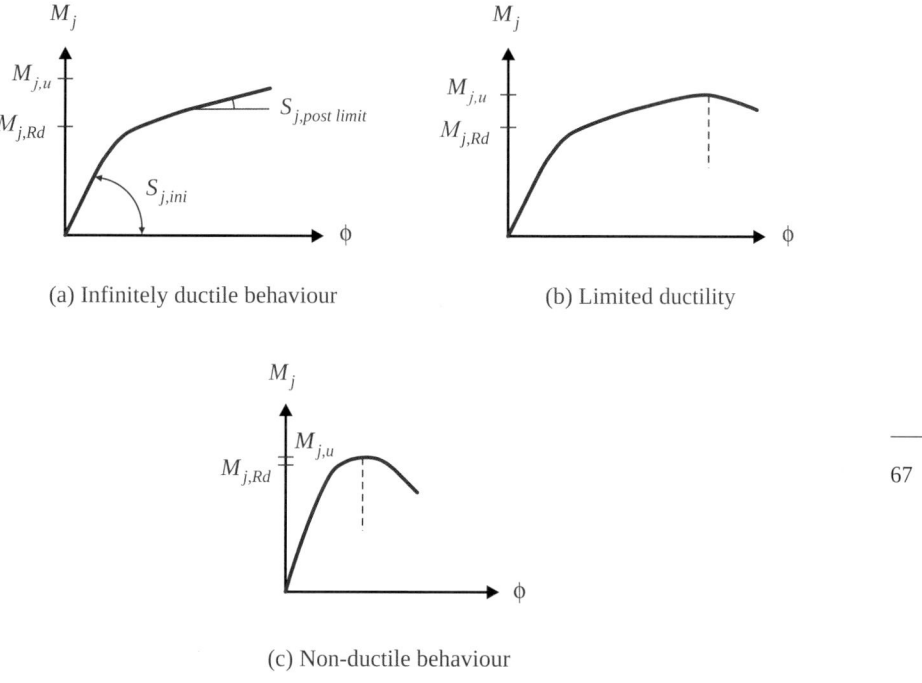

Figure 2.15 – Shape of joint $M - \phi$ characteristics

The development of plastic hinges during the loading of the frame and the corresponding redistribution of internal forces in the frame require, from the joints where hinges are likely to occur, a sufficient rotation capacity. In other words, there must be a sufficiently long yield plateau ϕ_{pl} (Figure 2.16) to allow the redistribution of internal forces to take place.

2. STRUCTURAL ANALYSIS AND DESIGN

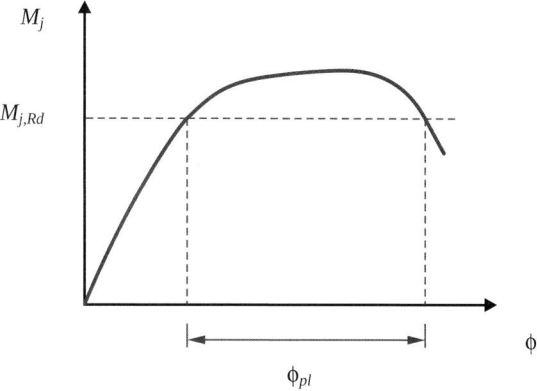

Figure 2.16 – Plastic rotation capacity

For beam and column sections, deemed-to-satisfy criteria allow the designer to determine the class of the sections and therefore the type of global frame analysis which can be contemplated.

A strong similarity exists for what regards structural joints. Moreover a similar classification may be referred to:

- Class 1 joints: $M_{j,Rd}$ is reached by full plastic redistribution of the internal forces within the joints and a sufficiently good rotation capacity is available to allow, without specific restrictions, a plastic frame analysis and design to be performed, if required;
- Class 2 joints: $M_{j,Rd}$ is reached by full plastic redistribution of the internal forces within the joints but the rotation capacity is limited. An elastic frame analysis possibly combined with a plastic verification of the joints has to be performed. A plastic frame analysis is also allowed as long as it does not result in a too high required rotation capacity in the joints where hinges are likely to occur. The available and required rotation capacities have therefore to be compared before validating the analysis;
- Class 3 joints: brittle failure (or instability) limits the moment resistance and does not allow a full redistribution of the internal forces in the joints. It is compulsory to perform an elastic verification of the joints.

As the moment design resistance $M_{j,Rd}$ is known whatever the collapse mode and the resistance level, no Class 4 has to be defined as it is done for member sections.

2.5.2 Requirements for classes of joints

In *Eurocode 3*, the procedure given for the evaluation of the design moment resistance of any joint provides the designer with other information such as:

- the collapse mode;
- the internal forces in the joint at collapse.

Through this procedure, the designer knows directly whether the full plastic redistribution of the forces within the joint has been reached - the joint is then Class 1 or 2 - or not - the joint is then classified as Class 3.

For Class 1 or 2 joints, the knowledge of the collapse mode, and more especially of the component leading to collapse, gives an indication about whether there is adequate rotation capacity for a global plastic analysis to be permitted. The related criteria are expressed in chapter 6.

Chapter 3

CONNECTIONS WITH MECHANICAL FASTENERS

3.1 MECHANICAL FASTENERS

The basic elements to compose joints in steel structures are mechanical fasteners like bolts or pins. Chapter 3 of Eurocode 3 Part 1-8 provides design rules for such mechanical fasteners. The designer will find here application rules for different types of mechanical fasteners: bolts, injections bolts, anchor bolts, rivets and pins. Beside these mechanical fasteners, other fasteners for more special applications can be used directly covered by Eurocode 3, for example flow drill bolts, HRC bolts (also called TC bolts) or nails. However, the present version of Eurocode 3 does not provide application rules for those types of mechanical fasteners. In some cases harmonised product standards may provide relevant data.

In steel construction, the most typical mechanical fasteners to connect plates or profiles are bolts, or more precisely: bolt assemblies (sets) including the bolt itself, a nut and one or more washers, see Figure 3.1. The bolts may be preloaded to improve serviceability performance or fatigue resistance. Joints made with preloaded bolts normally may have a slight higher stiffness, but this effect is not taken into account in the design rules. However, preloading requires a controlled tightening which leads to additional work during erection.

3. CONNECTIONS WITH MECHANICAL FASTENERS

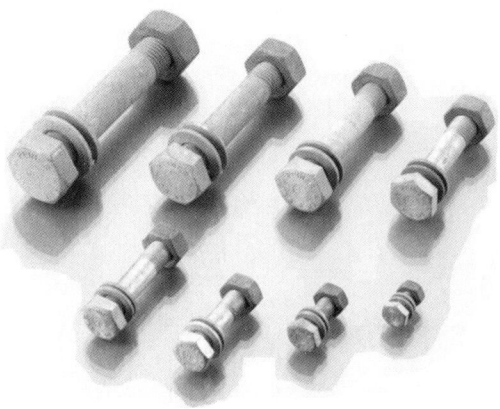

Figure 3.1 – Bolt assemblies

All bolts, nut and washers should comply with the standards listed in section 1.2.4 of Eurocode 3 Part 1-8. Most important standards are EN 15048 (Non-preloaded structural bolting assemblies) for non-preloaded bolts (CEN, 2007b) and EN 14399 (High strength structural bolting for preloading) for preloaded bolts (CEN, 2015). Mechanical properties are specified in EN ISO 898 (CEN, 2013). The design rules in Eurocode 3 Part 1-8 are valid for all bolt classes listed in Table 3.1, but the National Annex may exclude certain bolt classes.

Table 3.1 – Bolts classes and nominal values of yield strength f_{yb} and ultimate tensile strength f_{ub}

Bolt class	4.6	4.8	5.6	5.8	6.8	8.8	10.9
f_{yb} (N/mm²)	240	320	300	400	480	640	900
f_{ub} (N/mm²)	400	400	500	500	600	800	1000

The name of the bolt classes indicates the nominal values of the ultimate tensile strength f_{ub} and the yield strength f_{yb} as follows. The name consists of two numbers separated by a dot. The first number is the ultimate tensile strength f_{ub} divided by 100. The fractional part indicates the ratio f_{yb}/f_{ub}, for example for a 10.9 bolt, the ultimate tensile strength f_{ub} is $10 \times 100 = 1000$ (N/mm²) and the yield strength f_{yb} is obtained as $1000 \times 0.9 = 900$ (N/mm²).

For preloaded bolt, only bolt assemblies of classes 8.8 and 10.9 may be used. Requirements for controlled tightening are given in EN 1090 Part 2.

Bolt areas for common sizes of structural bolts are given in Table 3.2 where A is the gross section area and A_s is the tensile stress area (treaded portion of the bolt).

Table 3.2 – Bolt areas in accordance with EN ISO 898 (CEN, 2013)

d (mm)	10	12	14	16	18	20	22	24	27	30	36
A (mm^2)	78	113	154	201	254	314	380	452	573	707	1018
A_s (mm^2)	58	84	115	157	192	245	303	353	459	561	817

3.2 CATEGORIES OF CONNECTIONS

The very basic connection is composed of a bolt assembly connecting two (or more) steel plates as shown in Figure 3.2. For the design of these basic components (for example bolts in shear and/or tension, plate in bearing), see sub-chapter 3.4.

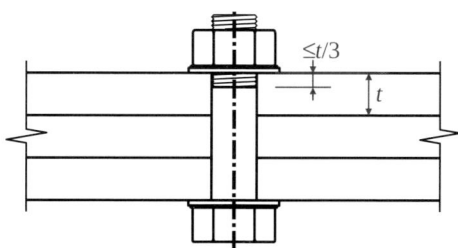

Figure 3.2 – Bolted connection

Eurocode 3 distinguishes different categories of bolted connections. The two main categories, dependent on the loading applied to the bolt, are shear connections and tension connections.

3.2.1 Shear connections

Shear connections (the bolt is subjected to shear) are subdivided into 3 sub-categories:

3. Connections with Mechanical Fasteners

a) Category A: Bearing type

In a bearing type connection, the applied load (design ultimate shear load $F_{v,Ed}$) will be transferred from the plate to the bolt (shank and/or threaded part) by bearing, while the bolt will be loaded in shear. Hence, the design resistance of a category A connection is the minimum of the design bearing resistance of the plate(s) $F_{v,Rd}$ and the design shear resistance of the bolt $F_{b,Rd}$, i.e. the following criteria should be checked:

$$F_{v,Ed} \leq F_{v,Rd}$$
$$F_{v,Ed} \leq F_{b,Rd} \quad (3.1)$$

All bolt classes from Table 3.1 may be used. Preloading or any special precision for the contact surfaces is not required. Depending on the number of plates, the shear load applied to the bolt may be divided into one or more shear planes.

b) Category B: Slip-resistant at serviceability limit state

With regards to the design resistance at ultimate limit state, a category B connection should meet the same requirement as a category A connection. In addition, at serviceability limit state, the connection should be designed so that no slip will occur. Therefore, preloaded bolts should be used. The design criteria are as follow:

$$F_{v,Ed} \leq F_{v,Rd}$$
$$F_{v,Ed} \leq F_{b,Rd} \quad (3.2)$$
$$F_{v,Ed,ser} \leq F_{s,Rd,ser}$$

where
  $F_{v,Ed,ser}$ is the design shear force per bolt for the serviceability limit state
  $F_{v,Rd,ser}$ is the design slip resistance per bolt at the serviceability limit state.

c) Category C: Slip-resistant at ultimate limit state

In this category, slip should not occur at the ultimate limit state. This will result in most severe design criteria for a shear connection. The design ultimate shear load will be transferred by friction between the connected plates only. This means, in theory, the connection only need to be checked for slip resistance. In addition, Eurocode 3 requires checking the bearing resistance. So, the following design criteria should be checked:

$$F_{v,Ed} \leq F_{s,Rd}$$
$$F_{v,Ed} \leq F_{b,Rd} \qquad (3.3)$$

where
$F_{s,Rd}$ is the design slip resistance per bolt at the ultimate limit state.

For category A and category B connections, if the connected plates are loaded in tension, the design tension resistance $N_{t,Rd}$ should be checked at ultimate limit state as well:

$$F_{v,Ed} \leq N_{t,Rd} \qquad (3.4)$$

For category C connections, a similar check has also to be achieved, but on the basis of the design plastic resistance of the net cross section at the bolt holes $N_{net,Rd}$:

$$F_{v,Ed} \leq N_{net,Rd} \qquad (3.5)$$

where
$N_{t,Rd}$ is the smaller of the design plastic resistance of the gross cross section $N_{pl,Rd}$ and the design ultimate resistance of the net cross section at holes for fasteners $N_{u,Rd}$
$N_{net,Rd}$ is design tension resistance of the net section at holes for fasteners

with

$$N_{pl,Rd} = \frac{A f_y}{\gamma_{M0}}, \quad N_{u,Rd} = \frac{0.9 A_{net} f_u}{\gamma_{M2}} \text{ and } N_{net,Rd} = \frac{A_{net} f_y}{\gamma_{M0}}.$$

3.2.2 Tension connections

In a tension connection the bolt is subjected to tension forces. Here, two sub-categories are defined:

3. CONNECTIONS WITH MECHANICAL FASTENERS

a) Category D: Non-preloaded

Pre-loading is not required and all bolt classes given in Table 3.1 may be used. This category may be used for connections subjected to predominantly static loading, for example wind loads acting on buildings. The design criteria to be checked are:

$$F_{t,Ed} \leq F_{t,Rd}$$
$$F_{t,Ed} \leq B_{p,Rd}$$
(3.6)

where

$F_{t,Rd}$ is the design tension resistance of the bolt
$B_{p,Rd}$ is the design punching resistance of the bolt head and the nut

b) Category E: Preloaded

In preloaded tension connections, only bolt classes 8.8 and 10.9 should be used. This category relates to connections which are frequently subjected to variations of loading, for example in crane supporting structures. The design criteria are the same as for category D connection (non-preloaded), see Eq. (3.6). Note that, for preloaded bolts, controlled tightening according to EN 1090 Part 2 is required.

When bolts are subjected to both shear and tension forces, interaction should also be checked, see 3.4.3.

If preloaded bolts are used in category A or category D connections where preloading is used to improve for example serviceability performance or durability or where preloading is required for execution purposes, but where preloading is not explicitly used in the design checks for slip resistance, then the level of preload can differ from the requirements in EN 1090 Part 2. In this case, the National Annex can specify less strict requirements.

3.3 POSITIONING OF BOLT HOLES

In bolted connections, independently of the type of loading, the bolt holes must fulfil certain requirements with respect to minimum and maximum spacing between the holes, end distances and edge distances. These requirements are given in Table 3.3 extracted from Table 3.3 of EN 1993-1-8. The symbols for the spacing and distances are defined in Figure 3.3. Note that these limits are valid for predominantly static loaded joints. For structures subjected to fatigue, requirements are given in EN 1993-1-9.

Table 3.3 – Minimum and maximum spacing, end and edge distances

Distances and spacings	Minimum	Maximum		
		Structures made from steels conforming to EN 10025 except steels conforming to EN 10025-5		Structures made from steels conforming to EN 10025-5
		Steel exposed to the weather or other corrosive influences	Steel not exposed to the weather or other corrosive influences	Steel used unprotected
End distance e_1	$1.2d_0$	$4t + 40$ mm		$\max(8t, 125 \text{ mm})$
Edge distance e_2	$1.2d_0$	$4t + 40$ mm		$\max(8t, 125 \text{ mm})$
Spacing p_1	$2.2d_0$	$\min(14t, 200 \text{ mm})$		$\min(14t_{min}, 175 \text{ mm})$
Spacing $p_{1,0}$		$\min(14t, 200 \text{ mm})$		
Spacing $p_{1,i}$		$\min(28t, 400 \text{ mm})$		
Spacing p_2	$2.4d_0$	$\min(14t, 200 \text{ mm})$		$\min(14t_{min}, 175 \text{ mm})$
where t is the thickness of the thinner outer connected part				

Maximum values are given to prevent corrosion in exposed members and to avoid local buckling between two fasteners in compression members. In other cases, or when no values are given, maximum values for the spacings, edge distances and end distances are unlimited. The local buckling resistance of the plate in compression between the fasteners should be calculated according to EN 1993-1-1 where a value of $0.6p_1$ should be used as buckling length. Local buckling between the fasteners need not to be checked if p_1/t is smaller than 9ε. The edge distance should not exceed the local buckling requirements for an outstand element in the compression members, see EN 1993-1-1. The end distance is not affected by this requirement.

3. CONNECTIONS WITH MECHANICAL FASTENERS

For slotted holes, minimum end distances and minimum edge distances are a little larger. For more information, see Table 3.3 of EN 1993-1-8.

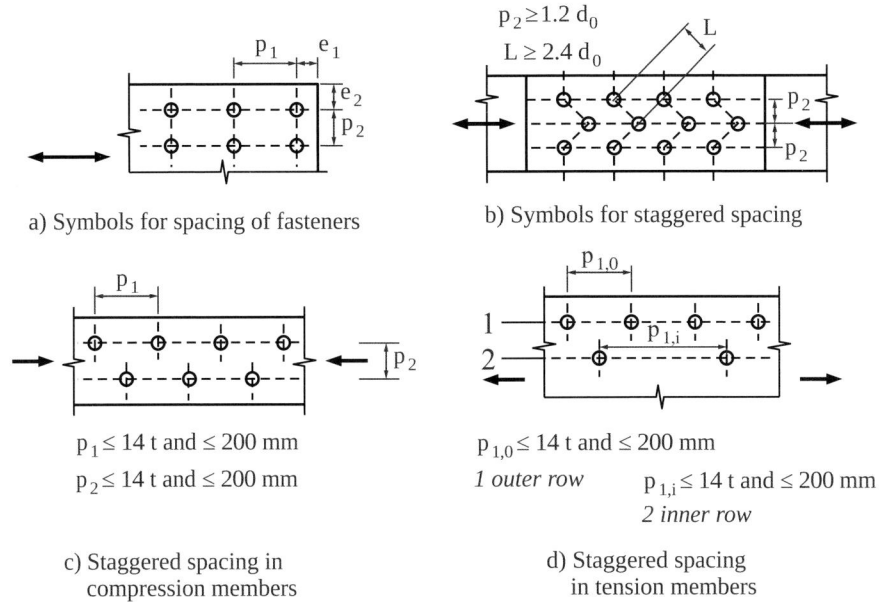

$p_1 \leq 14\,t$ and ≤ 200 mm
$p_2 \leq 14\,t$ and ≤ 200 mm

c) Staggered spacing in compression members

$p_{1,0} \leq 14\,t$ and ≤ 200 mm
1 outer row $p_{1,i} \leq 14\,t$ and ≤ 200 mm
2 inner row

d) Staggered spacing in tension members

Figure 3.3 – Symbols for end and edge distances and spacing of fasteners

3.4 DESIGN OF THE BASIC COMPONENTS

3.4.1 Bolts in shear

If the shear plane passes through the *threaded portion* of the bolt, the design resistance of bolts in shear $F_{v,Rd}$ is

$$F_{v,Rd} = \frac{\alpha_v f_{ub} A_s}{\gamma_{M2}} \qquad (3.7)$$

where
  α_v = 0.6 for (more ductile) bolt classes 4.6, 5.6 and 8.8
  α_v = 0.5 for (less ductile) bolt classes 4.8, 5.8 and 6.8 and 10.9
  A_s is the tensile stress area of the bolt

3.4 DESIGN OF THE BASIC COMPONENTS

If the shear plane passes through the *shank* (unthreaded portion) of the bolt, the design resistance of bolts in shear $F_{v,Rd}$ is

$$F_{v,Rd} = \frac{0.6 f_{ub} A}{\gamma_{M2}} \qquad (3.8)$$

where

A is the gross cross section area of bolt (the shank or unthreaded portion)

The design shear resistance according to Eq. (3.7) and Eq. (3.8) should only be used where the bolts are used in holes with nominal clearances as specified in EN 1090-2 (CEN, 2011), i.e. 1 mm for M12 or M14, 2 mm for M16 to M24 and 3 mm for M27 and thicker bolts. In Table 3.4 the design shear resistances $F_{v,Rd}$ is given for common bolt diameters.

Table 3.4 – Design shear resistances $F_{v,Rd}$ in kN (for $\gamma_{M2} = 1.25$)

Shear plane	Bolt grade	M12	M16	M20	M24	M27	M30	M36
in the shank	4.6	21.71	38.60	60.32	86.86	109.9	135.7	195.4
	5.6	27.14	48.25	75.40	108.6	137.4	169.6	244.3
	8.8	43.43	77.21	120.6	173.7	219.9	271.4	390.9
	10.9	54.29	96.51	150.8	217.1	274.8	339.3	488.6
in the thread	4.6	16.19	30.14	47.04	67.78	88.13	107.7	156.9
	5.6	20.23	37.68	58.80	84.72	110.2	134.6	196.1
	8.8	32.37	60.29	94.08	135.6	176.3	215.4	313.7
	10.9	33.72	62.80	98.00	141.2	183.6	224.4	326.8

The design shear resistance for *fit bolts*, where the nominal diameter of the hole is the same as the nominal diameter of the bolt, i.e. holes with no clearance, should be determined using Eq. (3.8). The thread of a fit bolt should not be included in the shear plane.

3.4.2 Bolts in tension

The design tension resistance of a (non-preloaded or preloaded) bolt is determined according to Eurocode 3 Part 1-8 as follows:

$$F_{t,Rd} = \frac{0.9 f_{ub} A_s}{\gamma_{M2}} \qquad (3.9)$$

In Table 3.5 the design tension resistances $F_{t,Rd}$ is given for common bolt diameters.

Table 3.5 – Design tension resistances $F_{t,Rd}$ in kN (for $\gamma_{M2} = 1.25$)

Bolt grade	M12	M16	M20	M24	M27	M30	M36
4.6	24.28	45.22	70.56	101.7	132.2	161.6	235.3
5.6	30.35	56.52	88.20	127.1	165.2	202.0	294.1
8.8	48.56	90.43	141.1	203.3	264.4	323.1	470.6
10.9	60.70	113.0	176.4	254.2	330.5	403.9	588.2

For countersunk bolts, in Eq. (3.9), a factor of 0.63 should be used instead of 0.9.

3.4.3 Bolts in shear and tension

When bolts are subjected to shear and tension forces, the following design criteria given in Table 3.4 of Eurocode 3 Part 1-8 should be satisfied:

$$\frac{F_{v,Ed}}{F_{v,Rd}} + \frac{F_{t,Ed}}{1.4 F_{t,Rd}} \leq 1.0 \qquad (3.10)$$

3.4 DESIGN OF THE BASIC COMPONENTS

Figure 3.4 shows the interaction diagram for combined shear and tension. If the shear load in a bolt does not exceed about 28% of its shear resistance, the design tension resistance of a bolt must not be reduced. In other words, the interaction check is not needed, if

$$F_{v,Ed} \leq (1-1/1.4)F_{v,Rd} \approx 0.286 F_{v,Rd} \tag{3.11}$$

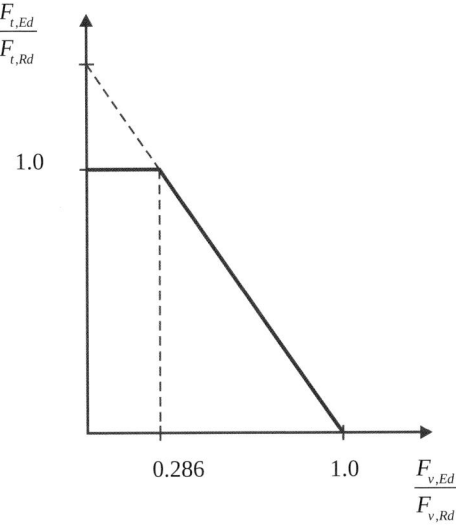

Figure 3.4 – Shear-tension interaction of bolts

3.4.4 Preloaded bolts

Dependent on the performance specification of a connection, bolts may be used as preloaded or non-preloaded bolts. In some particular cases, for example in connections subjected to dynamic loading, it is or it could be required that bolts be preloaded in order to ensure a minimum clamping force between connected plates. In such cases, a particular preparation of the surfaces as well as a controlled tightening is to be requested, see EN 1090-2. In these cases, the preload can be used in the design calculations, for example in slip resistant connections. The design preload $F_{p,Cd}$ is determined as follows:

$$F_{p,Cd} = 0.7 f_{ub} A_s / \gamma_{M7} \tag{3.12}$$

However, procedures to control preloading are cost-intensive. Also, controlled pre-loading of bolts will decrease the speed of fabrication and erection of the steel structures. So, it is obvious that preloaded bolts with controlled tightening should only be used when it is absolutely necessary. On the other hand, if bolts are preloaded to some extent, even if the level of preloading is not controlled by a particular tightening procedure, this will improve the stiffness performance of a bolted connection, for example by decreasing deformations (for example for execution purposes). However, in this case, the preload cannot be used in the design calculations.

a) Preloaded bolts in shear connections

The structural response in a connection is different depending of the load case. In a shear connection, as long as the bolts are not preloaded, the internal forces will be transferred by bearing between the plates and the bolt and by shear in the bolt shank respectively, see Figure 3.5. If the bolts are preloaded, a clamping pressure will develop between the connected parts. The internal forces are transferred directly between the connected parts. Friction will prevent slipping of the connection, see Figure 3.5. This is called a slip resistant connection. In this case the resulting deformation of the connection is significantly smaller that the deformation of a bearing type connection.

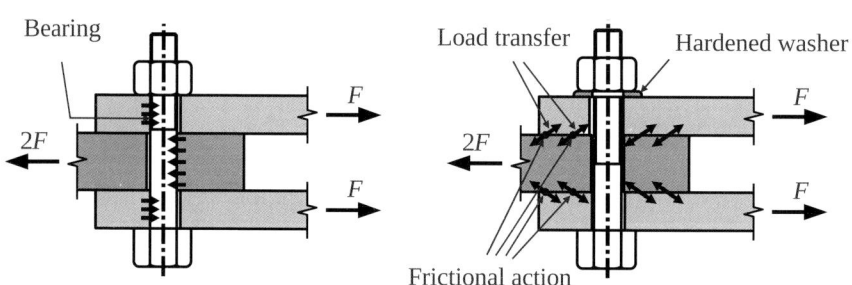

Figure 3.5 – Load transfer in a non-preloaded and a preloaded connection in a shear connection

The maximum load which can be transferred by friction is the design slip resistance at the ultimate limit state $F_{s,Rd}$ which is dependent on the preload of the bolts $F_{p,C}$ and on the friction coefficient or slip factor of the clamped plates. The design slip resistance of a preloaded class 8.8 or 10.9 bolt should be taken as:

3.4 DESIGN OF THE BASIC COMPONENTS

$$F_{s,Rd} = \frac{k_s n \mu}{\gamma_{M3}} F_{p,C} \qquad (3.13)$$

where

k_s is a reduction factor shape and size of the hole. For normal holes $k_s = 1$, for other cases, e.g. oversized or slotted holes, reference is made to EN 1993-1-8.

n is the number of the friction surfaces

μ is the slip factor. Values of the slip factor varies, depending on the treatment of friction surfaces, between 0.2 and 0.5. Slip factors μ and friction surface classes are defined in section 8.4 of EN 1090-2, see Table 3.7.

Values for the preload of the bolts $F_{p,C}$ are given in Table 3.6.

Table 3.6 – Nominal values of $F_{p,C}$ in kN

Bolt class	M12	M16	M20	M22	M24	M27	M30	M36
8.8	47	88	137	170	198	257	314	458
10.9	59	110	172	212	247	321	393	572

Table 3.7 – Friction surface classes and slip factors

Surface treatment	Class	Slip factor
Surfaces blasted with shot or grit with loose rust removed, not pitted	A	0.5
Surfaces blasted with shot or grit: a) spray-metallized with an aluminium or zinc based product; b) with alkali-zinc silicate paint with a thickness of 50 to 80 µm	B	0.4
Surfaces cleaned by wire-brushing or flame cleaning, with loose rust removed	C	0.3
Surfaces as rolled	D	0.2

If the applied load exceeds the design slip resistance $F_{s,Rd}$, the connection transforms into a bearing type connection.

In order to make erection easier, bolt holes have a clearance with respect to the bolt diameter. In bearing type connections, hole clearances cause slips which lead to an extra deformation as illustrated in Figure 3.6

- approximately half - to drive the threads up the helix, overcoming the friction between the mating screw surfaces and the resolved component of the axial force.

The tightening torque shall be applied continuously and smoothly. Tightening by the torque method comprises at least the two following steps:

- a first tightening step: the wrench shall be set to a torque value of about 75% of the torque reference values. This first step shall be completed for all bolts in one connection prior to commencement of the second step;
- a second tightening step: the wrench shall be set to a torque value of 110% of the torque reference values;

where the torque reference values must be declared by the fastener manufacturer in accordance with the relevant parts of EN 14399 (CEN, 2015) or determined by tests under site conditions according to the procedures given in Annex H of EN 1090-2.

– Combined method

Tightening by the combined method comprises two steps:

- a first tightening step, using a torque wrench offering a suitable operating range (see also *torque method* described before). The wrench shall be set to a torque value of about 75% of the torque reference values. This first step shall be completed for all bolts in one connection prior to commencement of the second step;
- a second tightening step in which a specified partial turn is applied to the turned part of the assembly. The position of the nut relative to the bolt threads shall be marked after the first step, using a marking crayon or marking paint, so that the final rotation of the nut relative to the thread in this second step can be easily determined. Further rotation to be applied, during the second step of tightening depends on total thickness of parts to be connected. Values are given in Table 3.8 extracted from EN 1090-2, see. Where the surface under the bolt head or nut (allowing for example for taper washers) is not perpendicular to the bolt axis, the required angle of rotation should be determined by testing.

3.4 DESIGN OF THE BASIC COMPONENTS

Table 3.8 – Additional rotation for the combined method

Total nominal thickness t of parts to be connected (including all packs and washers)	Further rotation to be applied, during the second step of tightening	
	Degrees	Part turn
$t < 2d$	60	1/6
$2d \leq t < 6d$	90	1/4
$6d \leq t \leq 10d$	120	1/3

Many other systems to control the preload in bolts have been tested. Most popular ones are "HRC systems" (bolt and nut assemblies with calibrated preload) and "direct tension indicators". Their use is very simple, but the scatter of the preload magnitude is large. They represent an alternative to the direct methods described before.

d) HRC systems

HRC assemblies have a torque control groove and a torque control spline at the end of the threaded part as seen in shown in Figure 3.8

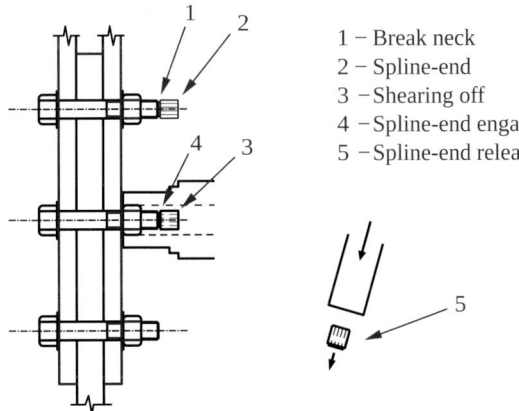

1 – Break neck
2 – Spline-end
3 – Shearing off
4 – Spline-end engagement
5 – Spline-end release

Figure 3.8 – HRC systems: Principle of tightening

When the torque reaches a definite value determined by the groove, the end of the screw breaks and the tightening is stopped. This type of bolt has advantages:

- it eliminates the need for a back-up man on the bolt screw.

3. Connections with Mechanical Fasteners

- it removes the possibility of operator error and checking is very rapid.
- the tool requires no calibration and tightening is not affected by field conditions, at least as long as the requirements in terms of slip factor are satisfied.

On the other hand, the reduction in preload could be rather large. To avoid such an unexpected event, a new European standard EN 14399-10 has been introduced where requirements for HRC systems are stated (CEN, 2009b).

e) Direct tension indicators

A direct tension indicator is an especially hardened washer with protrusions on one face; it is illustrated in Figure 3.9.

Figure 3.9 – Direct tension indicator

The protrusions bear against the underside of the bolt head leaving a gap. As the bolt is tightened the protrusions are flattened and the gap reduced. At a specified average gap, measured by feeler gauge, the induced shank tension will not be less than the minimum required by the standards. Figure 3.10 shows a standard assembly in place before and after tightening the bolt.

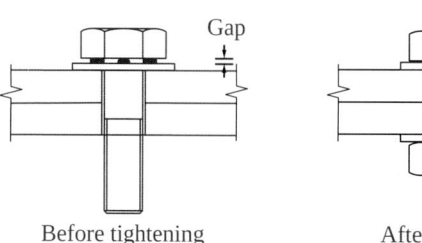

Before tightening   After tightening

Figure 3.10 – Principle of tightening with a direct tension indicator

All the bolts shall be tightened in accordance with the manufacturer's instructions and to the required tension indicator gap recommended by the manufacturer, as verified by calibration tests.

Tightening shall be carried out progressively from the middle of the joint to the free edges, to avoid loosening previously tightened fasteners.

Tightening to the required gap shall be carried out in two stages to ensure consistency of preloading. In the first stage 75% of the deformation of the protrusions of the tension indicator shall be reached. In the second stage of final tightening the required gap shall be reached.

Bolt and nut assemblies with direct tension indicators are standardised in EN 14399-9 (CEN, 2009a).

3.4.5 Plates in bearing

In bearing type connections, the internal force in the plate is transferred to the bolt by hole bearing. The design bearing resistance is determined as follows:

$$F_{b,Rd} = k_1 \alpha_b f_u dt / \gamma_{M2} \qquad (3.16)$$

where

k_1 = $\min(2.8 e_2/d_0 - 1.7; 1.4 p_2/d_0 - 1.7; 2.5)$ for edge bolts or $\min(1.4 p_2/d_0 - 1.7; 2.5)$ for inner bolts

α_b = $\min(\alpha_d; f_{ub}/f_u; 1.0)$
where $\alpha_d = e_1/3d_0$ for end bolts or $\alpha_d = p_1/3d_0 - 0.25$ for inner bolts

It should be noted that Eq. (3.16) covers two failure modes, see Figure 3.11. For small end distances or small bolt pitches in the direction of load transfer, the plate can fail by plate shear (Figure 3.11a). Otherwise it will fail by elongation of the hole (Figure 3.11b). Plate shear failure is taken into account by the factor α_d (Figure 3.11c). The factor k_1 considers that the bearing resistance is reduced if the edge distance e_2 or hole distance p_2 perpendicular to the direction of load transfer is small. If e_2 and p_2 are large enough, the full bearing capacity can develop.

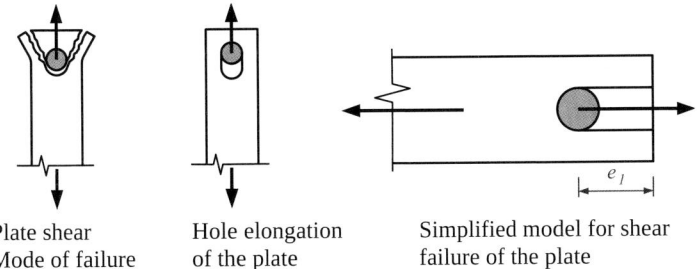

Plate shear
Mode of failure

Hole elongation
of the plate

Simplified model for shear
failure of the plate

Figure 3.11 – Failure modes for a plate in bearing

3. CONNECTIONS WITH MECHANICAL FASTENERS

The bearing resistance is calculated using the ultimate resistance of the plate. However, when the ultimate strength of the bolt is weaker than the ultimate strength of the plate, the bearing resistance is reduced to account for this situation.

Note that in EN 1993-1-8 this component is called "bolts in bearing on beam flange, column flange, end plate and cleat".

3.4.6 Block tearing

Block tearing failure, also called block shear failure can occur at a group of fastener holes, for example near the end of a beam web or a bracket. Figure 3.12 shows different situations of block tearing. It may be prevented by using appropriate hole spacing. Block tearing generally consists of tensile rupture along the line of fasteners on the tension face of the hole group and by yielding in shear at the row of fastener holes along the shear face of the hole group, see Figure 3.12.

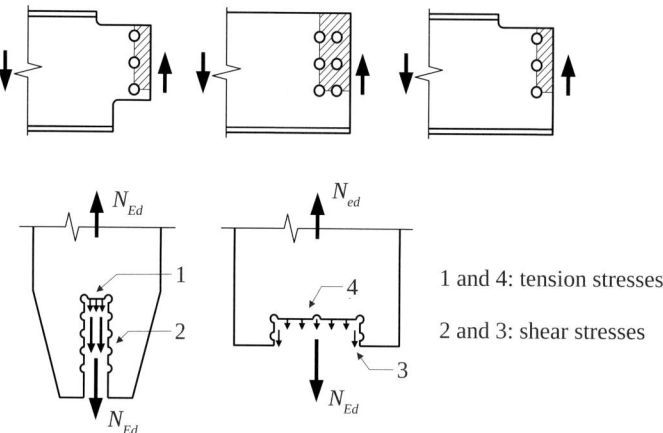

1 and 4: tension stresses
2 and 3: shear stresses

Figure 3.12 – Block tearing failure

The design block tearing resistance is for a symmetric bolt group
– subject to concentric loading:

$$V_{eff,1,Rd} = A_{nt}f_u/\gamma_{M2} + A_{nv}f_y/\sqrt{3}/\gamma_{M0} \qquad (3.17)$$

– subject to eccentric loading:

$$V_{eff,1,Rd} = 0.5A_{nt}f_u/\gamma_{M2} + A_{nv}f_y/\sqrt{3}/\gamma_{M0} \qquad (3.18)$$

where
- A_{nt} is net area subjected to tension
- A_{nv} is net area subjected to shear

3.4.7 Injection bolts

Injection bolts are bolts in which the cavity resulting from the clearance between the bolt shank and the wall of the plate is completely filled up with a two component resin (ECCS, 1994). After tightening the bolt, through a little hole in the head of the bolt, the resin in "injected" in the cavity, see Figure 3.13 (ECCS, 1994). The resin, after complete curing, has a defined minimum resistance and the connection becomes a slip resistant connection.

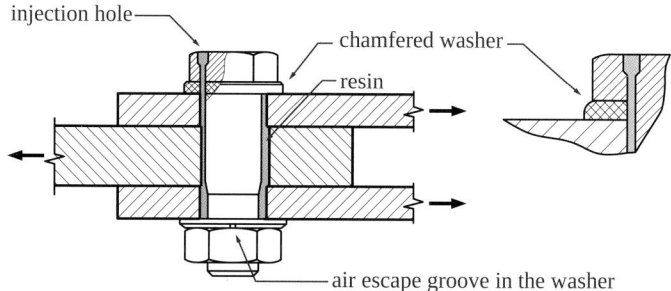

Figure 3.13 – Injection bolts in a double lap joint

Eurocode 3 Part 1-8 provides design rules for injection bolts. These ones should be of class 8.8 or 10.9 and may be non-preloaded or preloaded, i.e. bolts can be designed for Category A, B or C connections. For a connection with injection bolts, the design bearing resistance is taken as the design bearing resistance of the resin:

$$F_{b,Rd,resin} = k_t k_s d t_{b,resin} \beta f_{b,resin} / \gamma_{M4} \qquad (3.19)$$

where
- k_t is 1.0 for serviceability limit state (long duration) or 1.2 for ultimate limit state
- k_s is taken as 1.0 for holes with normal clearances, for oversized holes see 3.6.2.2 of EN 1993-1-8
- $t_{b,resin}$ is the effective bearing thickness of the resin, see Table 3.9

3. CONNECTIONS WITH MECHANICAL FASTENERS

β is a coefficient depending of the thickness ratio of the connected plates, see Table 3.9

$f_{b,resin}$ is the bearing strength of the resin. No values are given in EN 1993-1-8. Reference is made to EN 1090-2. Annex K of EN 1090-2 states that the design bearing strength of the resin should be determined similar to the procedure for the determination of the slip factor as specified in Annex G of EN 1090-2

Table 3.9 – Values for $t_{b,resin}$ and β

t_1/t_2	$t_{b,resin}$	β
≥ 2.0	$2t_2 \leq 1.5d$	1.0
$1.0 < t_1/t_2 < 2.0$	$t_1 \leq 1.5d$	$1.66 - 0.33(t_1/t_2)$
≤ 1.0	$t_1 \leq 1.5d$	1.33

Experimental tests (Gresnigt et al, 2000) have demonstrated that connections made with injection bolts have a very good long time behaviour. After a quite moderate relaxation of the resin, the deformation of bearing connections remains more or less constant. Even if the fabrication and erection of such connections requires a special attention, see Annex K of EN 1090-2, this type of fasteners could be an alternative of fitted bolts or rivets in order to build a slip resistant bolted connection without using preloaded bolts.

3.4.8 Pins

Pins are normally used in lap joints only subjected to shear forces, such as tension rods or pinned ended columns. A solid circular pin is placed through the holes of the connected elements. In contrast to a normal bolt, a pin has no head and no threaded part of its shank. It is only secured to ensure that the pin will not become loose. As the pin has no head or nut, the connected plate will not necessary be clamped together and a gap may remain between the connected plates. Therefore, a pin ended members can behave as a perfect hinge allowing for significant rotations. On the other side, bending moment may rise in the pin which needs to be taken into account in the design.

In pin-connected members the geometry of the unstiffened element that contains a hole for the pin should satisfy the dimensional requirements given in Figure 3.14.

Type A: Given thickness t

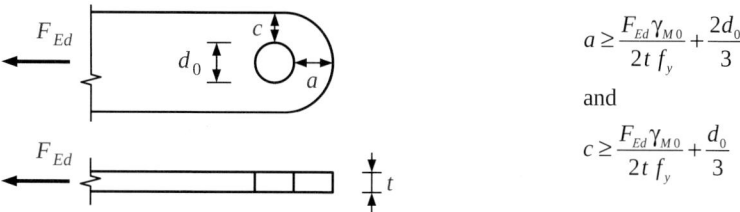

$$a \geq \frac{F_{Ed}\gamma_{M0}}{2t f_y} + \frac{2d_0}{3}$$

and

$$c \geq \frac{F_{Ed}\gamma_{M0}}{2t f_y} + \frac{d_0}{3}$$

Type B: Given geometry

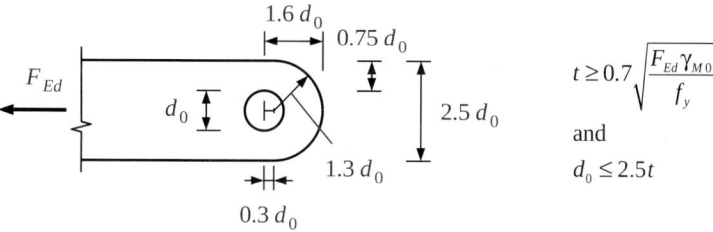

$$t \geq 0.7\sqrt{\frac{F_{Ed}\gamma_{M0}}{f_y}}$$

and

$$d_0 \leq 2.5t$$

Figure 3.14 – Geometrical requirements for pin ended members

Furthermore, pin connected members should be arranged such to avoid eccentricity and should be of sufficient size to distribute the load from the area of the member with the pin hole into the member away from the pin.

Design rules can be found in Section 3.13 of Eurocode 3 Part 1-8. They are repeated here below. As an only exception, Eurocode 3 allows also designing a pin as a single bolted connection as long as no significant rotation is required and the length of the pin is less than 3 times its diameter. Trusses constitute a potential application field for such a "bolt behaviour".

The design requirements for solid circular pins are given in Table 3.10. The moments in the pin are calculated on the basis that the connected parts form simple supports and it is generally assumed that the reactions

3. CONNECTIONS WITH MECHANICAL FASTENERS

between the pin and the connected parts are uniformly distributed along the length in contact on each part as indicated in Figure 3.15.

Table 3.10 – Design formulae for pin connections

Failure mode	Design requirements
Shear resistance of the pin	$F_{v,Rd} = 0.6 A f_{up} / \gamma_{M2} \geq F_{v,Ed}$
Bearing resistance of the plate and the pin	$F_{b,Rd} = 1.5 t d f_y / \gamma_{M0} \geq F_{b,Ed}$
If the pin is intended to be replaceable this requirement should also be satisfied.	$F_{b,Rd,ser} = 0.6 t d f_y / \gamma_{M6,ser} \geq F_{b,Ed,ser}$
Bending resistance of the pin	$M_{Rd} = 1.5 W_{el} f_{yp} / \gamma_{M0} \geq M_{Ed}$
If the pin is intended to be replaceable this requirement should also be satisfied.	$M_{Rd,ser} = 0.8 W_{el} f_{yp} / \gamma_{M6,ser} \geq M_{Ed,ser}$
Combined shear and bending resistance of the pin	$\left[\dfrac{M_{Ed}}{M_{Rd}}\right]^2 + \left[\dfrac{F_{v,Ed}}{F_{v,Rd}}\right]^2 \leq 1$

d is the diameter of the pin;
f_y is the lower of the design strengths of the pin and the connected part;
f_{up} is the ultimate tensile strength of the pin;
f_{yp} is the yield strength of the pin;
t is the thickness of the connected part;
A is cross-sectional area of the pin.

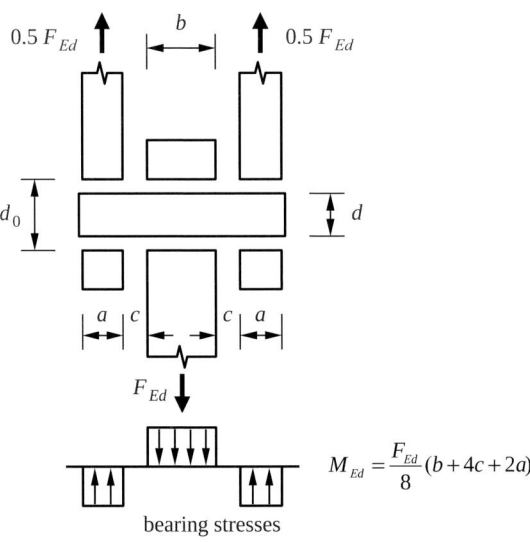

Figure 3.15 – Design bending moment M_{Ed} in a pin

3.4 DESIGN OF THE BASIC COMPONENTS

If the pin is intended to be replaceable, in addition to the provisions given before, the contact bearing stress should satisfy:

$$\sigma_{h,Ed} \leq f_{h,Rd} \quad (3.20)$$

with:

$$\sigma_{h,Ed} = 0.591\sqrt{\frac{EF_{Ed,ser}(d_0 - d)}{d^2 t}} \quad (3.21)$$

$$f_{h,Rd} = 2.5 \frac{f_y}{\gamma_{M6,ser}} \quad (3.22)$$

In these equations:
- d is the diameter of the pin;
- d_0 is the diameter of the pin hole;
- $F_{Ed,ser}$ is the design value of the force to be transferred in bearing, under the characteristic load combination for serviceability limit states.

3.4.9 Blind bolting

3.4.9.1 Flow drill blind bolting

The flow drill system is a special patented method for extruded holes. Flow drilling is a thermal drilling process to make a hole through a steel plate, for example the wall of hollow section as shown in Figure 3.16, by bringing a tungsten carbide bit into contact with steel plate and generating sufficient heat by friction to soften the steel. The bit moves through the wall, the metal flows to form an internal bush. In the subsequent step, the bush is threaded using a roll tap (Kurobane et al, 2004). Flow drill process is illustrated in Figure 3.17.

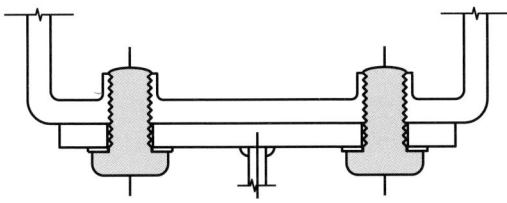

Figure 3.16 – Flow drill connection for joining end plates to RHS

3. CONNECTIONS WITH MECHANICAL FASTENERS

Note that flow drill blind bolting is not mentioned in Eurocode 3. It must be checked with the local building authorities if this type of bolting may be used when a structure is designed to Eurocode 3.

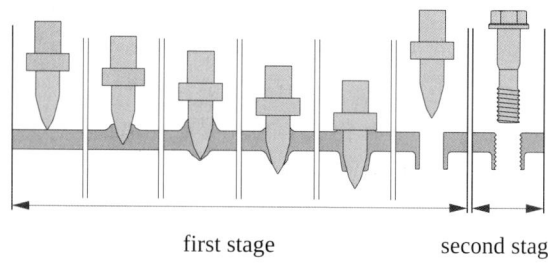

first stage　　　　　　second stage

Figure 3.17 – Flow drill process

(Kurobane *et al*, 2004) provides practical guidelines on flow drill blind bolting:

- wall thicknesses up to 12.5 mm can be recommended by using the flow drill method;
- threaded roll tapped holes for bolt sizes M16, M20 and M24 can be made;
- the full tension capacity of grade 8.8 bolts can be carried, provided that the plate thickness is equal to or greater than the minimum thickness shown in Table 3.11 and f_y is between 275 N/mm² and 355 N/mm²;

Table 3.11 – Minimum plate thickness

Bolt size	Minimum plate thickness [mm]
M16 grade 8.8	6.4
M20 grade 8.8	8.0
M24 grade 8.8	9.6

- the shear and bearing capacities of the hole can be calculated in the normal manner.
- In most applications in which the bolts are loaded in tension, the deformation or yielding of the steel plate will determine the connection capacity and not the capacity of each individual bolt.

For more details the reader may consult (Kurobane *et al*, 2004)

3.4.9.2 SHS Blind bolting connections

In recent years, other systems than the flow drill ones have been developed for connections with one side access only.

A number of patented blind bolting systems is available, e.g. Huck "Ultra Twist Blind Bolt", Lindapter "HolloFast", see Figure 3.18 (Wardenier *et al*, 2010) and "HolloBolt".

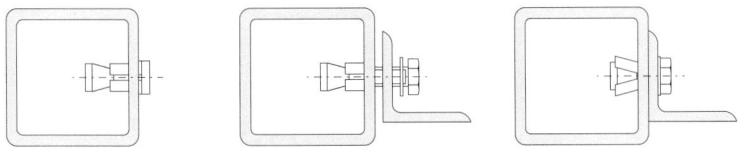

Figure 3.18 – Lindapter "HolloFast"

The systems are based on the principle that after bringing them in from one side the bolts are torqued and a "bolt head" forms on the inside of the connection.

The design rules for blind bolting systems are based on typical failure modes, i.e.:

– Punching shear of the fastener through the column face;
– Yielding of the column face;
– Bolt failure in shear, tension or a combination of both.

By means of these different systems, beams with end plates are directly connected to column faces by tension bolts. Sometimes the column walls have to be reinforced to prevent local distortion of the column walls or when full strength connections are required. The column walls may be partially thickened over the areas where the end plates are attached.

3.4.10 Nails

In some cases nails can be used to form reliable structural connections; for example, circular hollow sections (CHS) can be nailed together as illustrated in Figure 3.19 (Wardenier *et al*, 2010) for a splice connection between two co-axial tubes,. In such a connection, one tube can fit snugly inside the other.

3. CONNECTIONS WITH MECHANICAL FASTENERS

Figure 3.19 – Nailed connection

The observed failure modes are:

- Nail shear failure;
- Tube bearing failure;
- Net section fracture of the tube.

Simple design formulae, derived for bolted and riveted connections, are proposed in (Packer, 1996).

3.4.11 Eccentricity of angles

Joints are aimed at transferring the internal forces obtained by means of a structural frame analysis. Local eccentricities in the joints may however generate extra internal forces. The joints and members should be designed for the total resulting moments and forces, except in the case of particular types of structures where it has been demonstrated that it is not necessary; for instance in lattice girders, provided some conditions are met (see EN 1993-1-8).

One of the common situations where such eccentricities are to be considered is the case of an angle connected on one leg only by one or more lines of bolts; in-plane and out-of-plane eccentricities are here identified by considering the relative positions of the centroidal axis of the member and of the setting out line (see Figure 3.20).

3.4 DESIGN OF THE BASIC COMPONENTS

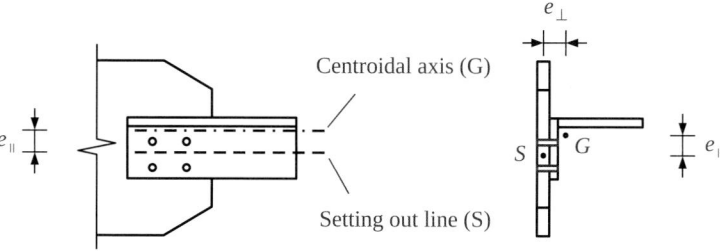

$e_{\parallel}$: eccentricity parallel to the gusset
$e_{\perp}$: eccentricity perpendicular to the gusset

Figure 3.20 – Local eccentricities in a bolted angle

For the specific case of angles connected through one line of bolts and subjected to an axial tension force only, EN 1993-1-8 provides a simplified design method which consists in virtually replacing the actual member by a concentrically loaded one (i.e. loaded along the centroidal axis), but characterised by an effective area. The design resistance of the member in the net section is then defined as follows (Figure 3.21):

one bolt:
$$N_{u,Rd} = \frac{2.0(e_2 - 0.5d_0)tf_u}{\gamma_{M2}} \qquad (3.23)$$

two bolts in line:
$$N_{u,Rd} = \frac{\beta_2 A_{net} f_u}{\gamma_{M2}} \qquad (3.24)$$

three bolts in line:
$$N_{u,Rd} = \frac{\beta_3 A_{net} f_u}{\gamma_{M2}} \qquad (3.25)$$

In these equations:

β_2, β_3 are reduction factors dependent on the pitch p_1 distance as given in Table 3.12. For intermediate values of p_1, the value of β may be determined by linear interpolation;

A_{net} is the net area of the angle. For an unequal-leg angle connected by its smaller leg, A_{net} should be taken as equal to the net section area of an equivalent equal-leg angle of leg size equal to that of the smaller leg.

3. Connections with Mechanical Fasteners

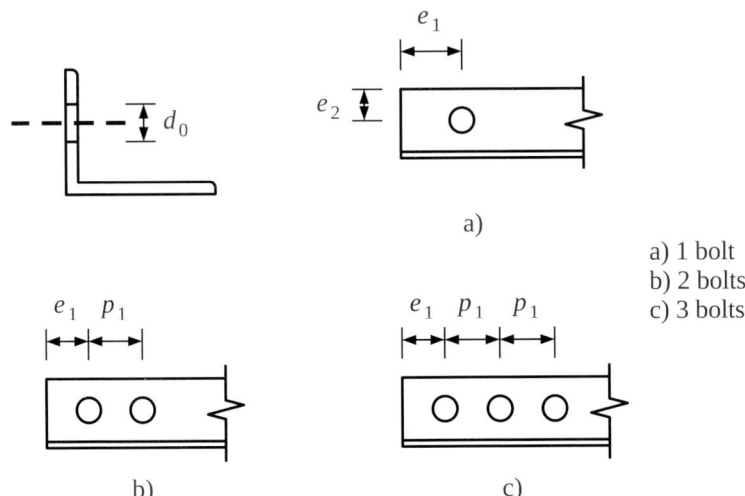

a) 1 bolt
b) 2 bolts
c) 3 bolts

Figure 3.21 – Angle connected by one leg (one bolt line)

Table 3.12 – Reduction factors β_2 and β_3

Pitch	p_1	$\leq 2.5 d_0$	$\geq 5.0 d_0$
2 bolts	β_2	0.4	0.7
3 bolts or more	β_3	0.5	0.7

3.5 DESIGN OF CONNECTIONS

3.5.1 Bolted lap joints

3.5.1.1 Introduction

The term "lap joint" designates a basic connection aimed at transferring forces between plates subjected to tension. One usually distinguishes single and double overlap joints (see Figure 3.22). To achieve this force transfer, various connection elements may be used; but in practice, bolts (preloaded or not) and welds (lateral and/or front ones) represent the more common connection means (see Figure 3.23).

3.5 DESIGN OF CONNECTIONS

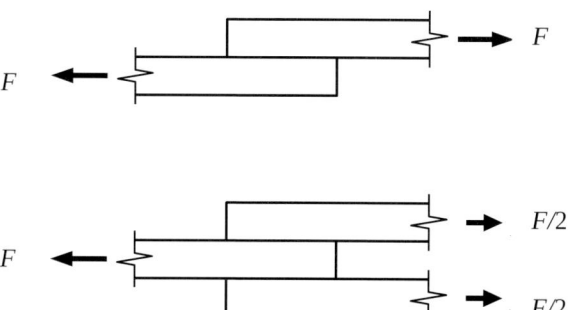

Figure 3.22 – Single and double overlap joints (lap joints)

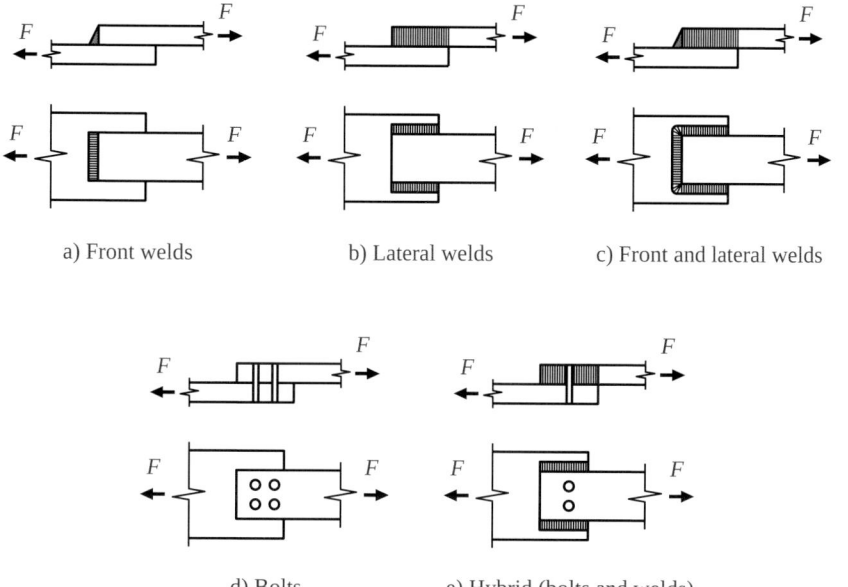

Figure 3.23 – Bolted and welded lap joints

In the next sections, the behaviour and the design of bolted lap joints are discussed. For welded and hybrid lap joints, reference should be respectively made to sub-chapter 4.4 and section 1.6.3.

3.5.1.2 Joints with non-preloaded bolts

In such joints, the forces are transferred from one plate to the other (the others in a double overlap joint) by bolt-plate contact (Figure 3.24).

3. CONNECTIONS WITH MECHANICAL FASTENERS

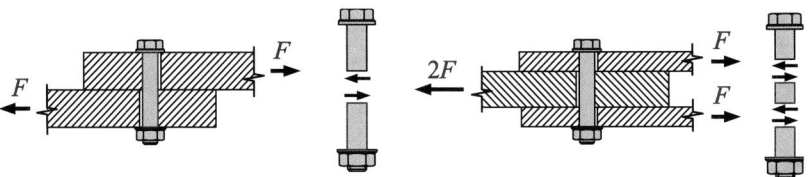

Figure 3.24 – Single and double bolted overlap joints

If the small friction effects between the plates and the bending forces in the bolts are neglected, the resistance of the joint may be limited by one of the following components:

- the bearing resistance of the plates and/or bolts;
- the shear resistance of the bolts (one or two shear planes, respectively for single or double overlap joints);
- the tension resistance of the plates.

The design resistance of a bolt in shear and of a plate/bolt in bearing has been defined in sections 3.4.1 and 3.4.5 respectively. For the design resistance of a plate in tension, reference is made to EN 1993-1-1 or to the ECCS Design Manual on "Design of Steel Structures" (Simões da Silva *et al*, 2010) where expressions are provided for the design of the gross and net sections.

Locally, i.e. around a bolt, the force which may be transferred can never exceed the design resistance of the weaker of the two first here-above mentioned components.

As far as the whole joint is now concerned (with more than one bolt-row), the force transferred between the plates is known to be non-uniformly distributed amongst the bolt-rows (see distribution of bolt forces C_i in Figure 3.25a) (Ju *et al*, 2004). If sufficient deformation is provided around each bolt zone, a full plastic redistribution of forces may be noticed, otherwise failure is reached by lack of ductility and the maximum external force to be transferred is lower than the one corresponding to a full plastic distribution. Schematically, the different stages of forces distribution in a shear bolted connection may be represented as in Figure 3.25.

3.5 DESIGN OF CONNECTIONS

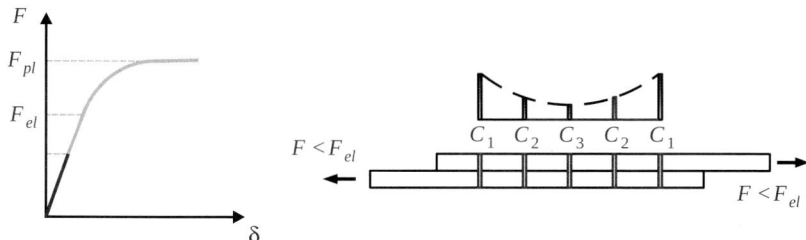

a) None of the bolt rows yields

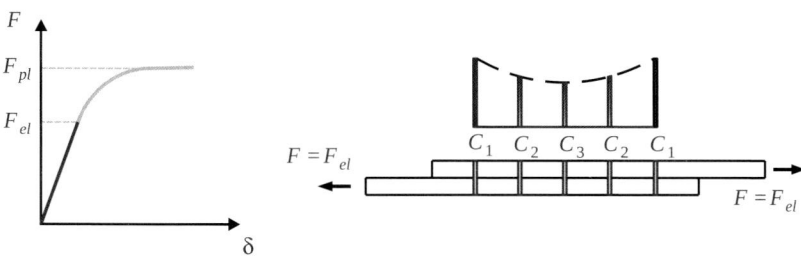

b) Outer bolt rows reach yielding (elastic resitance of the connection)

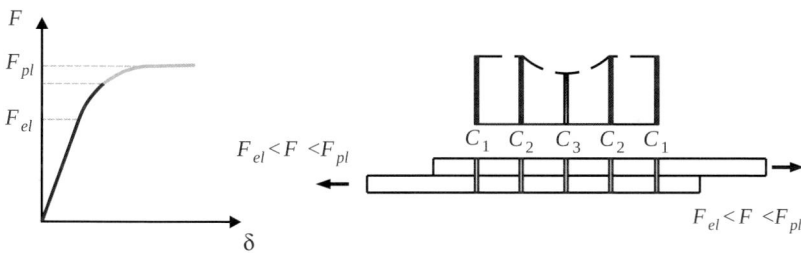

c) Progressive yielding of bolt rows from outside to inside (plastic redistribution)

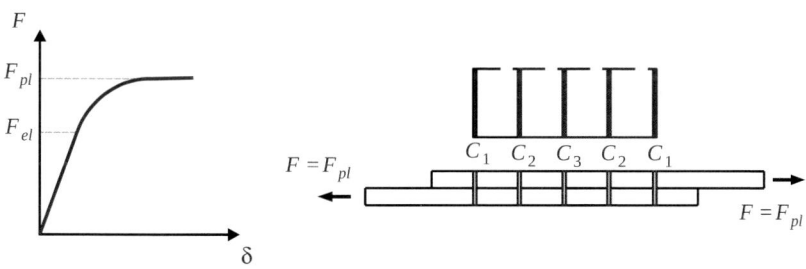

d) All bolt rows yield (full plastic resistance of the connection)

Figure 3.25 – Different stages of bolt force distribution in shear bolted connections

In Eurocode 3 Part 1-8, the design resistance of the lap joint is defined as the smaller of the two following values:

- the design resistance of the weakest plate(s) in tension
- the design resistance of the group of fasteners (the bolts)

The design resistance of the group of fasteners may be taken as the sum of the design bearing resistances $F_{b,Rd}$ of the individual fasteners provided that the design shear resistance $F_{v,Rd}$ of each individual fastener (rather brittle failure mode) is greater than or equal to the local design bearing resistance $F_{b,Rd}$ (rather ductile failure mode). This fulfilment of this criterion guarantees a sufficient deformation capacity to achieve a full redistribution of forces between the fasteners. The plastic resistance of the joint may so be reached.

Otherwise the design resistance of a group of fasteners should be taken as the number of fasteners multiplied by the smallest design resistance of any of the individual fasteners, what may be considered as a way to limit the plastic redistribution of yielding along the bolt rows.

Besides that, Eurocode 3 also prevents to profit from a full redistribution of plasticity in the case of so-called "long joints". This aspect is addressed in section 3.5.4.

3.5.1.3 Joints with preloaded bolts

Two categories of connections have here to be considered:
- slip resistance at ultimate limit state (Category C)
- slip resistance at serviceability limit state (Category B)

In the case of Category B connections, the design resistance of the lap joint is evaluated as specified in section 3.5.1.1.

For Category C, the design resistance identifies itself to the slip resistance; this one is simply obtained by multiplying the design slip resistance of a bolt by the number of bolts in the joints, see Eq. (3.13).

As already mentioned in section 3.2.1c), the resistance of the net cross section of the plates at the bolt holes has to be checked in a different way for category C than for the two others, see Eq. (3.5).

3.5.2 Bolted T-stubs

3.5.2.1 Generalities

The transfer of tensile forces in bolted connections is often achieved through the following basic components: column flange in bending, end plate in bending or cleat leg in bending, i.e. plates subjected to transverse forces.

The evaluation of the resistance of such plated components is based on a geometrical idealisation of the tension zone, known as the "T-stub" idealisation. A T-stub is, as the name indicates, a T profile made of a web in tension and a flange in bending; the latter is assumed to be bolted to a rigid foundation (Figure 3.26).

The reason why reference is made to such a model is illustrated in Figure 3.27 where the constitutive T-stubs of some bolted connections are visualised. In practice, the response of the actual plated components will be evaluated through the study of corresponding equivalent T-stubs; the equivalence will be reached through the definition of an appropriate length of the equivalent T-stubs, called l_{eff}.

The use of the T-stub has the advantage that the different above named plated components in bending may be studied with the same model and thus with the same design formulae, whatever the considered mechanical property (stiffness, resistance, ductility).

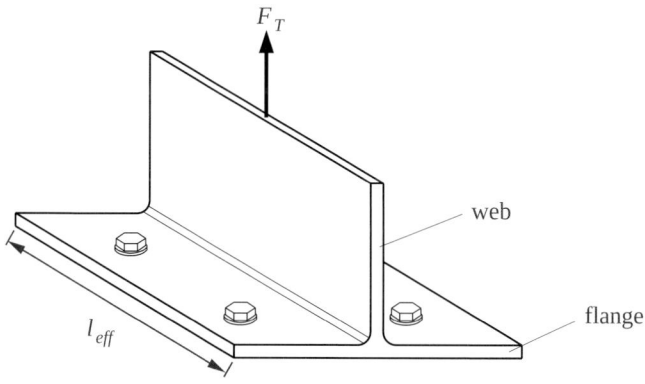

Figure 3.26 – T-stub geometry

3. Connections with Mechanical Fasteners

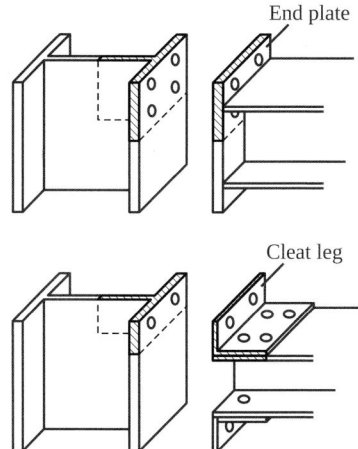

Figure 3.27 – Visualisation of equivalent T-stubs in bolted connections

3.5.2.2 Design resistance

Zoetemeijer (1974) developed initially the T-stub model for unstiffened plated components and then extended it to some stiffened configurations. Later on Jaspart (1991) applied the concept to cleated connections before extending it again further to various plate configurations (Jaspart, 1997).

In terms of resistance, three different failure modes may be contemplated according to the geometry and mechanical response of the constitutive plates and bolts, see Figure 3.28:

- Failure mode 1: Yielding of the flange

    The resistance is associated to the formation of a plastic yield mechanism in the flange. In such a case, the bolts are sufficiently strong to resist to the applied axial tension forces, including the prying forces Q.

- Failure mode 2: Failure of the bolts further to a partial yielding of the flange

    Mixed failure is achieved through the formation of yield lines in the flange (the full plastic mechanism being not reached) and the failure of the bolts in tension (again including prying effects).

- Failure mode 3: Failure of the bolts

    The resistance is linked to the failure of the bolts in tension. The deformation of the flange in bending is small, resulting in an absence of prying effects.

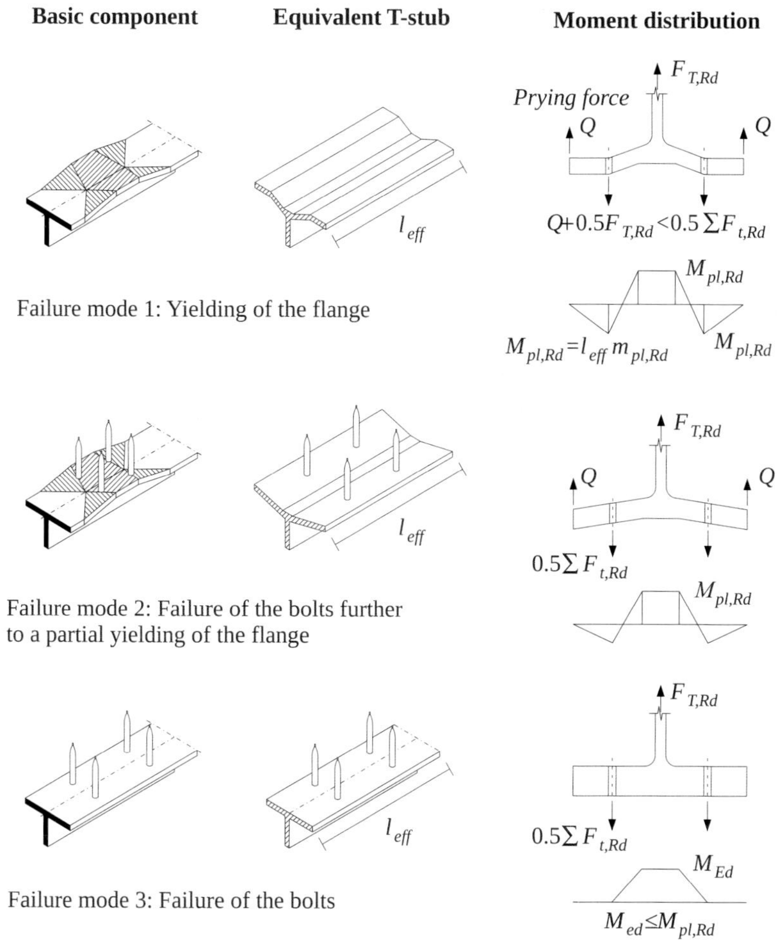

3.5 DESIGN OF CONNECTIONS

Basic component	Equivalent T-stub	Moment distribution
Failure mode 1: Yielding of the flange	l_{eff}	$F_{T,Rd}$, Prying force Q, $Q + 0.5F_{T,Rd} < 0.5\Sigma F_{t,Rd}$, $M_{pl,Rd} = l_{eff} m_{pl,Rd}$, $M_{pl,Rd}$
Failure mode 2: Failure of the bolts further to a partial yielding of the flange	l_{eff}	$F_{T,Rd}$, Q, $0.5\Sigma F_{t,Rd}$, $M_{pl,Rd}$
Failure mode 3: Failure of the bolts	l_{eff}	$F_{T,Rd}$, $0.5\Sigma F_{t,Rd}$, M_{Ed}, $M_{ed} \leq M_{pl,Rd}$

Figure 3.28 – Failure modes in the actual component and in the equivalent T-stubs

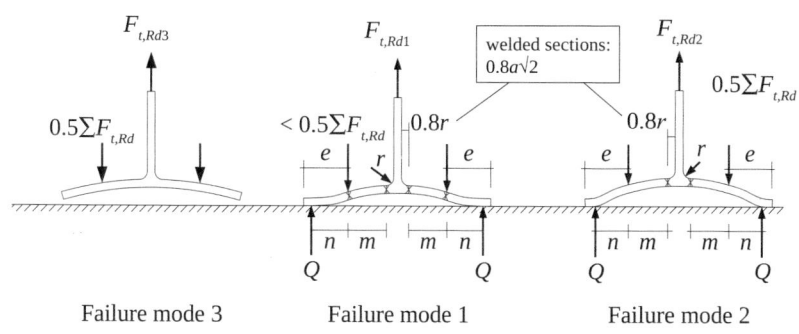

Figure 3.29 – Failure modes of an equivalent T-stub

3. Connections with Mechanical Fasteners

These failure modes occur similarly in both the actual component and in the equivalent T-stub. The effective length l_{eff} is defined such that the carrying capacity of the T-stub and of the actual component is the same. The evaluation of the resistance of the actual component may so be replaced by the study of the equivalent T-stub. As will be shown later, the T-stub model can also be used as a reference in determining the stiffness of the basic components.

Eurocode 3 Part 1-8 defines the design resistance of the T-stub, for each failure mode, as follows (Figure 3.29):

– Mode 1: Yielding of the flange

$$F_{T,Rd,1} = \frac{4\, l_{eff,1}\, m_{pl,Rd}}{m} \qquad (3.26)$$

– Mode 3: Failure of the bolts

$$F_{T,Rd,3} = \sum F_{t,Rd} \qquad (3.27)$$

– Mode 2: Failure of the bolts further to a partial yielding of the flange

$$F_{T,Rd,2} = \frac{2\, l_{eff,2}\, m_{pl,Rd} + \sum F_{t,Rd}}{m+n} \qquad (3.28)$$

where:

$m_{pl,Rd} = \dfrac{t^2 f_y}{4 \gamma_{M0}}$ is the plastic bending resistant of the T-stub flange per unit length

t is the thickness of the T-stub-flange

f_y is the yield strength of the T-stub-flange

γ_{M0} is the partial safety factor for the resistance of the flange (recommended value: $\gamma_{M0} = 1.0$)

m and e are geometrical dimensions characterising the position of the potential yield lines in the T-stub flange (see Figure 3.29)

$\sum F_{t,Rd}$ is the sum of the design resistances $F_{t,Rd} = \dfrac{0.9\, A_s\, f_{ub}}{\gamma_{M2}}$ of all the bolts in the T-stub

A_s is the tensile stress area of the bolts

3.5 DESIGN OF CONNECTIONS

f_{ub} is the yield strength of the bolts

γ_{M2} is the partial safety factor for the resistance of the bolts (recommended value $\gamma_{M0} = 1.25$)

n is a geometrical dimension characterising the position of the prying force: $n = e_{min}$, but with a limitation to $1.25\,m$

e_{min} is the smaller edge distance if two connected plates have different widths as show in Figure 3.30

$l_{eff,1}$, $l_{eff,2}$ minimum effective lengths of the yield lines for all the possible mechanisms in the T-stub, according to the failure modes.

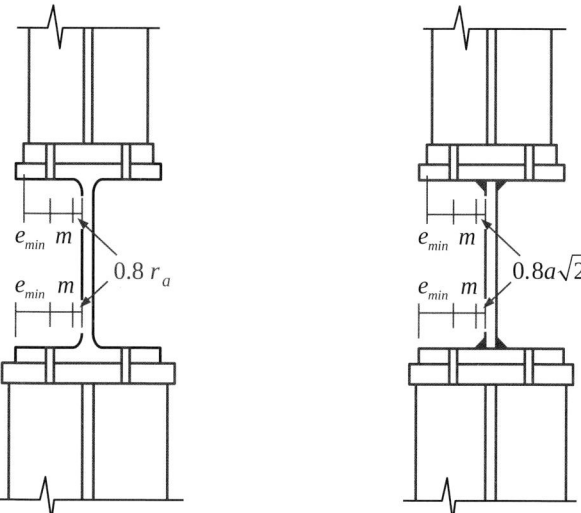

Figure 3.30 – Definition of e_{min} (for example in a beam-to-column joint)

In Figure 3.31a, so-called individual non-circular yield line patterns are illustrated. In other cases, depending on the geometry of the component, individual circular patterns could also appear (Figure 3.31b). Different values of effective lengths are associated to these two types of yield patterns.

The formation of non-circular yield patterns requires the development of prying forces in the T-stub. Failure modes 1, 2 or 3 may therefore occur, depending on the value of β, as shown in Figure 3.32. On the contrary, when circular patterns form, no prying forces develop in the T-stub and failure mode 2 can therefore not occur. In this case, a direct transition from failure mode 1 to failure mode 3 is contemplated (Figure 3.32).

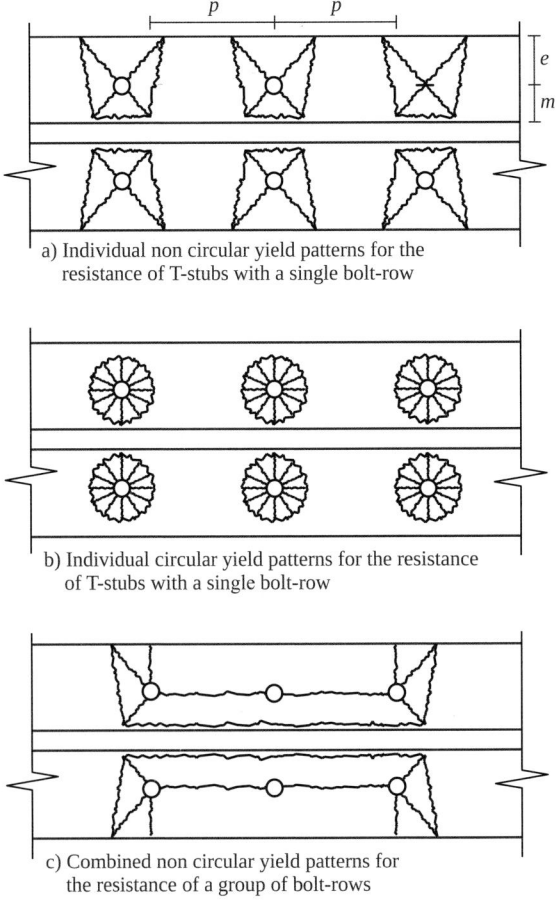

a) Individual non circular yield patterns for the resistance of T-stubs with a single bolt-row

b) Individual circular yield patterns for the resistance of T-stubs with a single bolt-row

c) Combined non circular yield patterns for the resistance of a group of bolt-rows

Figure 3.31 – Possible yield line mechanisms

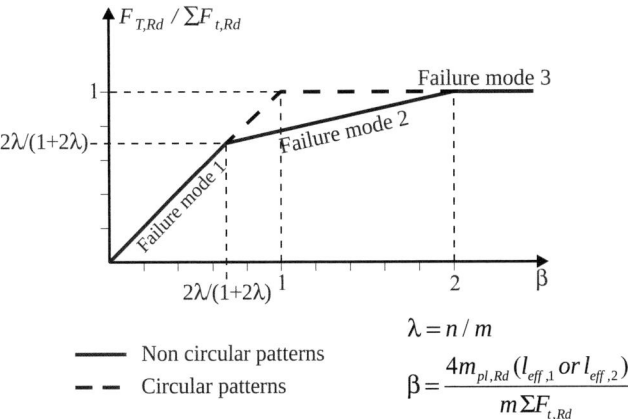

$\lambda = n/m$

— Non circular patterns
-- Circular patterns

$$\beta = \frac{4m_{pl,Rd}(l_{eff,1} \text{ or } l_{eff,2})}{m \Sigma F_{t,Rd}}$$

Figure 3.32 – Type of failure according to the geometry of the T-stub

The design resistance of the T-stub is the lower of the two following values:

- $F_{T,Rd,A} = \min\left(F_{T,Rd,1}; F_{T,Rd,2}; F_{T,Rd,3}\right)$ for non-circular patterns (3.29)
- $F_{T,Rd,B} = \min\left(F_{T,Rd,1}; F_{T,Rd,3}\right)$ for circular patterns (3.30)

Therefore:

$$F_{T,Rd} = \min\left(F_{T,Rd,A}; F_{T,Rd,B}\right) \quad (3.31)$$

In addition to considering T-stubs with a single bolt, as individual, the possibility exists to form yield line mechanism involving adjacent bolt-rows. Figure 3.31c illustrates the case of non-circular yield patterns extending over three bolt-rows. The same may also apply to circular patterns. In practice, the resistance of the T-stub involving a group of bolt-rows will always have to be compared to the sum of the design resistances of the individual T-stubs (involving one bolt-row only) and the lower of these two resistances will have to be kept as relevant.

Note that the resistance formulae in Eqs. (3.26) to (3.28) are only valid for a T-stub with two bolts per row. However, in practice, also configurations with four bolts per row are used. An extension of the design methods described in this section to a T-stub with four bolts per row can be found in Demonceau *et al* (2011).

3.5.2.3 Influence of the actual bolt dimensions on the design resistance

In Mode 1, a full plastic mechanism forms in the T-stub flange:

- two interior longitudinal yield lines at the toe of the root radius (or radius of fillet in a welded T-stub);
- two exterior longitudinal yield lines close to the bolts.

If the bolt force is assumed to be applied to the T-stub flange as a point load, as illustrated in Figure 3.33a, the exterior yield lines are straight and pass through the centre line of the bolts. Eq. (3.26) given above is fitted to cover this situation.

Should the bolts be infinitively stiff (Figure 3.33c), the two exterior yield lines would "move" towards the T-stub web, so resulting in a decrease of the *m* distance and therefore in an increase of the design

3. CONNECTIONS WITH MECHANICAL FASTENERS

resistance $F_{T,Rd,1}$. However the reality is somewhat different and Figure 3.33b shows that the actual yield lines will be conditioned by the geometrical dimensions of the screw head and washer, by the rigidity of the bolt and by the degree of bolt preloading. For the practical application of the T-stub model, the determination of the actual yield lines and the evaluation of the corresponding T-stub resistance would be too complicated and time-consuming. Eurocode 3 Part 1-8 therefore suggests an alternative method in which (Figure 3.33c):

- the assumption of straight yield lines passing through the bolt axes is kept;
- the positive local effect of the bolt head is considered by assuming that the bolt force is transferred to the T-stub flange in a distributed way, under the bolt head/washer (over a d_w width).

The distributed load on the T-stub, applied on a d_w width, may be further replaced by two concentrated forces acting at a e_w distance of the bolt line. The value of e_w is defined as follows:

$$e_w = \frac{d_w}{4} \tag{3.32}$$

where

d_w is the diameter of the washer, or the width across points of the head or nut, as relevant.

The refined alternative formula for Mode 1 writes then:

$$F_{T,Rd,1} = \frac{(8n - 2e_w) l_{eff,1}\, m_{pl,Rd}}{2mn - e_w(m+n)} \tag{3.33}$$

This one gives higher results than the one proposed in Eq. (3.26) above and has no specific limit of application. Jaspart (Jaspart, 1991) who developed the formulae also pointed out the possible benefit which would result, in addition, from the pre-stressing of the bolt. This extra benefit is presently not considered by Eurocode 3 Part 1-8.

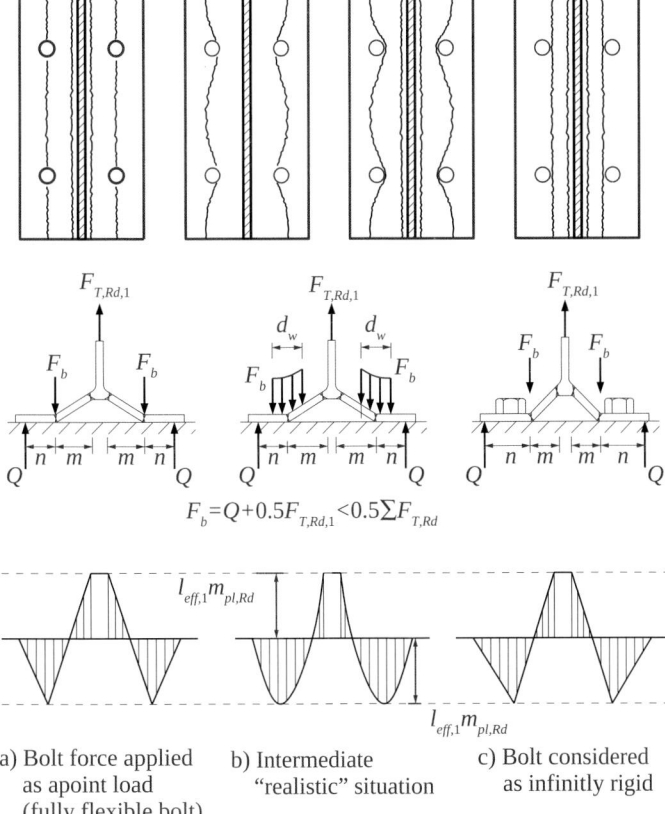

Figure 3.33 – Influence of the bolt geometry on the yield lines

3.5.2.4 Influence of the bolt/anchor length

In the previous sections, except for Mode 3 failure, prying forces develop at the interface between the T-stub flange and its foundation. This leads to overloading of the bolts in tension.

However in the case where the bolts would elongate to a significant extent, the prying forces could no more physically develop. In particular this may occur in column bases where the elongation of the anchor bolts could more than compensate the flexural deformation of the plate.

As a direct consequence, the validity of the design formulae presented in sections 3.5.2.2 and 3.5.2.3 becomes questionable. In EN 1993-1-8, a criterion has been included so as to identify situations where prying forces develop:

3. CONNECTIONS WITH MECHANICAL FASTENERS

$$L_b \leq \frac{8.8m^3 A_s n_b}{\sum l_{eff,1} t^3} \quad (3.34)$$

where

L_b is the bolt elongation length, taken equal to the grip length (total thickness of material and washers), plus half the sum of the height of the bolt head and the height of the nut

or

the anchor bolt elongation length, taken equal to the sum of 8 times the nominal bolt diameter, the grout layer, the plate thickness, the washer and half the height of the nut

n_b is the number of bolt rows (with 2 bolts per row)

When this criterion in not satisfied, Mode 1 and Mode 2 design formulae have to be substituted by a single design formula:

$$F_{T,1-2,Rd} = \frac{2M_{pl,1,Rd}}{m} \quad (3.35)$$

Note: Eq. (3.35) in the present version of EN 1993-1-8 is safe but not correct. In Table 6.2 of EN 1993-1-8, the value $M_{pl,1,Rd}$ is defined as

$$M_{pl,1,Rd} = 0.25 \sum \ell_{eff,1} t_f^2 f_y / \gamma_{M0} \quad (3.36)$$

where $\ell_{eff,1}$ is the effective length for Mode 1, i.e. the minimum of $\ell_{eff,cp}$ for circular patterns and $\ell_{eff,np}$ for non-circular patterns. While, in reality, only the patterns developing without prying forces need to be checked.

3.5.2.5 Direct applications to flange plate connections

3.5.2.5.1 RHS flange-plate connections in tension

A direct application of the T-stub model may be contemplated for various RHS flange-plate connections in tension.

Such a connection is shown in Figure 3.34a while, in Figure 3.34b, the way on how to idealise the connection as an equivalent T-stub is illustrated.

3.5 DESIGN OF CONNECTIONS

The evaluation of the tension resistance of the connection therefore simply resumes to the determination of the T-stub resistance.

A limitation however exists which may be expressed as follows: the distance d (Figure 3.34) between the two extreme bolts in a row should not exceed the width of the RHS profile. Would this condition not be fulfilled, specific yield mechanisms would appear in the plate corners, possibly involving the lateral not bolted sides of the plate. As a consequence, the T-stub, where only the bolted sides of the plate are considered, would not reflect the actual response of the connection.

Indication on how to overcome this "d distance" limitation may be found in (Demonceau *et al*, 2012)

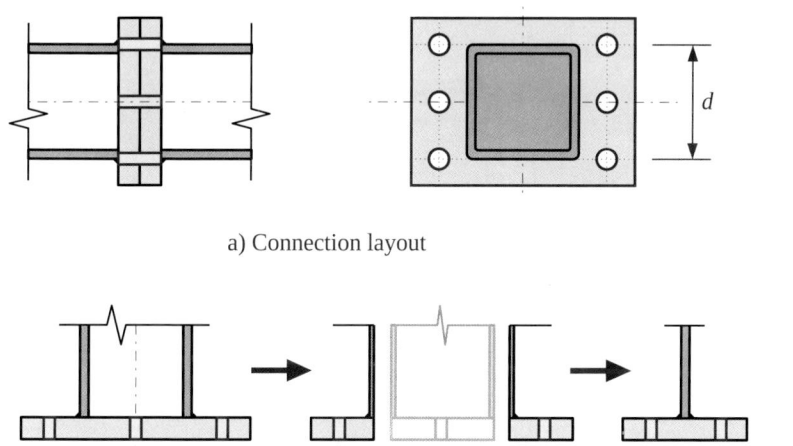

a) Connection layout

b) Substitution by an equivalent T-stub

Figure 3.34 – RHS flange-plate connection in tension (plate bolted on two sides)

Another example is provided in Figure 3.35a, where the plate is bolted on the four sides of the RHS profile. Similarly to what is said above, the connection may be modelled here by two equivalent T-stubs; and, in the case of a SHS profile, the resistance of the connection basically identifies to the sum of the resistance of the two T-stubs.

Should however the actual yield line patterns developing in the four extended parts of the plat merge into a single common one (see an example in Figure 3.35b, appropriate values of the effective lengths should be considered when evaluating the T-stub resistances.

3. CONNECTIONS WITH MECHANICAL FASTENERS

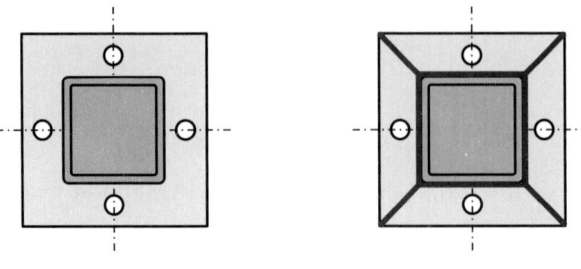

a) connection layout   b) Possible combined yield line pattern

Figure 3.35 – SHS flange-plate connection in tension (plate bolted on four sides)

3.5.2.5.2 CHS flange-plate connections in tension

In fact, Eurocode 3 Part 1-8 is not directly proposing recommendations for the design of bolted joints with circular flanges (Figure 3.36).

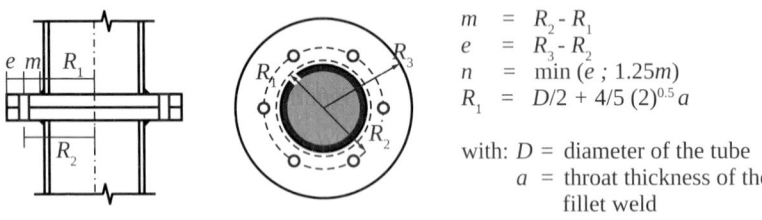

$$m = R_2 - R_1$$
$$e = R_3 - R_2$$
$$n = \min(e\,;\,1.25m)$$
$$R_1 = D/2 + 4/5\,(2)^{0.5}\,a$$

with: D = diameter of the tube
a = throat thickness of the fillet weld

Figure 3.36 – Bolted CHS flange-plate connection

However, in Hoang et al (2013), the concept of T-stub has been extended so to cover such joint configurations:

– Mode 1 (thin flanges – full plastic mechanism in the flange):

$$F_{T,Rd,1} = 2\pi m_{pl,Rd}\left[1 + \frac{R_1 + R_2}{m}\right] \quad (3.37)$$

– Mode 2 (intermediate flanges – yield lines in the flange + bolt failure):

$$F_{T,Rd,2} = \frac{2\pi m_{pl,Rd} + n\Sigma F_{t,Rd}}{m + n} \quad (3.38)$$

– Mode 3 (thick flanges – bolt failure):

$$F_{T,Rd,3} = F_{t,Rd} \quad (3.39)$$

3.5 DESIGN OF CONNECTIONS

These expressions are valid in the following field of application:

$$R_2 \leq N \cdot \min(2\pi m; 4m+1.25n)/2\pi \qquad (3.40)$$

In these formulae, the various parameters are defined according to Figure 3.36 and section 3.5.2.2; N is equal to the number of regularly spaced connecting bolts.

The failure mode of the connection is defined as the one corresponding to the minimum of three above defined design resistances.

3.5.3 Gusset plates

Gusset plates are commonly used in steel construction. The connected members are either bolted or welded to the gusset plate. But in many cases, both types of connectors are selected, as shown in Figure 3.37.

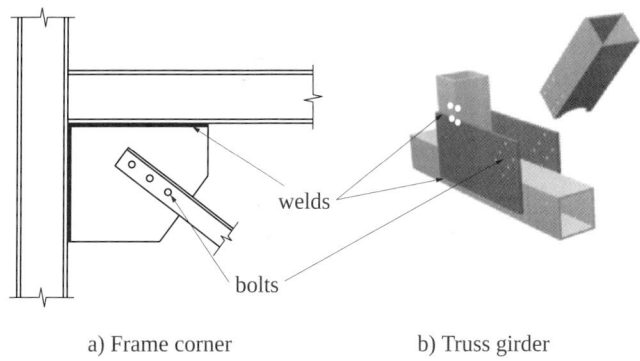

a) Frame corner　　　　　　b) Truss girder

Figure 3.37 – Examples of gusset plate connections using welds and bolts

In the vicinity of the gusset, several failure modes may occur at different locations:

- in the supported element(s);
- in the bolted or welded connection(s) between the supported member(s) and the gusset plate;
- in the gusset plate;

3. CONNECTIONS WITH MECHANICAL FASTENERS

- in the bolted or welded connection(s) between the gusset plate and the supported member(s);
- in the supported member(s).

In the present section, only the way on how the gusset plate resists to the applied forces is contemplated as all the other failure modes are covered elsewhere in chapter 3 and chapter 4.

In fact, two categories of failure modes are likely to develop in the gusset plate, independently of the kind of connectors used (bolts or welds):

- local failure in the gusset plate under the forces transferred by the supported member(s);
- global cross section failure of the gusset plate under the resultant of forces transferred by more than one of the supported member(s).

These ones are illustrated in Figure 3.38 in the particular case of a gusset plate connecting two diagonals and a main chord in a truss girder.

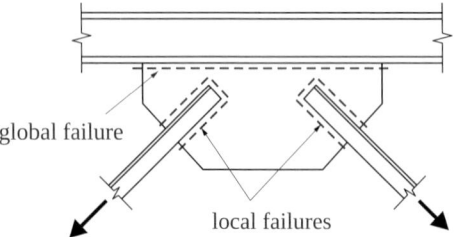

Figure 3.38 – Failure modes in a gusset plate

In EN 1993-1-8, no specific guidelines are provided for the local and global verifications of the gusset plate, or at least no full consistent set of design formulae.

The following recommendations may however be proposed:

a) Local failure (supported member in tension)

A block tearing type failure has to be checked, both for welded and bolted connections, as illustrated in Figure 3.39. To achieve it, reference may be simply made to corresponding Eurocode verification formulae (see section 3.4.6).

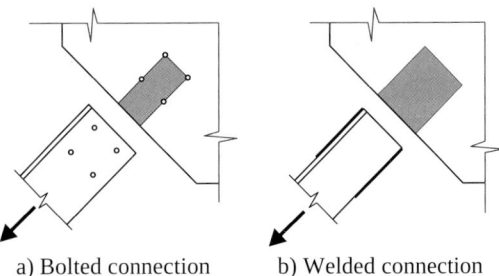

a) Bolted connection b) Welded connection

Figure 3.39 – Block tearing failure

b) Local failure (supported member in tension)

For supported members in compression, gusset local plate instability may occur.

In this context, good practice indicates that the region of unsupported gusset plate at the ends of the members should be kept to a minimum. End and edge distance requirements for the bolts should however be strictly followed.

No information is provided in Eurocode 3 Part 1-8 for such a stability check. In Disque and Robert (1984), the authors recommend to assimilate the local plate instability of the gusset to the flexural buckling of a column:

- with a buckling length defined as the average of l_1, l_2 and l_3 (see Figure 3.40) multiplied by a buckling length coefficient K assumed to be equal to 0.65, or even 0.5 by some authors (Gross, 1990);
- with a unit width;
- subjected to compressive stresses equal to those defined on the Whitmore section (Whitmore, 1952), i.e. on an effective width defined as specified in Figure 3.41.

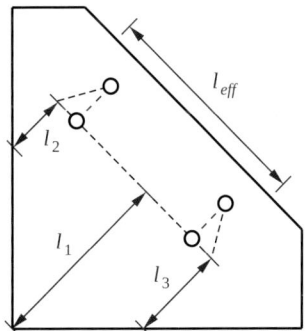

Figure 3.40 – Buckling lengths

3. CONNECTIONS WITH MECHANICAL FASTENERS

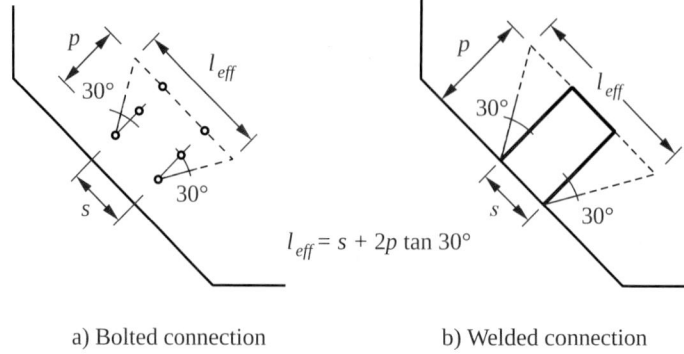

a) Bolted connection   b) Welded connection

Figure 3.41 – Gusset plate yielding

A safe application of this concept, assuming that the Whitmore stress equals the yield strength of the material, leads to an easy-to-apply design criterion:

$$K \cdot l \leq 5.4 t\varepsilon \qquad (3.41)$$

with:

$$\varepsilon = \sqrt{\frac{235}{f_{y,g}}} \quad \text{coefficient dependent on } f_{y,g} \text{ (expressed in N/mm}^2\text{)}$$

c) Global failure (supported member in tension)

Reference is here made to section 1.6.1 where the "static approach" is introduced.

The verification of the global resistance therefore resumes to the checking of the resistance of all potential cross sections in the gusset plates; in these ones the statically and plastically admissible character of the assumed stress distributions will have to be checked (see, as an example, Figure 3.42). Elastic or plastic stress distributions may be considered under the applied forces.

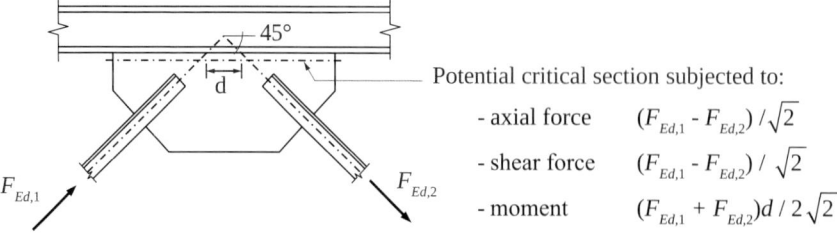

Figure 3.42 – Example of verification for the global failure mode

3.5.4 Long joints

The evaluation of the design resistance of bolted lap joints using preloaded and non-preloaded bolts has been discussed in section 3.5.1 and design formulae have been provided.

In Category A (bearing type) and Category B (slip resistance at serviceability limit state) connections, a reduction factor has however possibly to be applied for so-called "long joints", i.e. joints with a significant distance between the two extreme bolt rows. In such long joints, and contrarily to what has been shown for short joints in section 3.5.1, the deformation capacity in the end bolt-rows is too limited to allow a full plastic redistribution of the internal forces to take place between the bolt-rows, even if the design shear resistance $F_{v,Rd}$ of each individual fastener is greater than or equal to the local design bearing resistance $F_{b,Rd}$.

Practically, EN 1993-1-8 recommends to reduce the design resistance of the joint by a factor β_{Lf} where the distance L_j between the centres of the end fasteners in a joint, measured in the direction of force transfer (see Figure 3.43), is more than $15d$. d is the nominal diameter of the bolts while β_{Lf} is defined as:

$$\beta_{Lf} = 1 - \frac{L_j - 15d}{200d} \qquad (3.42)$$

but with $\beta_{Lf} \geq 0.75$ and $\beta_{Lf} \leq 1.0$.

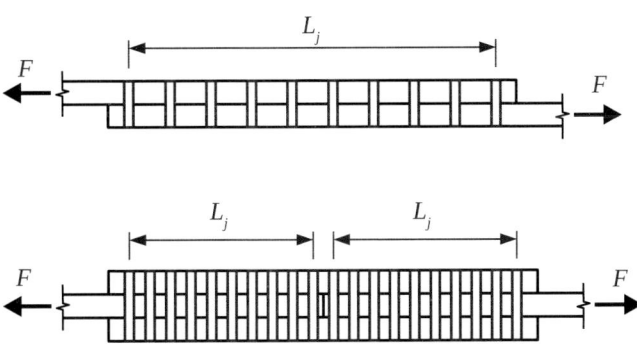

Figure 3.43 – Lap joint length

3. Connections with Mechanical Fasteners

It has to be specified that this reduction factor never applies where there is a uniform transfer of forces over the length of the joint, e.g. the transfer of longitudinal shear forces between the web and the flange of a section in a beam in bending.

Chapter 4

WELDED CONNECTIONS

4.1 TYPE OF WELDS

In the construction field, fillet welds and butt welds are widely used. One speaks about 80% of fillet welds and 15% of butt welds. For the five remaining per cent, plug and fillet all round welds are mostly used. In the following paragraphs, these different weld types are briefly introduced, in accordance with ESDEP Lecture 6 (ESDEP).

4.1.1 Butt welds

Butt welds are applied within the cross section of the abutting plates in so-called butt and tee-joints (Figure 4.1a).

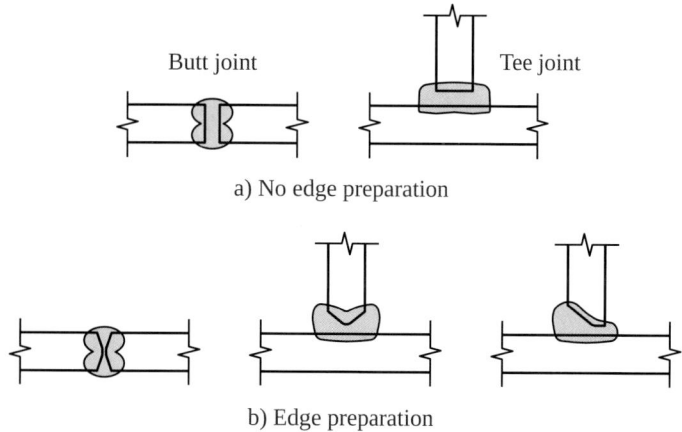

Figure 4.1 – Butt welds with full penetration

4. WELDED CONNECTIONS

The preparation of the plates, before welding, usually appears as a prerequisite, except for thin plates, less than about 5 mm, where it can be avoided. In all the other cases, bevelled plate edges will have to be realised, as seen in Figure 4.1b and Figure 4.2. This may have an impact on the economy of the project.

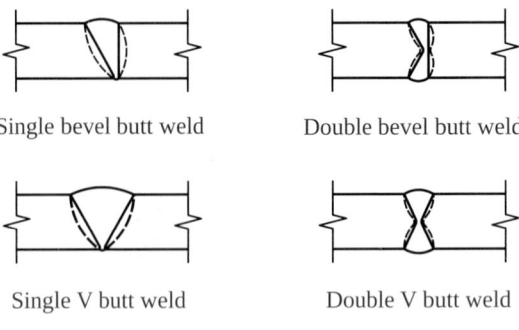

Figure 4.2 – Examples of types of bevelled edges

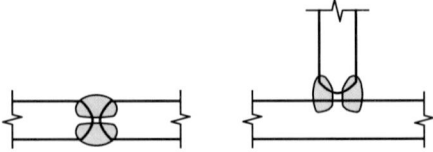

Figure 4.3 – Butt welds with partial penetration

According to the penetration of the weld, i.e. the degree of fusion of the weld and parent materials throughout the thickness of the connected plates, butt welds with complete penetration (equal to the plate thickness, see Figure 4.1) or partial penetration (less than the plate thickness, see Figure 4.3) are distinguished.

4.1.2 Fillet welds

Contrarily to a butt weld, a fillet weld requires no preparation of the plates to connect. This is an important feature in terms of fabrication costs.

Fillet welds are applied to the surface of the connected plates; they exhibit an approximately triangular shape.

Lap joints, tee or cruciform joints and corner joints are shown in Figure 4.4. In tee joints, the two plates are not necessarily welded perpendicularly each to another.

4.1 TYPE OF WELDS

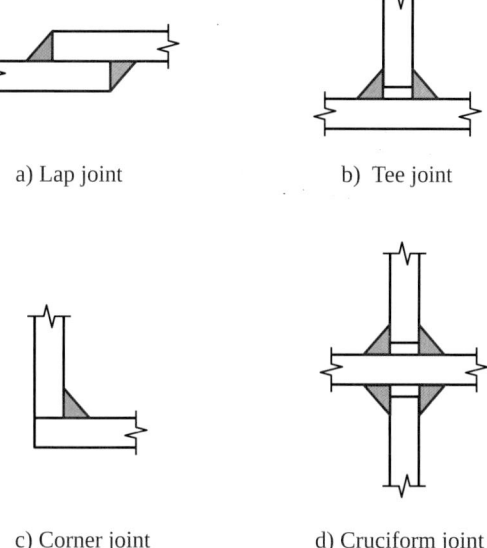

a) Lap joint   b) Tee joint

c) Corner joint   d) Cruciform joint

Figure 4.4 – Schematic representation of various fillet weld joint configurations

As explained later in section 4.4.2.4, the one-sided character of the fillet weld in the corner joint represented in Figure Figure 4.4c is likely to negatively influence the joint response. As a way of improvement, the outer corner could be butt welded (Figure 4.5).

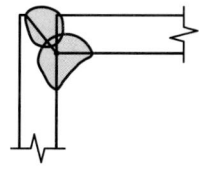

Figure 4.5 – Improved corner joint

Fillet welds are either continuous or intermittent along the whole joint, Figure 4.6. The latter should not be used in corrosive conditions.

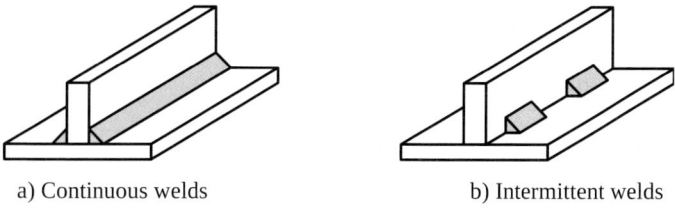

a) Continuous welds   b) Intermittent welds

Figure 4.6 – Continuous and intermittent fillet welds

4. WELDED CONNECTIONS

4.1.3 Fillet welds all round

Fillet welds all round (Figure 4.7) designate fillet welds in circular or elongated holes; such welds may be used only to transmit shear or to prevent the buckling or separation of lapped parts.

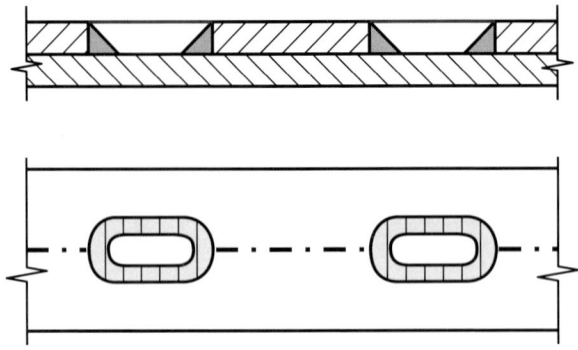

Figure 4.7 – Fillet welds all round

4.1.4 Plug welds

As seen in Figure 4.8, the holes (round or elongated) realised in one of the two plates to assemble are here filled with welding material.
Such welds may be used:

– to transmit shear,
– to prevent the buckling or separation of lapped parts, and
– to inter-connect the components of built-up members

but should not be used to resist externally applied tension.

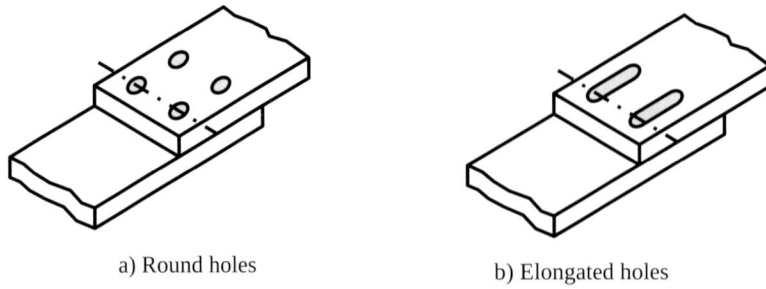

a) Round holes   b) Elongated holes

Figure 4.8 – Plug welds

4.2 CONSTRUCTIVE CONSTRAINTS

Welding is a quite convenient assembly technique, but its reliability is highly dependent on the respect of strict requirements in terms of properties of parent and consumable materials, welding process, tolerances, preparation of surfaces, storage and handling of consumables, weather protection, welding sequence, preheating, operator qualification, acceptance criteria, etc.

These aspects, extensively covered in EN 1090-2, require from the manufacturers high competence on the engineering and operating sides.

In the present chapter, focus is made on design issues (sub-chapter 4.3) and therefore these technological aspects will simply be briefly addressed below.

4.2.1 Mechanical properties of materials

4.2.1.1 Parent material

The design provisions proposed in sub-chapter 4.3 apply to weldable structural steels conforming to EN 1993-1-1 and to material thicknesses of 4 mm and over. The provisions also apply to joints in which the mechanical properties of the weld metal are compatible with those of the parent metal, see section 4.2.1.2.

For welds in thinner material reference should be made to EN 1993 Part 1-3 and, for welds in structural hollow sections in material thicknesses of 2.5 mm and over, guidance is given in EN 1993 Part 1-8.

Quality level C according to EN ISO 5817 (CEN, 2014) is usually required, if not otherwise specified.

The choice of material as well as the verification for lamellar tearing should be performed according to EN 1993-1-10 (CEN, 2005e).

4.2.1.2 Welding consumables

All welding consumables should conform to the relevant standards specified in EN 1090-2. The specified yield strength, ultimate tensile strength, elongation at failure and minimum Charpy V-notch energy value of the filler metal, should be equivalent to, or better than that specified for the parent material.

Generally it is safe to use electrodes that are overmatched with regard to the steel grades up to S460. For higher steel grades up to S700 undermatching need to be accounted for according to the rules given in EN 1993-1-12 (CEN, 2007a).

4.2.2 Welding processes, preparation of welds and weld quality

One of the following arc welding processes may be used:
- metal arc-welding with covered electrodes
- flux-cored arc welding
- submerged arc welding
- MIG (metal inert gas) welding
- MAG (metal active gas) welding.

All these processes can be used in the workshop. Usually, only bolting or metal arc welding with covered electrodes is employed in the erection phase on the building site. With the metal arc welding process, welds can be made in all positions. The various weld positions are shown in Figure 4.9, where the arrows give the arc direction during the welding operation. It is clear that welding in the flat position is easily carried out, allowing a greater rate of metal deposition than the other positions; by welding in this favourable position, the maximum size of weld run can be obtained. With ordinary welding consumables and favourable welding conditions, a fillet weld with a throat thickness of 6mm can be produced with only one run. For welds of greater thickness, more than one run is necessary. In this case, the welding sequence must be carefully planned, see Figure 4.10.

The welding conditions, particularly the current limitation of the welding equipment, constitute a limit to the depth of penetration into the parent plate. For example, if a closed butt joint (no gap between the two plates) is welded with one run on each side, the penetration may not be complete and the central part of the joint will remain unfused (Figure 4.11a). With a gap between the two parts of the joint, full penetration can be achieved with the same welding equipment. The limitation is then set by the thickness of the plates to be joined. In practice, the limit for butt welds with square edges, i.e. without preparation, is 10mm plate thickness with a 5mm gap. When the plate thickness exceeds this value, bevelled edges permit full penetration by several runs, see Figure 4.11b.

4.2 CONSTRUCTIVE CONSTRAINTS

Edge preparation consists essentially of cutting and bevelling the edges of the plates which are to be welded. These operations can be done by thermal cutting, by machining or by chipping or grinding. The resulting surfaces of the bevelled edges should be smooth, uniform, free from cracks and without corrosion. If thermal cutting or another process which hardens the material is used, the approved welding procedure must take account of this weld preparation process. Different bevel geometries are shown in Figure 4.2.

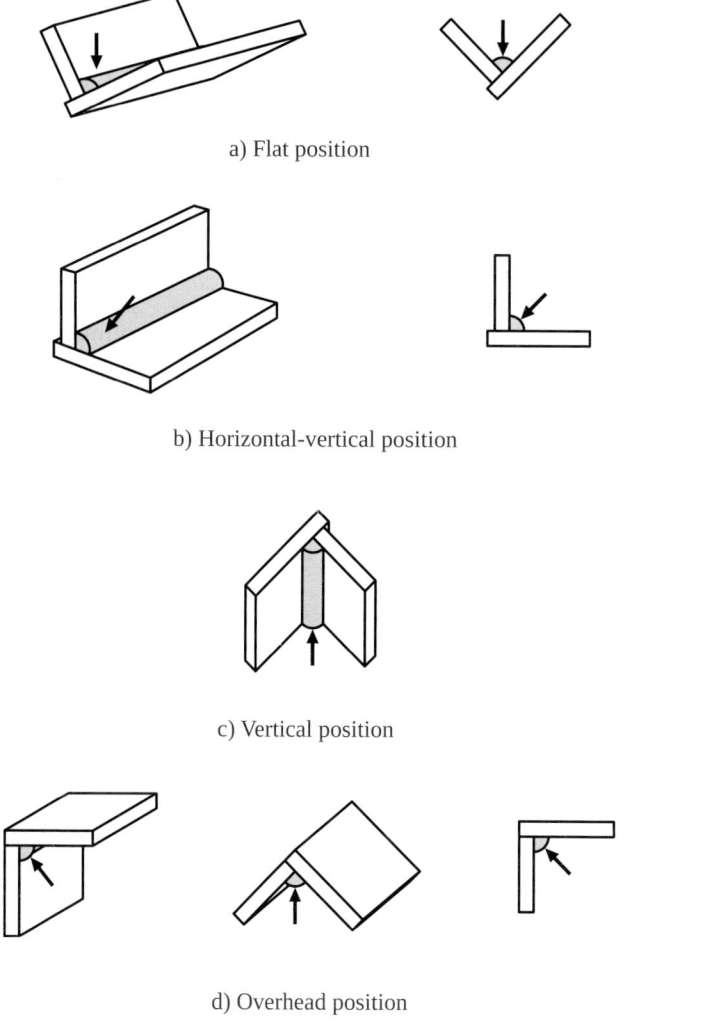

a) Flat position

b) Horizontal-vertical position

c) Vertical position

d) Overhead position

Figure 4.9 – Weld positions

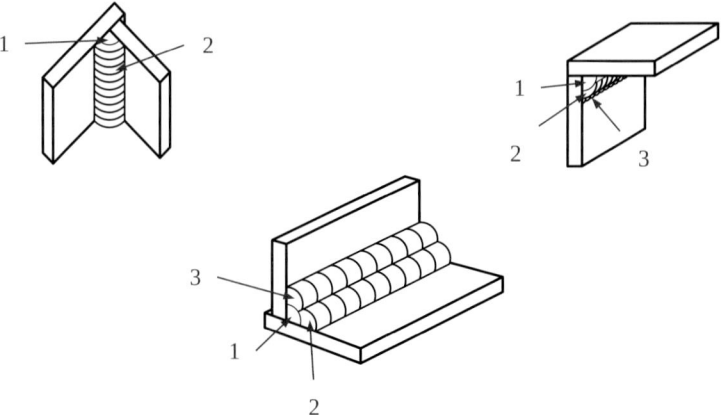

Figure 4.10 – Welds with successive runs

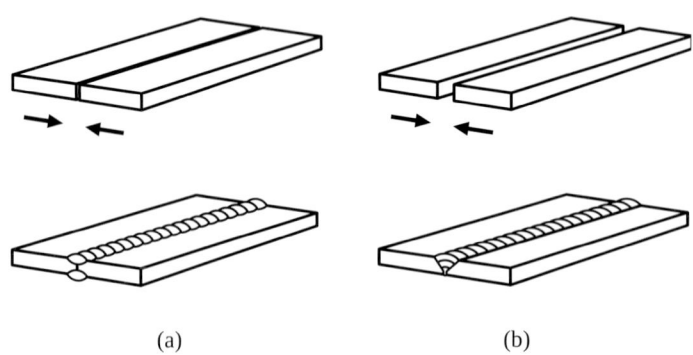

Figure 4.11 – Effects of the gap on weld penetration

Practical recommendations, for example EN 1993-1-9 for fatigue (CEN, 2005d) or EN 1993-2 for steel bridges (CEN, 2006e), give some tolerance values for various weld types.

Before welding, surfaces and edges adjacent to the weld location must be cleaned to remove oil, grease, paint or any other contaminants, which can affect the quality of the weld and the weld strength.

The appropriate welding method and procedure are defined in a project specification provided by the designer. Relative information may be found in EN 1090-2.

Quality control is an important part of industrial activity. The term quality includes all the characteristics of a product which affect its ability to serve its purpose. Here, attention is drawn to quality control applied to welding, including the qualification of firms and the procedure qualification for welding tests.

Normally, all welded structures undergo weld inspection. The type and the extent for the inspection, as well as the choice of welds to be inspected are selected in accordance with the project specification.

The principal purpose of weld inspection is to discover possible weld defects. Examples on weld defects are, see Figure 4.12:

- Undercut. The thickness of the parent metal is reduced near the weld toe.
- Porosity or gas inclusions. Air or gas bubbles are incorporated in the melted metal, where they remain after cooling.
- Insufficient throat. The throat thickness is smaller than the design thickness. The resistance of the joint might be insufficient.
- Incomplete penetration. The throat thickness is smaller than the design thickness. The resistance of the joint might be insufficient.

All these defects can be measured. Codes of practice specify the allowable tolerances for each defect, see EN ISO 5817 (CEN, 2014).

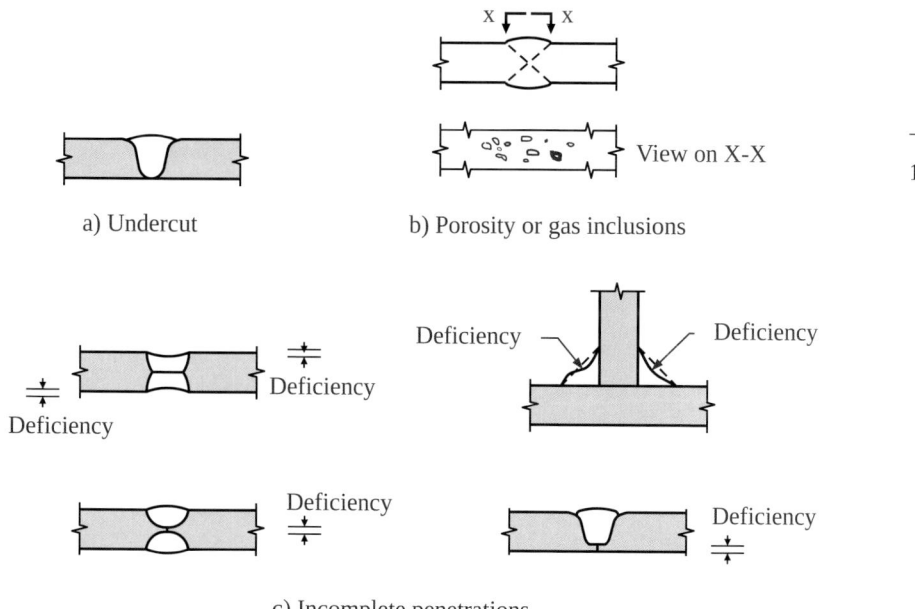

Figure 4.12 – Examples of weld defects

4. WELDED CONNECTIONS

4.2.3 Geometry and dimensions of welds

4.2.3.1 Fillet welds

In tee-joints, fillet welds may be used to connect parts where the fusion faces form an angle of between 60° and 120°. Angles smaller than 60° are also permitted; however, in such cases, the weld should be considered to be a partial penetration butt weld. For angles greater than 120° the resistance of fillet welds should be determined by testing in accordance with EN 1990 Annex D: Design by testing (CEN, 2002).

Fillet welds finishing at the ends or sides of parts should be returned continuously, full size, around the corner for a distance of at least twice the leg length of the weld, unless access or the configuration of the joint renders this impracticable. In the case of intermittent welds this rule applies only to the last intermittent fillet weld at corners. End returns should be indicated on the drawings.

4.2.3.2 Intermittent fillet welds

As already said, intermittent fillet welds should not be used in corrosive conditions.

The gaps (L_1 or L_2) between the ends of each length of weld L_w should fulfil the requirement given in Figure 4.13. They should be taken as the smaller of the distances between the ends of the welds on opposite sides and the distance between the ends of the welds on the same side.

In any run of intermittent fillet weld there should always be a length of weld at each end of the part connected.

In a built-up member in which plates are connected by means of intermittent fillet welds, a continuous fillet weld should be provided on each side of the plate for a length at each end equal to at least three-quarters of the width of the narrower plate concerned.

4.2 CONSTRUCTIVE CONSTRAINTS

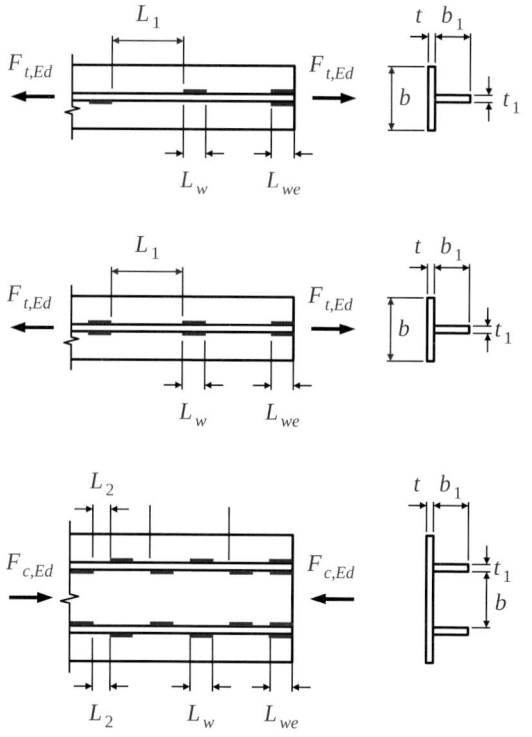

Requirements for the gaps (L_1, L_2):   min ($L_{we} \geq 0.75\ b$; $0.75\ b_1$)

For build-up members in tension: min ($L_1 \leq 16\ t$; $16\ t_1$; 200 mm)

For build-up members in compression or shear:

$$\min (L_2 \leq 12\ t\ ;\ 12\ t_1;\ 025\ b\ ;200\ \text{mm})$$

Figure 4.13 – Geometry of intermittent fillet welds

4.2.3.3 Fillet all round

As specified earlier, fillet welds all round may be used only to transmit shear or to prevent the buckling or separation of lapped parts.

The diameter of a circular hole, or width of an elongated hole, for a fillet weld all round should not be less than four times the thickness of the part containing it.

The ends of elongated holes should be semi-circular, except for those ends which extend to the edge of the part concerned.

The centre to centre spacing of fillet welds all round should not exceed the value necessary to prevent local buckling. To satisfy this condition, reference may be made to Table 3.3 in sub-chapter 3.3.

4.2.3.4 Butt welds

Full penetration and partial penetration welds may be contemplated, as defined in section 4.3.4. Intermittent butt welds should not be used.

4.2.3.5 Plug welds

It has been said before that plug welds may be used:
- to transmit shear,
- to prevent the buckling or separation of lapped parts, and
- to inter-connect the components of built-up members

but should not be used to resist externally applied tension.

The diameter of a circular hole, or width of an elongated hole, for a plug weld should be at least 8 mm more than the thickness of the part containing it. The ends of elongated holes should either be semi-circular or else should have corners which are rounded to a radius of not less than the thickness of the part containing the slot, except for those ends which extend to the edge of the part concerned.

The thickness of a plug weld in parent material up to 16 mm thick should be equal to the thickness of the parent material. The thickness of a plug weld in parent material over 16 mm thick should be at least half the thickness of the parent material and not less than 16 mm.

As for fillet welds all round, the centre to centre spacing of fillet welds all round should not exceed the value necessary to prevent local buckling. To satisfy this condition, reference may again be made to Table 3.3 in sub-chapter 3.3.

4.2.3.6 Welding in cold-formed zones

Welding may be carried out within a length $5t$ either side of a cold-formed zone (see Table 4.1), provided that one of the following conditions is fulfilled:

- the cold-formed zones are normalized after cold-forming but before welding;
- the r/t-ratio satisfy the relevant value obtained from Table 4.1.

Table 4.1 – Conditions for welding cold-formed zones and adjacent material

r/t	Strain due to cold forming (%)	Maximum thickness (mm)		
		Generally		Fully killed Aluminium-killed steel (Al ≥ 0.02%)
		Predominantly static loading	Where fatigue predominates	
≥ 25	≤ 2	any	any	any
≥ 10	≤ 5	any	16	any
≥ 3.0	≤ 14	24	12	24
≥ 2.0	≤ 20	12	10	12
≥ 1.5	≤ 25	8	8	10
≥ 1.0	≤ 33	4	4	6

4.3 DESIGN OF WELDS

4.3.1 Generalities

For weld design, three fundamental assumptions are usually made, as recalled in (ESDEP):

- The welds are homogeneous and isotropic.
- The connected elements are rigid and their deformations are negligible.
- Only nominal stresses due to external loads are considered. Effects of residual stresses, stress concentrations and shape of the welds are neglected in static design.

These assumptions will be briefly commented in sub-chapter 4.4.

4. WELDED CONNECTIONS

4.3.2 Fillet welds

A fillet weld is characterised by two main geometrical properties: its effective length and its effective throat thickness. These ones are defined in sections 4.3.2.1 and 4.3.2.2. On the basis of these two values and of the weld and base material grades, the design resistance of a fillet weld may then be derived. How to proceed is explained in section 4.3.2.3.

4.3.2.1 Effective length

To the actual physical length L_w is usually substituted, for the design, an effective length of a fillet weld l_{eff}. This one should be taken as the length over which the fillet is full-size, which is defined as the overall length of the weld reduced by twice the effective throat thickness a (see 4.3.2.2).

Obviously, if the weld is full size throughout its length, including starts and terminations, no reduction in effective length need be made for either the start or the termination of the weld.

A weld which a length smaller than 30 mm or less than six times its throat thickness, whichever is larger, should not be designed to carry loads.

4.3.2.2 Effective throat thickness

The effective throat thickness, a, of a fillet weld should be taken as the height of the largest triangle (with equal or unequal legs) that can be inscribed within the fusion faces and the weld surface, measured perpendicular to the outer side of this triangle (see Figure 4.14).

In practice, the effective throat thickness of a fillet weld should not be less than 3 mm.

In the case of a deep penetration fillet weld, a a value higher than the one defined before may be taken, provided that preliminary tests show that the required penetration can consistently be achieved by the welder. This higher throat thickness is illustrated in Figure 4.15.

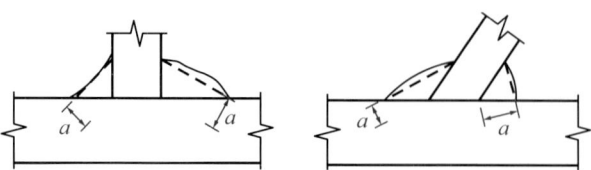

Figure 4.14 – Throat thickness of a fillet weld

4.3 DESIGN OF WELDS

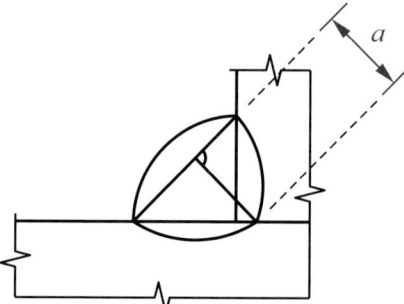

Figure 4.15 – Throat thickness of a deep penetration fillet weld

4.3.2.3 Design resistance

Two procedures for the evaluation of the design resistance of a fillet weld are provided in EN 1993-1-1. They are respectively known as the directional and simplified methods.

The most economical one, i.e. the one providing the higher resistance level, is the directional method. In the case of side welds (i.e. weld parallel to the applied force) subjected to shear forces along the weld axis, equal results are obtained from both approaches.

4.3.2.3.1 Directional method

In this method, the forces transmitted by a unit length of weld are resolved into four components. These ones are respectively parallel and transverse to the longitudinal axis of the weld and normal and transverse to the plane of the throat:

$\sigma_\perp$ = normal stress perpendicular to the throat,

$\sigma_\parallel$ = normal stress parallel to the axis of the weld,

$\tau_\perp$ = shear stress (in the plane of the throat) perpendicular to the axis of the weld,

$\tau_\parallel$ = shear stress (in the plane of the throat) parallel to the axis of the weld.

These stresses, shown in Figure 4.16, are assumed to be uniformly distributed on the throat section $A_w = 1 \cdot a$.

4. WELDED CONNECTIONS

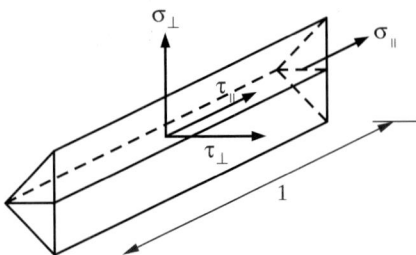

Figure 4.16 – Stresses on the throat section of a fillet weld (unit length)

The design resistance of the fillet weld will be sufficient if the following equations are both satisfied:

$$\left[\sigma_\perp^2 + 3\left(\tau_\perp^2 + \tau_\parallel^2\right)\right]^{0,5} \leq \frac{f_u}{\beta_w \gamma_{M2}} \quad (4.1)$$

$$\sigma_\perp \leq 0.9 f_u / \gamma_{M2}$$

where:

f_u = nominal ultimate tensile strength of the weaker part joined;
β_w = appropriate correlation factor taken from Table 4.2.

It has to be mentioned that the value of $\sigma_\parallel$ is not influencing the resistance of the weld.

When the connected members exhibit different material strength grades, the design of the weld should be based on the mechanical properties of the material with the lower strength grade.

Table 4.2 – Correlation factor β_w for fillet wed design

Standard and steel grade			Correlation factor
EN 10025	EN 10210	EN 10219	β_w
S 235 S 235 W	S 235 H	S 235 H	0.8
S 275 S 275 N/NL S 275 M/ML	S 275 H S 275 NH/NLH	S 275 H S 275 NH/NLH S 275 MH/MLH	0.85
S 355 S 355 N/NL S 355 M/ML S 355 W	S 355 H S 355 NH/NLH	S 355 H S 355 NH/NLH S 355 MH/MLH	0.9

Table 4.2 – Correlation factor β_w for fillet wed design (continuation)

Standard and steel grade			Correlation factor β_w
EN 10025	EN 10210	EN 10219	
S 420 N/NL S 420 M/ML		S 420 MH/MLH	1.0
S 460 N/NL S 460 M/ML S 460 Q/QL/QL1	S 460 NH/NLH	S 460 NH/NLH S 460 MH/MLH	1.0

4.3.2.3.2 Simplified method

As an alternative to section 4.3.2.3.1, the design resistance of a fillet weld may be assumed to be adequate if, at every point along its length, the resultant of all the forces per unit length transmitted by the weld satisfy the following criterion:

$$F_{w,Ed} \leq F_{w,Rd} \qquad (4.2)$$

where:
$F_{w,Ed}$ = design value of the weld force per unit length;
$F_{w,Rd}$ = design weld resistance per unit length.

The resistance of the weld per unit length $F_{w,Rd}$ is defined independently of the orientation of the weld throat plane relatively to the applied force, as follows:

$$F_{w,Rd} = f_{vw,d} \cdot a \qquad (4.3)$$

where the design shear strength of the weld $f_{vw,d}$ is defined as:

$$f_{vw,d} = \frac{f_u/\sqrt{3}}{\beta_w \gamma_{M2}} a \qquad (4.4)$$

The values of f_u and β_w are defined in section 4.3.2.3.1.

4. WELDED CONNECTIONS

4.3.3 Fillet welds all round

The design resistance of a fillet weld all round is evaluated through the same expressions than those presented in section 4.3.2.

4.3.4 Butt welds

Providing the welding process has been correctly carried out, the butt weld filler metal may be considered as parent metal. Hence, to determine the resistance of the joint, the calculation is based on the throat area, i.e. the penetration area. Depending on the penetration, two kinds of butt welds are defined: full and partial penetration welds.

For a full penetration butt weld (in usual overmatching situations), calculation is not necessary because the filler metal strength is at least as high as the parent metal strength of the weaker connected element and the throat thickness of the weld is equal to the thickness of the plate, see Figure 4.17. Thus the butt weld may effectively be regarded simply as replacing the parent material.

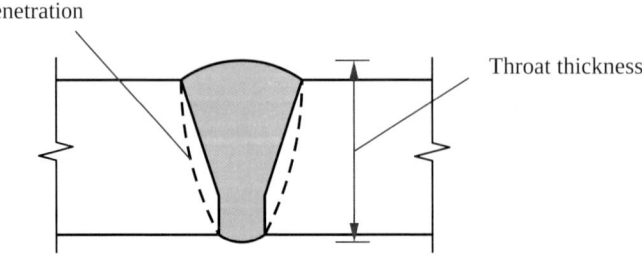

Figure 4.17 – Butt weld with full penetration

The evaluation of the design resistance of a partial penetration butt weld (Figure 4.18) is similar to the one prescribed for a deep penetration fillet weld. The throat thickness of a partial penetration butt weld should not be greater than the depth of penetration that can be consistently achieved.

4.3 DESIGN OF WELDS

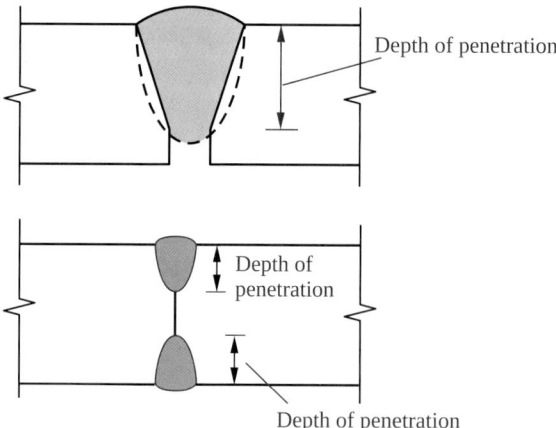

Figure 4.18 – Butt weld with partial penetration

A partial penetration tee-butt joint with superimposed fillet welds may be considered as a full penetration butt weld, if the total throat thickness is greater than the material thickness and the gap dimension meets certain conditions (Figure 4.19). If it is not the case, the resistance should be determined using the method for a fillet weld or a deep penetration fillet weld, depending on the amount of penetration. The throat thickness should be determined in conformity with the provisions for fillet welds or partial penetration butt welds, as relevant.

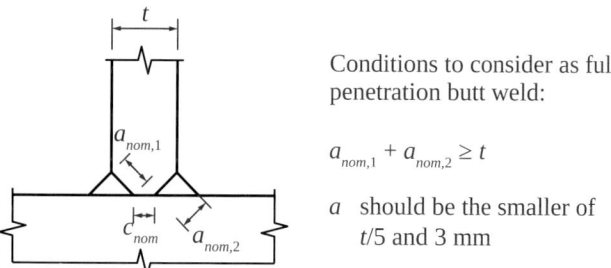

Conditions to consider as full penetration butt weld:

$a_{nom,1} + a_{nom,2} \geq t$

a should be the smaller of $t/5$ and 3 mm

Figure 4.19 – Tee-butt joint with superimposed fillet welds

4.3.5 Plug welds

The design resistance $F_{w,Rd}$ of a plug weld is assessed as follows:

$$F_{w,Rd} = f_{vw,d} \cdot A_w \qquad (4.5)$$

where:
- $f_{vw,d}$ is the design shear strength of a weld given by Eq. (4.4);
- A_w is the design throat area defined as the area of the hole.

4.3.6 Concept of full strength fillet weld

Fillet welds must be designed according to sections 4.3.2.3.1 or 4.3.2.3.2. In the case of relatively small loads in relation to the capacity of the weld, Part 1-8 design rules may lead to rather thin welds. If the rupture strength of those thin welds is lower than the yield strength of the weakest of the connected parts, the connection has so little deformation capacity that it is usually not sufficient to accommodate effects due for instance to imposed deformations. In such a case the connection will behave in a brittle way.

To avoid this, the welds should be designed "full strength". The rupture strength of full strength welds would then be greater than the rupture strength of the connected elements; so, in the case of overloading, the elements would fail before the welds.

A similar requirement is also regularly expressed in connections where global ductility is required. This aspect will be addressed in some different further chapters of the present book; for instance in sub-chapter 4.4 for welded joints, but also in chapter 5 on simple joints or chapter 8 on joints under various "other than static" loading situations.

The "full-strength" character of the welds may be achieved by expressing that the design resistance of the weld is higher or equal than the design resistance of the weakest connected element.

As an example, the derivation of "full strength" throat thicknesses is illustrated here below in the case of double fillet end welds (i.e. weld perpendicular to the applied force). If reference is made to the directional method, the design resistance of a fillet weld is checked as follows:

$$\sigma_c = \sqrt{\sigma_\perp^2 + 3\tau_\perp^2 + 3\tau_{//}^2} \leq \frac{f_u}{\beta_w \gamma_{M2}} \quad \text{and} \quad \sigma_\perp \leq 0.9 \frac{f_u}{\gamma_{M2}} \qquad (4.6)$$

4.3 DESIGN OF WELDS

where:

- f_u = nominal ultimate tensile strength of the weaker part joined
- γ_{M2} = partial safety factor for welded connections
- β_w = correlation factor (β_w = 1.0 for steel grades S420 and S460, see Table 4.3)

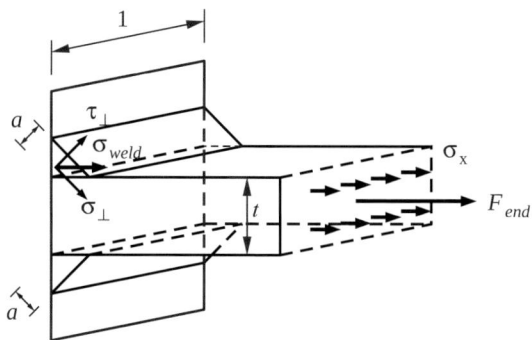

Figure 4.20 – End fillet and side fillet welds

For end fillet welds (Figure 4.20):

$$\sigma_\perp = \tau_\perp = \frac{\sigma_{weld}}{\sqrt{2}} \qquad (4.7)$$

$$\tau_\parallel = 0 \qquad (4.8)$$

From the first formula reported above (more restrictive here than the second one), it follows:

$$\sigma_c = \sqrt{\left(\frac{\sigma_{weld}}{\sqrt{2}}\right)^2 + 3\left(\frac{\sigma_{weld}}{\sqrt{2}}\right)^2} \leq \frac{f_u}{\beta_w \gamma_{M2}} \qquad (4.9)$$

and so:

$$\sigma_{weld} \leq f_{w,u,end} = \frac{f_u}{\beta_w \gamma_{M2} \sqrt{2}} \qquad (4.10)$$

For double end fillet welds, the above-expressed "full strength" criterion writes (Figure 4.20):

4. WELDED CONNECTIONS

$$2al\sigma_{weld} \geq_y tlf_y/\gamma_{M0} \tag{4.11}$$

where:
- f_y = nominal yield strength of the weaker part joined
- γ_{M0} = partial safety factor for steel material
- l = length of the weld
- a = throat thickness of the weld

This last equation is obtained by assuming, in Figure 4.20, that $F_{end} = \sigma_x tl$ with $\sigma_x = f_y/\gamma_{M0}$.

The minimum weld size to satisfy the full strength design requirement is therefore expressed as:

$$a \geq \frac{f_y t}{2 f_{w,u,end} \gamma_{M0}} \tag{4.12}$$

and finally:

$$a \geq \frac{f_y \beta_w \gamma_{M2}}{\sqrt{2} f_u \gamma_{M0}} t \tag{4.13}$$

Minimum throat thicknesses required for a full strength end weld design are reported in Table 4.3 for various steel grades.

Table 4.3 – Values of β_w and $f_{w,u,end}$ for steels according to EN 10025 and minimum "full strength" required weld thickness in case of double fillet end welds (plate thickness smaller than 40 mm; $\gamma_{M0} = 1.0$ and $\gamma_{M2} = 1.25$)

Steel grade	f_y	f_u	β_w	$f_{w,u,end}$	Full strength double fillet welds
	N/mm²	N/mm²		N/mm²	
S235	235	360	0.80	255	$a \geq 0.46\, t$
S275	275	430	0.85	286	$a \geq 0.48\, t$
S355	355	510	0.90	321	$a \geq 0.55\, t$
S420 M	420	520	1.00	294	$a \geq 0.71\, t$
S420 N	420	550	1.00	311	$a \geq 0.68\, t$
S460 M	460	550	1.00	311	$a \geq 0.74\, t$
S460 N	460	580	1.00	328	$a \geq 0.70\, t$

4.4 DISTRIBUTION OF FORCES IN A WELDED JOINT

4.4.1 Generalities

This topic has been addressed in chapter 1 (see section 1.6.1) in which reference is made to the so-called "static approach". For welded joints, however, the limited ductility of the welds may lead to a limitation of its application even if, according to Eurocode 3, "the distribution of forces in a welded connection may be calculated on the assumption of either elastic or plastic behaviour".

In reality:

– Welds are characterised by the presence of residual stresses as well as of geometrical imperfections; the latter in local stress concentrations. So even before a loading is applied, variations of internal stresses and strains are observed along the weld. Locally, the residual stresses could even tend to become higher than the material strength and so provoke failure; but thanks to the limited but anyway available ductility of the weld material, plastic redistribution takes place and the stresses are just limited locally to the material yield strength.

– When the welds are then externally loaded, new plastic redistributions have to take place to allow, at the end, to reach a uniform distribution of stresses equal to the yield strength along the design throat area. To reach such a plastic distribution of internal stresses, further local ductility is obviously required.

In section 4.3.1, it has been assumed that the effect of these various imperfections may be neglected for static design. But this does not mean that the ductility of the weld is sufficient to rely in all cases on a full plastic redistribution of internal forces along the weld length. Eurocode 3 Part 1-8 is not explicitly expressing what is allowed or not, even if some clauses may be interpreted at a mean to limit plastic redistributions.

In the following lines, an attempt is made to clarify the point.

When a weld is subjected to pure shear forces or axial forces, the ductility of the weld is assumed to be sufficient to reach a full plastic redistribution of stresses along the weld, but only as long as some geometrical requirements or layouts are strictly fulfilled (see 4.4.2.1).

4. WELDED CONNECTIONS

In other cases, as a general statement, it may be recommended to adopt an elastic distribution of forces as long as partial-strength welds are used. Doing so reduces the risk of brittle failure resulting from a local yielding of a weld segment. Consequently, a plastic distribution of forces is seen to be basically limited to cases where full-strength welds are adopted (the local ductility being so provided by the parent material).

In Figure 4.21, both situations are illustrated.

Obviously, some exceptions may be contemplated as soon as the plastic redistribution of forces only requires a quite limited deformation capacity of the weld (see Figure 4.22).

Besides that, it is worthwhile recalling that residual stresses and stresses not subjected to transfer of load (as long as they remain rather limited) need not to be included when checking the resistance of a weld. This applies specifically to the normal stress parallel to the axis of a weld.

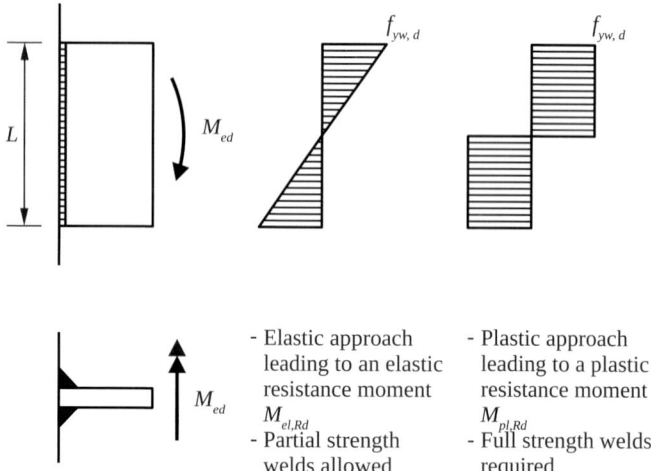

- Elastic approach leading to an elastic resistance moment $M_{el,Rd}$
- Partial strength welds allowed

- Plastic approach leading to a plastic resistance moment $M_{pl,Rd}$
- Full strength welds required

Figure 4.21 – Elastic and plastic distributions of forces

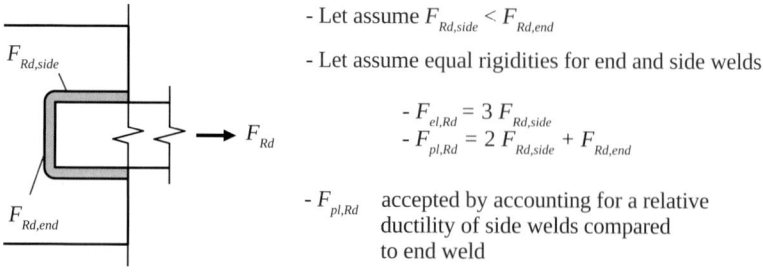

- Let assume $F_{Rd,side} < F_{Rd,end}$
- Let assume equal rigidities for end and side welds
- $F_{el,Rd} = 3\, F_{Rd,side}$
- $F_{pl,Rd} = 2\, F_{Rd,side} + F_{Rd,end}$
- $F_{pl,Rd}$ accepted by accounting for a relative ductility of side welds compared to end weld

Figure 4.22 – Example of elastic and plastic distributions

4.4 DISTRIBUTION OF FORCES IN A WELDED JOINT

As Eurocode 3 Part 1-8 is not providing detailed recommendations in this matter, a safe but anyway not systematically too strict attitude is expected from the designer. In section 6.3.5, a particular application of the static theorem to beam-to-end plate connections will be presented.

4.4.2 Particular situations

4.4.2.1 Long lap joints with side fillet welds

In lap joints, the distribution of stresses in the side fillet welds is usually assumed to be uniformly distributed along the weld length (see Figure 4.23a) and the design resistance is evaluated accordingly. The actual distribution may however differ significantly from this assumption, as larger stresses occur at the ends of the connection (see Figure 4.23b). This has similarly been observed in section 3.5.4 for long bolted joints. At the ultimate state, just before failure, plastic deformation near the ends may contribute to a more uniform shear stress in the welds. However, in long connections, the limited available ductility in the welds resulting from local plasticity does not allow for a full plastic redistribution and the resulting stress distribution is not fully uniform.

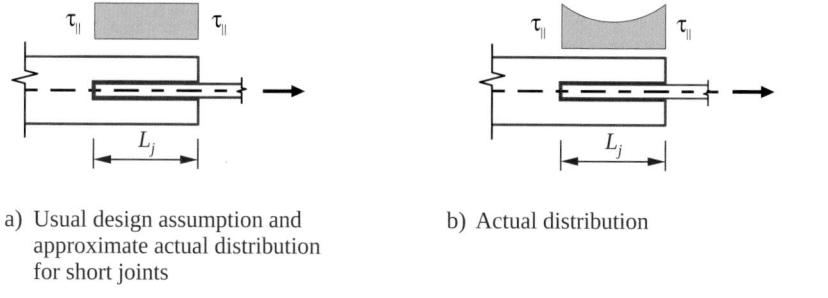

a) Usual design assumption and approximate actual distribution for short joints

b) Actual distribution

Figure 4.23 – Elastic distribution of shear stresses along the welds

EN 1993 Part 1-8 specifies that the design resistance of lap welded joints may be evaluated on the basis of an assumed uniform distribution of shear stresses for joints with a length lower or equal to $150a$. For longer joints, the so-evaluated resistance should be multiplied by a reduction factor β_{Lw} ($\beta_{Lw,1}$ or $\beta_{Lw,2}$) to allow for the effects of non-uniform stress distribution:

4. WELDED CONNECTIONS

a) Lap joints longer than $150a$:

$$\beta_{Lw,1} = 1.2 - \frac{0.2L_j}{150a} \leq 1.0 \qquad (4.14)$$

L_j = overall length of the lap in the direction of the force transfer

b) For fillet welds longer than 1.7 metres connecting transverse stiffeners in plated members

$$\beta_{Lw,2} = 1.1 - \frac{L_w}{17} \quad \text{but } 0.6 \leq \beta_{Lw,2} \leq 1.0 \qquad (4.15)$$

L_w = length of the weld (in metres)

4.4.2.2 Welds to unstiffened flanges

If a plate is welded to an unstiffened flange of an I, H or a box section, loading will tend to deform the flange or the box side unequally along the breadth. The result is that the parts of the weld near the web will be more heavily loaded than the other parts, see Figure 4.24. Therefore a reduced effective width shall be taken into account both for the parent material and for the weld design.

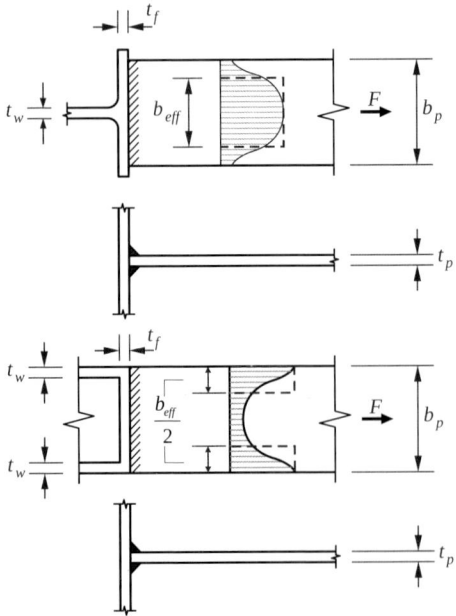

Figure 4.24 – Effective width for unstiffened tee-joints

4.4 DISTRIBUTION OF FORCES IN A WELDED JOINT

a) For an I or H section the effective width b_{eff} should be taken as:

$$b_{eff} = t_w + 2s + 7kt_f \qquad (4.16)$$

where:

$$k = \left(\frac{t_f}{t_p}\right)\left(\frac{f_{y,f}}{f_{y,p}}\right) \text{ but } \leq 1.0 \qquad (4.17)$$

$f_{y,f}$ = design yield strength of the flange of the I or H section
$f_{y,p}$ = design yield strength of the welded plate

The geometrical parameters t_w, t_f and t_p are shown in Figure 4.24 while s is obtained from:

– for a rolled I or H section: $s = r$ (root radius)
– for a welded I or H section: $s = \sqrt{2}a$

However, for the unstiffened flange of an I or H section, the following criterion should be satisfied:

$$b_{eff} \geq \left(\frac{f_{yp}}{f_{up}}\right) b_p \qquad (4.18)$$

where:
$f_{u,p}$ = ultimate strength of the plate welded to the I or H section;
b_p = width of the plate welded to the I or H section.

Should this requirement not be satisfied than the supporting member should be stiffened.

b) For other sections such as box sections or channel sections where the width of the connected plate is similar to the width of the flange, the effective width b_{eff} should be obtained from:

$$b_{eff} = 2t_w + 5t_f \quad \text{but } b_{eff} \leq 2t_w + 5kt_f \qquad (4.19)$$

4. WELDED CONNECTIONS

4.4.2.3 Intermittent welds

As an extra rule, Part 1-8 indicates that if the design resistance of an intermittent weld is determined by using the total length l_{tot}, the weld shear force per unit length $f_{w,Ed}$ should be multiplied by the factor $(e+l)/l$ (see Figure 4.25).

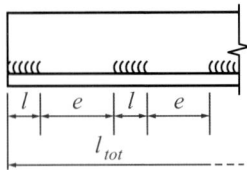

Figure 4.25 – Geometry of intermittent welds

4.4.2.4 Eccentrically loaded single fillet or single-sided partial penetration butt welds

Local eccentricity should be avoided whenever it is possible through appropriate conceptual design. When present, local eccentricities (relative to the line of action of the force to be resisted) should be taken into account in the following cases:
- where a bending moment transmitted about the longitudinal axis of the weld produces tension at the root of the weld (Figure 4.26a);
- where a tensile force transmitted perpendicular to the longitudinal axis of the weld produces a bending moment, resulting in a tension force at the root of the weld (Figure 4.26b).

Obviously, local eccentricity has not to be taken into account if a weld is used as part of a weld group around the perimeter of a structural hollow section.

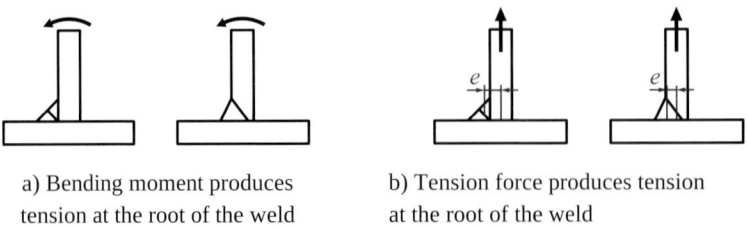

a) Bending moment produces tension at the root of the weld

b) Tension force produces tension at the root of the weld

Figure 4.26 – Single fillet or single-sided partial penetration butt welds

4.4.2.5 Angles connected by one leg

In section 3.4.11, attention has been drawn to the presence of eccentricities in angles connected, for instance to a gusset plate, by bolts located on one leg only. The same occurs for similar welded joints. Here again a simplified solution may be found by adopting an effective cross sectional area and then treating the member as concentrically loaded.

For an equal-leg angle, or an unequal-leg angle connected by its larger leg, the effective area may be taken as equal to the gross area.

For an unequal-leg angle connected by its smaller leg, the effective area should be taken as equal to the gross cross sectional area of an equivalent equal-leg angle of leg size equal to that of the smaller leg, when determining the design resistance of the cross section. However, when determining the design buckling resistance of a compression member, the actual gross cross sectional area should be used.

4.4.2.6 Gusset plates

When gusset plates are used in a joint, they may be either connected by bolts or welds to each of the connected members. For the design of the weld, no further aspects need to be addressed here while, for the gusset plate, design rules have already been proposed in section 3.5.3.

Chapter 5

SIMPLE JOINTS

5.1 INTRODUCTION

Simple joints are assumed to transfer no bending moments and should be designed accordingly. In plane structures, these ones are so intended to resist to axial and shear forces.

In the following figures, various joint configurations where simple joints are likely to be used are illustrated. They are grouped in five categories:

- Beam-to-column joint configurations (Figure 5.1)
- Beam-to-beam joint configurations (Figure 5.2)
- Beam splice joint configurations (Figure 5.3)
- Bracing joint configurations (Figure 5.4)
- Column base joint configurations (Figure 5.5)

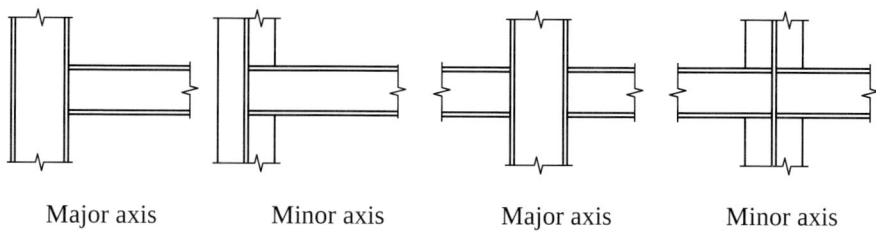

Major axis   Minor axis   Major axis   Minor axis

a) Single-sided joint configurations   b) Double-sided joint configurations

Figure 5.1 – Beam-to-column joint configurations

5. Simple Joints

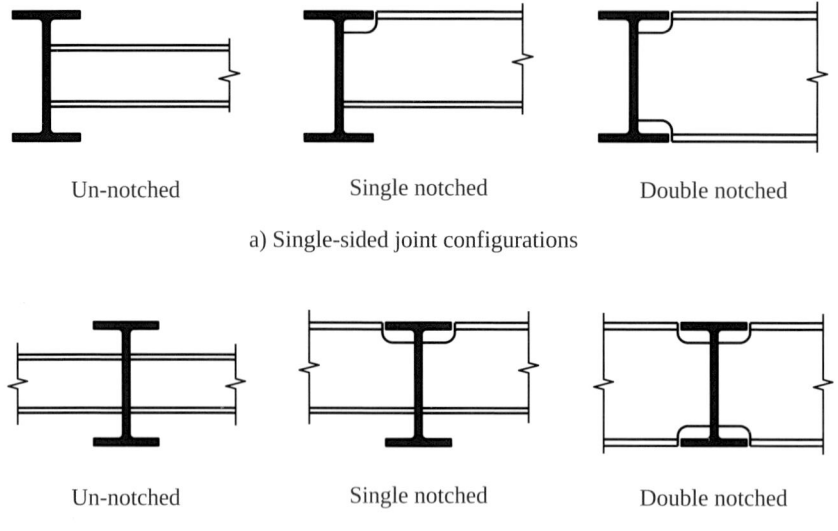

a) Single-sided joint configurations

b) Double-sided joint configurations

Figure 5.2 – Beam-to-beam joint configurations

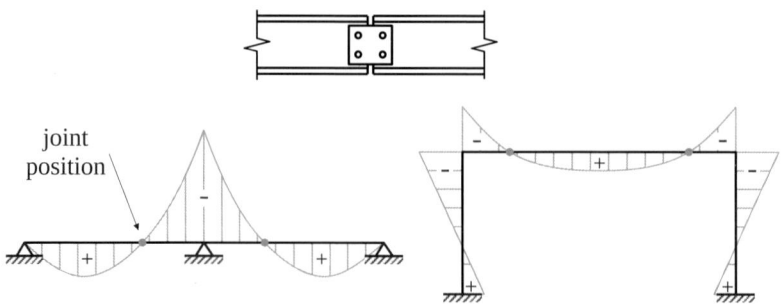

Figure 5.3 – Beam splices and possible locations of simple beam splice joints

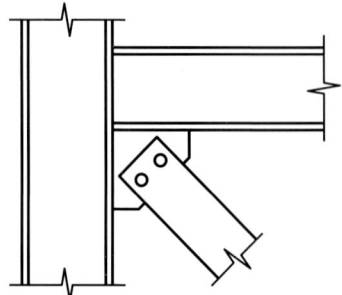

Figure 5.4 – Bracing configuration

5.2 STEEL JOINTS

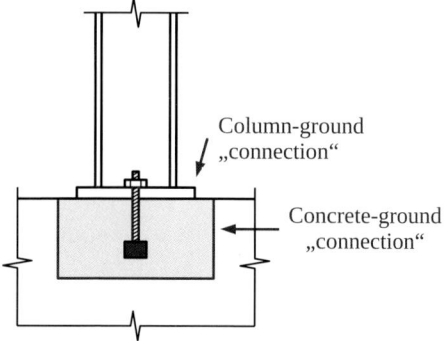

Figure 5.5 – Column base joint configuration

The connected members may be steel ones, or composite ones. Design rules for beam-to-column, beam-to-beam configurations and column bases are provided in the present chapter. As a straightforward extension of these rules to beam splices may be contemplated, this specific configuration will not be explicitly addressed. The same applies to connections of bracings, the design of which is covered by rules already provided in chapter 3 and chapter 4.

5.2 STEEL JOINTS

5.2.1 Introduction

In Eurocode 3 Part 1-8, no specific design rules are provided for simple joints. For sure, formulae are provided for some joint components, but no full consistent set of design rules is made available. On the other hand, in some European countries, national rules for simple structural joints were existing, often since many years. Unfortunately, these recommendations did not cover all the possible types of failure and were sometimes given significantly different design rules for a typical failure mode.

In a first step, a comparative study (Guillaume, 2000) of such available design rules for simple connections has been performed. In this work, reference has been made to different normative documents or design recommendations:

– Eurocode 3 (CEN, 2005a) and its Part 1-8 (CEN, 2005c);

5. Simple Joints

- BS 5950 (BSI, 2000) and BCSA-SCI recommendations (BCSA/SCI Connections Group, 1993, 1992)
- NEN 6770 (NEN, 1997; Dutch Commissie SG/TC-10a (Verbindingen), 1998);
- German "Ringbuch" (Sedlacek *et al*, 2000)

Each of these documents covers its own application field, in which a limited number of possible failure modes will potentially occur. So, the comparison between them was difficult.

With the aim of establishing a full design approach according to the general design principles stated in Eurocode 3, design sheets for header plate and fin plate connections were so prepared at the University of Liège and discussed at several meetings of Technical Committee 10 "Connections" of the European Convention for Constructional Steelwork (ECCS). The present section present these design rules, as finally agreed by ECCS in the form of "European Design Recommendations for the Design of Simple Joints in Steel Structures" (Jaspart *et al*, 2009).

The practical design recommendations presented in the aforementioned publication or in its eventual revised version could possibly one day be implemented in EN 1993-1-8, as it is already the case in the last published version of the BCSA/SCI recommendations (Moreno *et al*, 2011) and the German "Ringbuch" (Weynand *et al*, 2013).

5.2.2 Scope and field of application

5.2.2.1 Types of structure

Simple structural joints are commonly met in steel framed buildings but they can be used also in other types of structures to connect steel elements (for example in bridges).

5.2.2.2 Types of connected elements

The shapes of the structural connected elements which are considered are:

- I or H beams;
- I or H columns (with a possible easy extension to RHS and CHS).

5.2.2.3 Types of loading

The design methods are intended for joints subject to predominantly static or quasi-static loading. Fatigue aspects are not considered.

The resistance of the joints is checked under shear forces, what corresponds to usual loading conditions of the structure during its life;

5.2.2.4 Types of fasteners

a) Bolts

There are two classes of bolts: normal bolts and high strength bolts. The second class is characterised by a slip-type resistance mode in shear.

Here under, only non-preloaded bolts are explicitly covered.

b) Welds

In Eurocode 3, various types of welds are considered: fillet welds, fillet welds all round, butt welds, plug welds and flare groove welds. Only fillet welds are explicitly considered here.

5.2.2.5 Types of connections

Three connections types, used in the present design recommendations to connect a beam to a column or a beam to a beam, are specified below.

a) Header plate connections

The main components of a header plate connection are shown in Figure 5.6: a steel plate, a fillet weld on both sides of the supported beam web and two single or two double vertical bolt lines. The plate is welded to the supported member and bolted to a supporting element such as a steel beam or column. Its height does not exceed the clear depth of the supported beam web. The end of the supported steel beam may be un-notched, single notched or double notched.

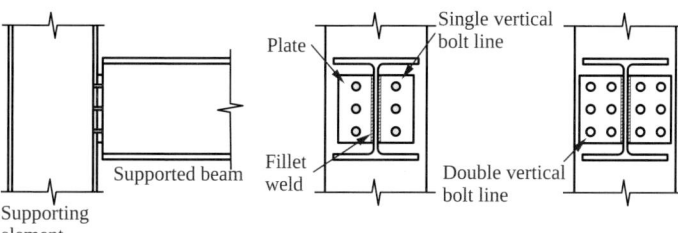

Figure 5.6 – Header plate connection

5. SIMPLE JOINTS

b) Fin plate connections

The main components of a fin plate connection are shown in Figure 5.7: a fin plate, a fillet weld on both sides of the plate and a single or double vertical bolt line. The plate is welded to a supporting member such as a steel beam or column and bolted to the web of the supported beam. The end of the supported steel beam may be un-notched, single notched or double notched.

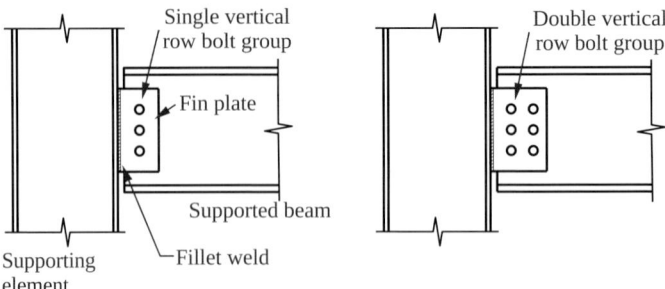

Figure 5.7 – Fin plate connection

c) Web cleat connections

A web cleat connection is characterised (see Figure 5.8) by two web cleats and three single or double vertical bolt lines (two on the supporting element and one on the supported member). The cleats are bolted to the supporting and supported members. Un-notched, single notched or double notched supported beams may be considered.

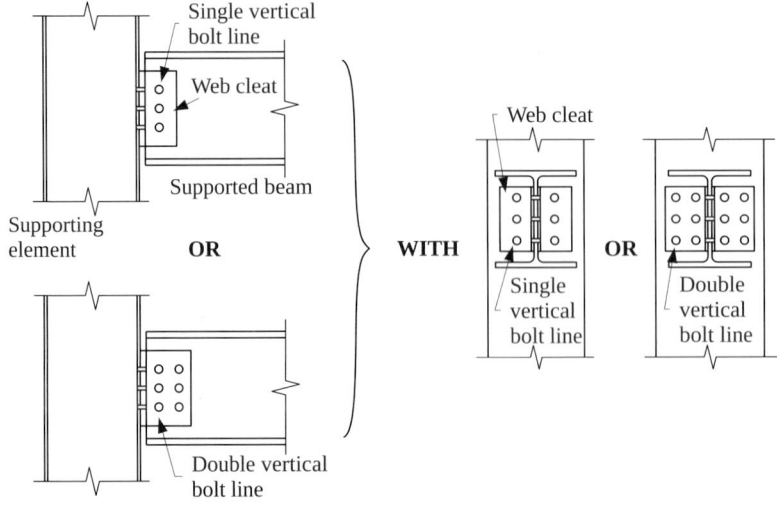

Figure 5.8 – Web cleat connection

Traditionally, other types of beam-to-column connections are sometimes considered as hinges. But nowadays Eurocode 3 Part 1-8 classifies them as semi-rigid. Two examples are given in Figure 5.9.

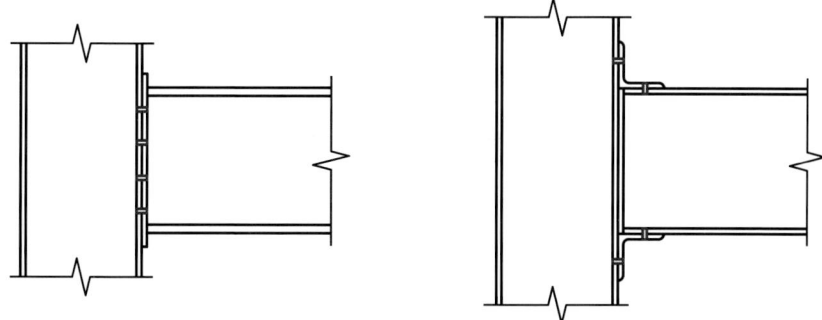

Figure 5.9 – Other simple connections

5.2.2.6 Reference code

The design rules presented in this chapter are based on the resistance formulae provided by Eurocode 3 Part 1-8, at least as far as information is available. When this is not the case, the basic design principles prescribed by Eurocode 3 are followed.

5.2.3 Joint modelling for frame analysis and design requirements

5.2.3.1 General

As explained in sub-chapter 2.2, the effects of the actual response of the joints on the distribution of internal forces and moments within a structure, and on the overall deformations, should generally be taken into account; but when these effects are sufficiently small, they may be neglected.

To identify whether the effects of joint behaviour on the analysis need be taken into account, a distinction should be made between the three previously defined types of joint modelling: simple, continuous and semi-continuous.

The appropriate type of joint modelling depends on the classification of the joint (see sub-chapter 2.4) and on the selected procedure for structural analysis and design.

5. SIMPLE JOINTS

5.2.3.2 Simple joint modelling

The joints considered in this chapter are assumed not to transmit bending moments. Thus, they should be modelled by hinges. Unfortunately, many joints which are traditionally considered as a hinge do not fulfil the stiffness and/or strength limitations required by Eurocode 3 for nominally pinned joints.

Two different attitudes may be adopted in such a case:

- According to the Eurocode 3 requirements, the joint is modelled by a rotational spring and is therefore considered as semi-rigid (what it is in reality). Its rotational stiffness, design bending resistance and shear resistance have to be evaluated and the actual properties of the joint have to be explicitly taken into consideration in the structural analysis and in the design phase. This approach is the more scientifically correct one but it needs more complex calculations as far as the global analysis and joint design are concerned.
- Despite its actual properties, the joint is considered as a hinge and the design rules presented in this chapter for simple joints can be applied, but under some strict conditions which ensure the safe character of the approach. The global analysis and the joint design are simpler in this case as they are based on a more traditional hinged (simple) approach.

If the second option is chosen, the joint is assumed not to transfer bending moments even if it is not the truth. As a consequence, bending moments actually develop in the joints although they are designed to resist only shear forces. This is potentially unsafe and at first sight is not basically acceptable.

But a careful examination of this problem leads to the conclusion that the "hinge assumption" is safe if the two following requirements are fulfilled:

- the joint possesses a sufficient rotation capacity;
- the joint possesses a sufficient ductility.

The first requirement relates to the rotational capacity that the joint should have, in order to "rotate" as a hinge, without developing too high internal bending moments.

The second requirement is there to ensure that the development of combined shear and bending forces into the joint is not leading to brittle failure modes (for instance, because of the rupture of a bolt or a weld). In other words, the design of the joint should exhibit internal plastic deformations instead of brittle phenomena.

If these two requirements (sufficient rotation capacity and ductility) are fulfilled, it can be demonstrated that to consider an actually semi-rigid joint as a nominally pinned one is safe for design purposes and, in particular, for the evaluation of:

- *the frame displacements:*

    the stiffness of the actual structure is always greater than that of the hinged one, and all the actual displacements are therefore lower than the calculated ones;

- *the plastic failure loading:*

    as the actual bending strength of the joint is higher than the considered one (equal to zero), the first order plastic resistance of the frame is higher than the one evaluated on the basis of a hinge behaviour;

- *the critical loading of linear elastic instability:*

    the transversal stiffness of the actual structure is larger than the one of the structure with nominally pinned joints, and the rotational restraints at the end of the columns in the actual structure are higher than these calculated with a hinge assumption; this ensures the safe character of the hinge assumption as far as global and local elastic instability are concerned;

- *the elastic-plastic phenomena of instability:*

    the actual stiffness of the structure is greater than the considered one but the actual loading conditions are more important than those acting on the structure with nominally pinned joints; nevertheless, various studies (Owens, Cheal, 1989; Gibbons *et al*, 1993; Gaboriau, 1995) show that the "hinged" approach is safe.

For further explanations, see (Renkin, 2003).

In this chapter, the design recommendations for simple joints relate to the "hinge model". Specific design requirements ensuring safety are presented for each of the connection types considered.

5. SIMPLE JOINTS

5.2.3.3 Summary of design requirements

As said before, the internal forces in the joint are here determined by a structural analysis based on simple joint modelling. The hinge is assumed to be located at the intersection of the axes of the connected elements. As a result of this structural analysis, the maximum applied shear force and rotation in the joint, respectively V_{Ed} and $\phi_{required}$ are obtained.

From the geometrical properties of the joint and the mechanical properties of its constitutive materials, the available rotation capacity of the joint $\phi_{available}$ can be estimated, as well as its design shear resistance V_{Rd}. To ensure the validity of this approach, some ductility requirements have to be satisfied and the available rotation of the joint has to be higher than the required one. Finally, the joint will be considered as acceptable if the applied shear force does not exceed the design shear resistance.

5.2.4 Practical ways to satisfy the ductility and rotation requirements

The interested reader will find detailed information on the ways to satisfy the ductility and rotation requirements in (Jaspart et al, 2009). In the next sections, these requirements are illustrated for every connection type.

5.2.4.1 Header plate connection

5.2.4.1.1 Design requirements for sufficient rotation capacity

To enable rotation without increasing too much the bending moment which develops into the joint, contact between the lower beam flange and the supporting member has to be strictly avoided. So, it is imperative that the height h_p of the plate is less than that of the supported beam web (Figure 5.10):

$$h_p \leq d_b \tag{5.1}$$

where d_b is the clear depth of the supported beam web.

If such a contact takes place, a compression force develops at the place of contact; it is equilibrated by tension forces in the bolts and a significant bending moment develops (Figure 5.10).

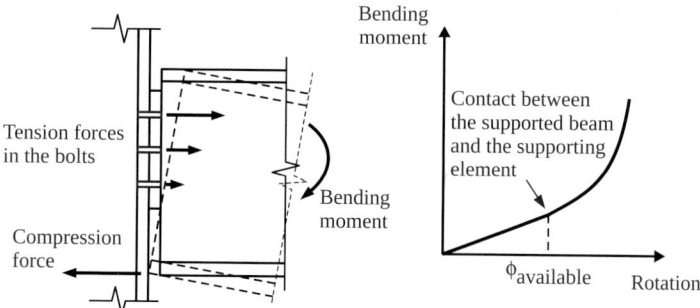

Figure 5.10 – Contact and evolution of the bending moment

The level of rotation at which the contact occurs is obviously dependent on the geometrical characteristics of the beam and of the header plate, but also on the actual deformations of the joint components.

In order to derive a simple criterion that the user could apply, before any calculation, to check whether the risk of contact may be disregarded, the following rough assumptions are made (see Figure 5.12):

- the supporting element remains un-deformed;
- the centre of rotation of the beam is located at the lower extremity of the header plate.

On the basis of such assumptions, a safe estimation (i.e. a lower bound) of the so-called "available rotation of the joint" $\phi_{available}$ may be easily derived:

$$\phi_{available} = \frac{t_p}{h_e} \qquad (5.2)$$

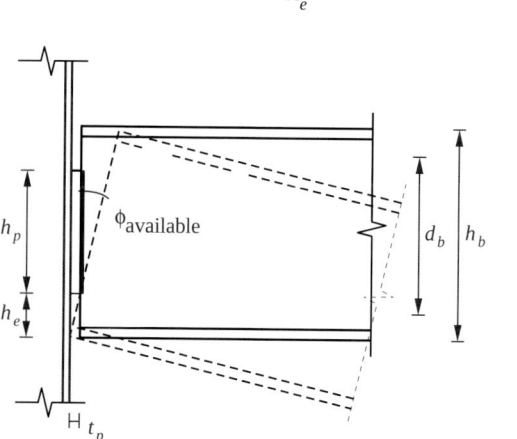

Figure 5.11 – Geometrical characteristics of the joint and illustration of contact between the beam and the supporting element

5. SIMPLE JOINTS

This available rotation has to be greater than the "required rotation capacity" which varies according to the structural system and loading. A simple criterion ensuring the sufficient joint rotation capacity may be written as:

$$\phi_{available} > \phi_{required} \qquad (5.3)$$

For instance, the required rotation capacity, for a beam (length L and inertia I) simply supported at its extremities and subjected to an uniformly distributed load (factored load γp at ULS), is given by:

$$\phi_{required} = \frac{\gamma p L^3}{24EI} \qquad (5.4)$$

By expressing that $\phi_{available} > \phi_{required}$, a simple criterion ensuring a sufficient joint rotation capacity may be derived:

$$\frac{t}{h_e} > \frac{\gamma p L^3}{24EI} \qquad (5.5)$$

Similar criteria may be derived for other load cases (Annex A).

5.2.4.1.2 Design requirements for sufficient joint ductility

As bending moments develop in the joint, the bolts and the welds are subjected to tension forces in addition to shear forces. Premature failure of those elements which exhibit a brittle failure and which are more heavily loaded in reality than in the calculation model has therefore to be strictly avoided. Simple related criteria should therefore be proposed.

a) Criterion to avoid premature bolt failure because of tension forces

In Eurocode 3, a criterion based on the T-stub approach ensures that a yield lines mechanism develops in the plate before the strength of the bolts is exhausted (see EN 1993-1-8 (CEN, 2005c)); its background is given in (Jaspart, 1997). This criterion, initially developed for end plates and column flanges, is here safely extended to column (weak axis beam-to-column joints) or beam (beam-to-beam joint configurations) webs.

According to this criterion, at least one of the two following inequalities (5.6) to (5.8) has to be satisfied:

– for header plate:

$$\frac{d}{t_p} \geq 2.8 \sqrt{\frac{f_{yp}}{f_{ub}}} \qquad (5.6)$$

– for a supporting column flange:

$$\frac{d}{t_{cf}} \geq 2.8 \sqrt{\frac{f_{ycf}}{f_{ub}}} \qquad (5.7)$$

– for a supporting column or beam web (or faces of hollow sections):

$$\frac{d}{t_w} \geq 2.8 \sqrt{\frac{f_{yw}}{f_{ub}}} \qquad (5.8)$$

where:
- d is the nominal diameter of the bolt shank;
- t_p is the thickness of the header plate;
- t_{cf} is the thickness of the supporting column flange;
- t_w is the thickness of the supporting column or beam web;
- f_{yp} is the yield strength of the steel constituting the header plate;
- f_{ycf} is the yield strength of the steel constituting the supporting column flange;
- f_{yw} is the yield strength of the steel constituting the supporting column or beam web;
- f_{ub} is the ultimate strength of the bolt.

Such a criterion does not ensure that the whole shear capacity of the bolt may be considered when evaluating the shear resistance of the joint. In fact, when this requirement is satisfied, it may be demonstrated:

– that the tension force in the bolts may amount up to $0.5 B_{t,Rd}$, i.e. 50% of the design tension resistance $B_{t,Rd}$ of the bolts, but no more;

5. SIMPLE JOINTS

- that, for such a maximum tension force, the actual shear resistance of the bolt only amounts to 64% of the full shear resistance of the bolts (according to the EC 3 resistance formula for bolts in shear and tension).

This looks at first sight to be disappointing: the user tries to maximise the shear resistance of the joint but should limit the resistance of the bolts in shear by 36%. It may be argued though that only the bolts located in the upper half of the header plane are affected by such a reduction, as the others are located in a compression zone, and are therefore not subjected to tension forces.

So finally a reduction is taken into consideration by multiplying the total resistance of the bolts in shear by a factor 0.8 (i.e. a reduction factor of 0.64 for half of the bolts located in the upper half of the header plate $-0.5[1+0.64] \approx 0.8$).

b) *Criterion to avoid premature weld failure because of tension or shear forces*

The welds must be designed according to EN 1993-1-8. In the case of relatively small loads in relation to the capacity of the web, application of the rules in 4.5.3.2 of EN 1993-1-8 may lead to rather thin welds. If the rupture strength of those thin welds is lower than the yield strength of the weakest of the connected parts, the connection has so little deformation capacity that is usually not sufficient to accommodate imposed deformations etc. In such a case the connection will behave in a brittle way.

To avoid this, the welds should be designed "full strength". The rupture strength of full strength welds would then be greater than the rupture strength of the adjacent plate; so, in the case of overloading, the plate would yield before the welds fail.

In section 4.3.6, the concept of "full strength welds" has been introduced and related design criteria have been derived. These criteria may be expressed separately for welds in shear and welds in tension. But in the present case, the welds are subjected simultaneously to shear and tension forces; as a result, the worst situation in terms of weld loading is therefore considered for design: end welds, i.e. welds subjected to pure tension.

By applying clause 4.5.3.2 of Eurocode 3 Part 1-8, using the directional method, minimum throat thicknesses required for a full strength weld design are obtained; these ones are again reported here below in Table 5.1 for various steel grades (values extracted from Table 4.3 in section 4.3.6).

Table 5.1 – Minimum required "full strength" weld thickness a in case of double fillet welds. Plate thickness smaller than 40 mm; $\gamma_{M0} = 1.0$ and $\gamma_{M2} = 1.25$

Steel grade	S235	S275	S355	S420 M	S420 N	S460 M	S460 N
Full strength double fillet welds	$a \geq$ 0.46t	$a \geq$ 0.48t	$a \geq$ 0.55t	$a \geq$ 0.71t	$a \geq$ 0.68t	$a \geq$ 0.74t	$a \geq$ 0.70t

5.2.4.2 Fin plate connection

5.2.4.2.1 Design requirements for sufficient rotation capacity

So as to permit a rotation without increasing too much the bending moment which develops into the joint, contact between the lower beam flange and the supporting member has to be strictly avoided. To achieve it, the height h_p of the fin plate should be lower than that of the supported beam web (Figure 5.12):

$$h_p \leq d_b \tag{5.9}$$

where d_b is the clear depth of the supported beam web.

If such a contact takes place, a compression force develops at the place of contact; it is equilibrated by tension forces in the welds and in the plate, and additional shear forces in the bolts.

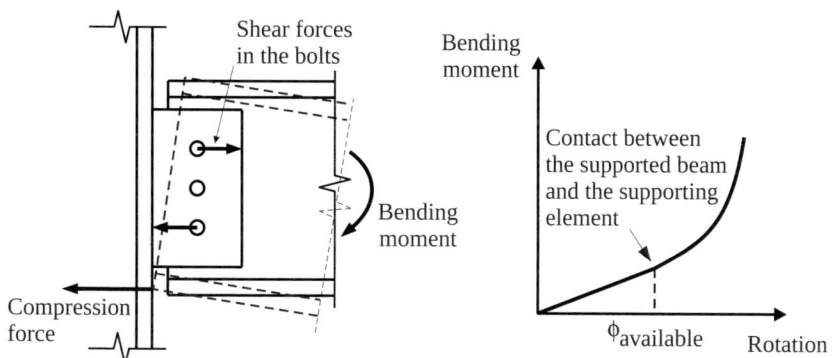

Figure 5.12 – Contact and evolution of the bending moment

5. SIMPLE JOINTS

The level of rotation at which the contact occurs is obviously dependent on the geometrical characteristics of the beam and of the fin plate, but also on the actual deformations of the joint components.

In order to derive a simple criterion that the user could apply, before any calculation, to check whether the risk of contact may be disregarded, the following rough assumptions are made (see Figure 5.13):

- the supporting element and the fin plate remain un-deformed;
- the centre of rotation of the beam is located at the centre of gravity of the bolt group.

On the basis of such assumptions, a safe estimation (i.e. a lower bound) of the so-called "available rotation of the joint" $\phi_{available}$ may be easily derived:

$$\text{if } z > \sqrt{(z-g_h)^2 + \left(\frac{h_p}{2}+h_e\right)^2} \; :$$

$$\phi_{available} = \infty \tag{5.10}$$

else:

$$\phi_{available} = \arcsin\left(\frac{z}{\sqrt{(z-g_h)^2 + \left(\frac{h_p}{2}+h_e\right)^2}}\right) - \arctg\left(\frac{z-g_h}{\frac{h_p}{2}+h_e}\right) \tag{5.11}$$

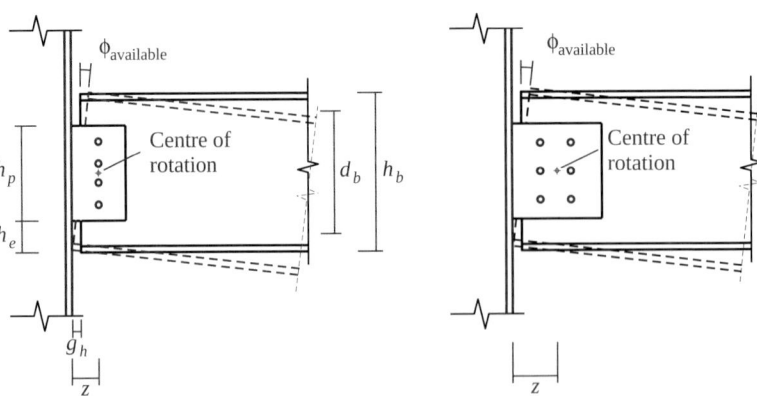

Figure 5.13 – Geometrical characteristics of the joint and illustration of the contact between the beam and the supporting element

This available rotation has to be greater than the "required rotation capacity" which varies according to the structural system and loading. A simple criterion ensuring the sufficient joint rotation capacity may be written as:

$$\phi_{available} > \phi_{required} \qquad (5.12)$$

Expressions for $\phi_{required}$ are given in 5.2.4.1.1 and Annex A.

5.2.4.2.2 Design requirements for sufficient joint ductility

As previously explained, the design shear resistance of the joint may be reached, as a result of a plastic redistribution of internal forces amongst the different constitutive components. This requires that no local brittle failure modes or instabilities develop during this redistribution. The failure modes which could prevent redistribution of internal forces to take place are, for fin plate connections: the bolts and the welds in shear on account of their brittle nature, and the buckling of the fin plate which is assumed to be non-ductile in terms of plastic redistribution.

a) Criterion to avoid premature weld failure because of tension forces

The same full strength criterion as the one established for the header plate connection may be considered (see 5.2.4.1.2). Related minimum weld throat thicknesses may so again be obtained from Table 5.1.

b) Criterion to permit a plastic redistribution of internal forces between the "actual" and "design" loading situations

1. First of all, the design shear resistance of the connection should be associated with a ductile mode. Failure by bolts in shear or by buckling of the fin plate is therefore excluded. A first criterion can be written as:

$$\min(V_{Rd1}; V_{Rd7}) > V_{Rd} \qquad (5.13)$$

where:

V_{Rd1} is the shear resistance of the bolts;
V_{Rd7} is the buckling resistance of the fin plate;
V_{Rd} is the design shear resistance of the connection.

5. SIMPLE JOINTS

2. Secondly, the component which yields under the "actual" loading in the connection has also to be ductile (so, no bolts in shear or buckling of the fin plate). To ensure this, different criteria have to be fulfilled dependent on the failure mode obtained through treating the connections as "hinged":

– Failures by bolts in shear or buckling of the fin plate:

Excluded by the first criterion (1), just above.

– All other failure modes:

For one vertical bolt row, at least one of the following two inequalities has to be satisfied:

For the beam web:

$$F_{b,hor,Rd} \leq \min\left(F_{v,Rd}; V_{Rd\,7}\,\beta\right) \qquad (5.14)$$

For the fin plate:

$$F_{b,hor,Rd} \leq \min\left(F_{v,Rd}; V_{Rd\,7}\,\beta\right) \qquad (5.15)$$

For two vertical bolt rows, at least one of the following three inequalities has to be satisfied:

For the beam web:

$$\max\left(\frac{1}{F_{v,Rd}^2}\left(\alpha^2+\beta^2\right); \frac{1}{V_{Rd\,7}^2}\right) \leq \left(\frac{\alpha}{F_{b,ver,Rd}}\right)^2 + \left(\frac{\beta}{F_{b,hor,Rd}}\right)^2 \qquad (5.16)$$

For the fin plate:

$$\max\left(\frac{1}{F_{v,Rd}^2}\left(\alpha^2+\beta^2\right); \frac{1}{V_{Rd\,7}^2}\right) \leq \left(\frac{\alpha}{F_{b,ver,Rd}}\right)^2 + \left(\frac{\beta}{F_{b,hor,Rd}}\right)^2 \qquad (5.17)$$

And:

$$V_{Rd\,6} \leq \min\left(\frac{2}{3\sqrt{\alpha^2+\beta^2}}F_{v,Rd}; \frac{2}{3}V_{Rd\,7}\right) \qquad (5.18)$$

3. Lastly, during the redistribution process, the "bolts in shear" failure mode should not be met. To avoid that, simple criteria can be written that again depends on the failure mode resulting from treating the connection as a "hinge":

   – Failure by bolts in shear or buckling of the fin plate:

   Excluded by the first criterion (1).

   – Failure by fin plate or beam web in bearing:

   If the two first criteria (1) and (2) are fulfilled, no additional criterion is necessary.

   – All other failure modes:

$$V_{Rd1} > \min\left(V_{Rd2}; V_{Rd8}\right) \tag{5.19}$$

where:

V_{Rd1} is the shear resistance of the bolts;
V_{Rd2} is the bearing resistance of the fin plate;
V_{Rd8} is the bearing resistance of the beam web.

Notation used in the above requirements is given in section 5.2.5. The criteria (1), (2) and (3) can be only checked after the evaluation of the design shear resistance of the joint. For further explanations about the derivation of these requirements, see (Renkin, 2003).

5.2.4.3 Web cleat connection

In (Jaspart *et al*, 2009), the web cleat connections are just briefly addressed as it is stated that "the components to be considered are simply those met in header plate (cleat leg connected to the supporting member) and fin plate (cleat leg connected to the supported member) connections". It is so concluded that the design rules and requirements for a safe approach may be simply deduced from those established for the two previous connection types.

In (Weynand *et al*, 2012) however it is pointed out that the previous statement has sense for all components active in web cleat connection, except for the "header plate in bending" one which required further consideration. The conclusion drawn in (Weynand *et al*, 2012) are summarised below.

In (Jaspart *et al*, 2009), the header plate is modelled as a simple single-span beam with elastic stress distribution over its rectangular cross section. This

5. SIMPLE JOINTS

model is not applicable for web cleats. In a double web cleated connection, the eccentricity of the bolts connected to the supporting member lead to a moment M_y, see Figure 5.14, resulting in compression forces in the upper part and horizontal tension forces in the lower part of the cleat. It is assumed that the compression forces will be carried by contact of the cleats to the web of the beam (Figure 5.15). The width of the compression zone b_{com} is derived by assuming an angle $\alpha = 60°$. The height of the compression zone h_{com} results from the criterion that the compression force should not exceed the yield strength of the plates (Figure 5.16).

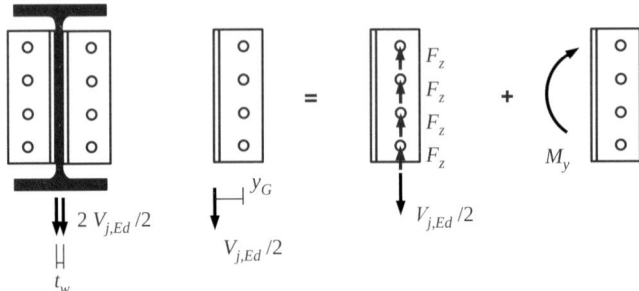

Figure 5.14 – Forces at supporting member side

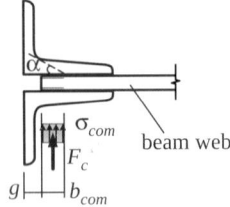

Figure 5.15 – Compression zone

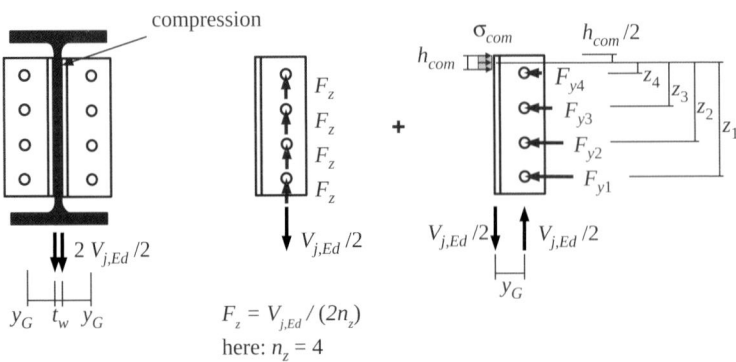

Figure 5.16 – Forces on the supporting element side

The tension forces are taken by additional (horizontal) forces in the bolts of the leg connected to the supporting member (Figure 5.16), leading to a reduction in the vertical component of the shear force in the bolt.

The centre of compression is assumed mid height of the compression zone. An elastic distribution of the resulting horizontal forces is assumed. Hence, horizontal forces in the bolts increase linear (Figure 5.16). With these assumptions and an initial guess for the height of the compression zone h_{com}, all forces may be determined and a more accurate value of h_{com} can be found by a short iteration.

Because of the short distance from the compression zone to the supporting element side, the horizontal forces are directly absorbed by the bolts at the supporting element side (Figure 5.15). Torsion in the cleat and additional tying forces in the bolts on the supported element side do not arise.

In order to avoid notched beams, a configuration with double web cleated connections with "long" legs may be used as shown in Figure 5.17. To check the "bending in the cleat" connected to the supporting member, for such a configuration, a compression zone as shown in Figure 5.15 may not be assumed. Hence, the cleat can rotate more freely and the moment M_y, see Figure 5.14, must be carried by the bolt pattern of the cleat legs on the supporting member. Resulting forces are shown in Figure 5.17. If these "bolts in shear" are the decisive component of the joint, the presence of the horizontal forces due to M_y will obviously lead to a reduction of the shear resistance of the joint.

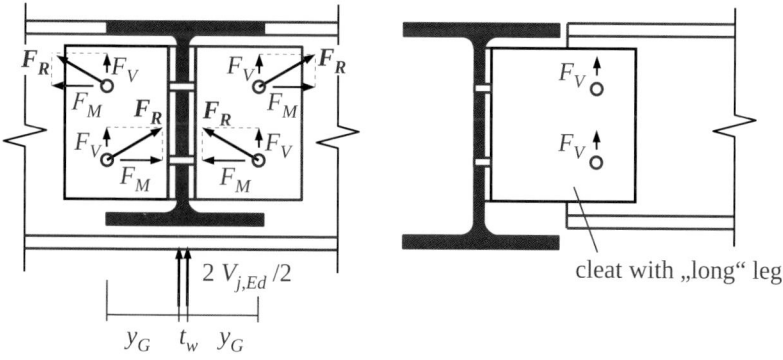

Figure 5.17 – Forces on the supporting element side for cleats with long legs

5. SIMPLE JOINTS

5.2.5 Design rules for joint characterisation

As stated before, provided that ductility and rotation requirements are satisfied, the design of the simple joints reduces to the limitation of the applied shear force to the shear design resistance. The evaluation of the latter is provided below for three types of connections, namely header plate, fin plate and web cleat connections. The forces applied to joints at the ultimate limit state result from a structural analysis and shall be determined according to the principles given in EN 1993-1-1. The resistance of the joint is determined on the basis of the resistances of the individual fasteners, welds and other components, as shown below.

5.2.5.1 Connections with a header plate

5.2.5.1.1 Notations

Specific notations used in the following sections for header plates are shown in Figure 5.18.

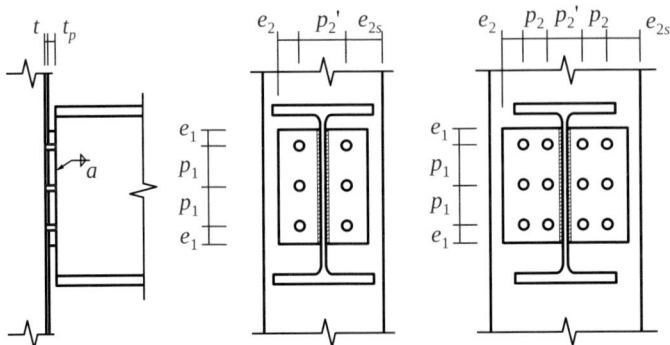

Figure 5.18 – Header plate notations

5.2.5.1.2 Requirements to ensure the safety of the approach

To apply the design rules presented in section 5.2.5.1.3, all the following inequalities have to be satisfied.

1. $h_p \leq d_b$

2. $\dfrac{t_p}{h_e} > \phi_{required}$

3. If the supporting element is a beam or column web:

$$\frac{d}{t_p} \geq 2.8\sqrt{\frac{f_{yp}}{f_{ub}}} \quad \text{or} \quad \frac{d}{t_w} \geq 2.8\sqrt{\frac{f_{yw}}{f_{ub}}}$$

If the supporting element is a column flange:

$$\frac{d}{t_p} \geq 2.8\sqrt{\frac{f_{yp}}{f_{ub}}} \quad \text{or} \quad \frac{d}{t_{cf}} \geq 2.8\sqrt{\frac{f_{ycf}}{f_{ub}}}$$

4. $\quad a \geq \dfrac{\beta_w f_{ybw} \gamma_{M2}}{\sqrt{2} f_{ubw} \gamma_{M0}} t_{bw}$

(β_w is given in Table 4.2)

5.2.5.1.3 Resistance to shear forces

To apply the design rules presented hereafter, the criterion given in section 5.2.5.1.2 have to be satisfied. The shear resistance of a header plate connection may be determined by considering the eight components:

a) Bolts in shear

$$V_{Rd1} = 0.8 n F_{v,Rd}$$

According Table 3.4 in EN 1993 Part 1-8:

$$F_{v,Rd} = \frac{\alpha_v f_{ub} A}{\gamma_{M2}}$$

- where the shear plane passes through the threaded portion of the bolt:
  $A = A_s$ is the tensile stress area of the bolt
  $\alpha_v = 0.6$ for 4.6, 5.6 and 8.8 bolt grades
  $\alpha_v = 0.5$ for 4.8, 5.8, 6.8 and 10.9 bolt grades

- where the shear plane passes through the unthreaded portion of the bolt:
  A is the gross cross area of the bolt
  $\alpha_v = 0.6$

5. SIMPLE JOINTS

b) Header plate in bearing

$$V_{Rd2} = nF_{b,Rd}$$

According Table 3.4 in EN 1993 Part 1-8:

$$F_{b,Rd} = \frac{k_1 \alpha_b f_{up} d t_p}{\gamma_{M2}}$$

where

$$k_1 = \min\left(2.8\frac{e_2}{d_0} - 1.7 \,;\, 1.4\frac{p_2}{d_0} - 1.7 \,;\, 2.5\right)$$

$$\alpha_b = \min\left(\frac{e_1}{3d_0} \,;\, \frac{p_1}{3d_0} - \frac{1}{4} \,;\, \frac{f_{ub}}{f_{up}} \text{ or } 1.0\right)$$

c) Supporting member in bearing

$$V_{Rd3} = nF_{b,Rd}$$

$$F_{b,Rd} = \frac{k_1 \alpha_b f_u d t}{\gamma_{M2}}$$

— where the supporting element is a column flange:

$$k_1 = \min\left(2.8\frac{e_{2s}}{d_0} - 1.7 \,;\, 1.4\frac{p_2}{d_0} - 1.7 \,;\, 2.5\right)$$

$$\alpha_b = \min\left(\frac{p_1}{3d_0} - \frac{1}{4} \,;\, \frac{f_{ub}}{f_u} \text{ or } 1.0\right)$$

$$f_u = f_{ucf}$$

$$t = t_{cf}$$

— where the supporting element is a column web:

$$k_1 = \min\left(1.4\frac{p_2}{d_0} - 1.7 \,;\, 2.5\right)$$

$$\alpha_b = \min\left(\frac{p_1}{3d_0} - \frac{1}{4}; \frac{f_{ub}}{f_u} \text{ or } 1.0\right)$$

$$f_u = f_{ucw}$$

$$t = t_{cw}$$

- where the supporting element is a beam web:

$$k_1 = \min\left(1.4\frac{p_2}{d_0} - 1.7; 2.5\right)$$

$$\alpha_b = \min\left(\frac{p_1}{3d_0} - \frac{1}{4}; \frac{f_{ub}}{f_u} \text{ or } 1.0\right)$$

$$f_u = f_{ubw}$$

$$t = t_{bw}$$

Formulae as written here apply to major axis beam-to-column joints (connection to a column flange), to single-sided minor axis joints and to single-sided beam-to-beam joint configurations. In the other cases, the bearing forces result from both the left and right connected members, with the added problem that the number of connecting bolts may differ for the left and right connections. The calculation procedure may cover such cases without any particular difficulty. It could just bring some more complexity in the final presentation of the design sheet.

d) Header plate in shear: Gross section

$$V_{Rd\,4} = \frac{2h_p t_p}{1.27} \frac{f_{yp}}{\sqrt{3}\gamma_{M0}} \quad \text{(2 resisting sections)}$$

e) Header plate in shear: Net section

$$V_{Rd\,5} = 2A_{v,net} \frac{f_{up}}{\sqrt{3}\gamma_{M2}} \quad \text{(2 resisting sections)}$$

$$\text{with } A_{v,net} = t_p\left(h_p - n_1 d_0\right)$$

5. SIMPLE JOINTS

f) Header plate in shear: Shear block

$$V_{Rd6} = 2F_{eff,Rd} \quad \text{(2 resisting sections)}$$

- If $h_p < 1.36 p_{22}$ and $n_1 > 1$:

$$F_{eff,Rd} = F_{eff,2,Rd} = 0.5 \frac{f_{up} A_{nt}}{\gamma_{M2}} + \frac{1}{\sqrt{3}} f_{yp} \frac{A_{nv}}{\gamma_{M0}}$$

- else:

$$F_{eff,Rd} = F_{eff,1,Rd} = \frac{f_{up} A_{nt}}{\gamma_{M2}} + \frac{1}{\sqrt{3}} f_{yp} \frac{A_{nv}}{\gamma_{M0}}$$

with

$p_{22} = p_2'$ for $n_2 = 2$

$p_{22} = p_2' + p_2$ for $n_2 = 4$

A_{nt} = net area subjected to tension

- for one vertical bolt row ($n_2 = 2$):

$$A_{nt} = t_p \left(e_2 - d_0/2 \right)$$

- for two vertical bolt rows ($n_2 = 4$):

$$A_{nt} = t_p \left(p_2 + e_2 - 3d_0/2 \right)$$

A_{nv} = net area subjected to shear

$$A_{nv} = t_p \left(h_p - e_1 - (n_1 - 0.5) d_0 \right)$$

(see clause 3.10.2 in EN 1993 Part 1-8)

g) Header plate in bending

- If $h_p \geq 1.36 p_{22}$: $V_{Rd7} = \infty$

- else: $V_{Rd7} = \dfrac{2 W_{el}}{\dfrac{(p_{22} - t_w)}{2}} \dfrac{f_{yp}}{\gamma_{M_0}}$

with:

$p_{22} = p_2'$ for $n_2 = 2$

$p_{22} = p_2' + p_2$ for $n_2 = 4$

5.2 STEEL JOINTS

$$W_{el} = \frac{t_p h_p^2}{6}$$

h) Beam web in shear

$$V_{Rd\,8} = t_{bw} h_p \frac{f_{ybw}}{\gamma_{M0} \sqrt{3}} \qquad \text{(clause 5.4.6 in Eurocode 3)}$$

i) Shear resistance of the joint

$$V_{Rd} = \min_{i=1}^{8}\{V_{Rd\,i}\}$$

Note: The design shear resistance of the joint can only be considered if all the requirements given in section 5.2.5.1.2 are satisfied.

5.2.5.2 Connections with a fin plate

5.2.5.2.1 Notations

Specific notations for fin plate connections used in the following sections are shown in Figure 5.19.

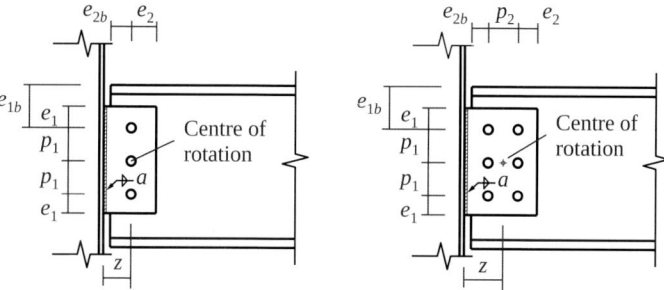

Figure 5.19 – Fin plate notations

5.2.5.2.2 Requirements to ensure sufficient rotation capacity

The two following inequalities have to be satisfied.

1. $h_p \leq d_b$

2. $\phi_{available} > \phi_{required}$

5. SIMPLE JOINTS

where:

if $z > \sqrt{(z-g_h)^2 + \left(\dfrac{h_p}{2}+h_e\right)^2}$

$\phi_{available} = \infty$

else:

$\phi_{available} = \arcsin\left(\dfrac{z}{\sqrt{(z-g_h)^2 + \left(\dfrac{h_p}{2}+h_e\right)^2}}\right) - \arctg\left(\dfrac{z-g_h}{\dfrac{h_p}{2}+h_e}\right)$

5.2.5.2.3 Requirements to avoid premature weld failure

The following inequality has to be fulfilled.

1. $a \geq \dfrac{\beta_w\, f_{yp}\, \gamma_{M2}}{\sqrt{2}\, f_{up}\, \gamma_{M0}} t_p$

(β_w is given in Table 4.2)

5.2.5.2.4 Resistance to shear forces

a) Bolts in shear

$V_{Rd1} = \dfrac{n\, F_{v,Rd}}{\sqrt{1 + \left(\dfrac{6z}{(n+1)\, p_1}\right)^2}}$ for $n_2 = 1$

$V_{Rd1} = \dfrac{F_{v,Rd}}{\sqrt{\left(\dfrac{z\, p_2}{2I} + \dfrac{1}{n}\right)^2 + \left(\dfrac{z\, p_1}{2I}(n_1-1)\right)^2}}$ for $n_2 = 1$

with:

$I = \dfrac{n_1}{2} p_2^2 + \dfrac{1}{6} n_1 (n_1^2 - 1) p_1^2$

where:

According Table 3.4 in EN 1993 Part 1-8:

$$F_{v,Rd} = \frac{\alpha_v f_{ub} A}{\gamma_{M2}}$$

- where the shear plane passes through the threaded portion of the bolt:

  $A = A_s$ tensile stress area of the bolt

  $\alpha_v = 0.6$ for 4.6, 5.6 and 8.8 bolt grades

  $\alpha_v = 0.5$ for 4.8, 5.8, 6.8 and 10.9 bolt grades

- where the shear plane passes through the unthreaded portion of the bolt:

  A gross cross area of the bolt)

  $\alpha_v = 0.6$

b) Fin plate in bearing

$$V_{Rd2} = \frac{1}{\sqrt{\left(\dfrac{\dfrac{1}{n}+\alpha}{F_{b,ver,Rd}}\right)^2 + \left(\dfrac{\beta}{F_{b,hor,Rd}}\right)^2}}$$

- for $n_2 = 1$:

  $\alpha = 0$

  $$\beta = \frac{6z}{p_1 n(n+1)}$$

- for $n_2 = 2$:

  $$\alpha = \frac{z}{I}\frac{p_2}{2}$$

  $$\beta = \frac{z}{I}\frac{n_1 - 1}{2} p_1$$

5. SIMPLE JOINTS

with $I = \dfrac{n_1}{2} p_2^2 + \dfrac{1}{6} n_1 \left(n_1^2 - 1\right) p_1^2$

According Table 3.4 in EN 1993 Part 1-8:

- bearing resistance in vertical direction:

$$F_{b,ver,Rd} = \dfrac{k_1 \alpha_b f_{up} d t_p}{\gamma_{M2}}$$

where:

$$k_1 = \min\left(2.8\dfrac{e_2}{d_0} - 1.7 \,;\, 1.4\dfrac{p_2}{d_0} - 1.7 \,;\, 2.5\right)$$

$$\alpha_b = \min\left(\dfrac{e_1}{3 d_0} \,;\, \dfrac{p_1}{3 d_0} - \dfrac{1}{4} \,;\, \dfrac{f_{ub}}{f_{up}} \text{ or } 1.0\right)$$

- bearing resistance in horizontal direction:

$$F_{b,hor,Rd} = \dfrac{k_1 \alpha_b f_{up} d t_p}{\gamma_{M2}}$$

where:

$$k_1 = \min\left(2.8\dfrac{e_1}{d_0} - 1.7 \,;\, 1.4\dfrac{p_1}{d_0} - 1.7 \,;\, 2.5\right)$$

$$\alpha_b = \min\left(\dfrac{e_2}{3 d_0} \,;\, \dfrac{p_2}{3 d_0} - \dfrac{1}{4} \,;\, \dfrac{f_{ub}}{f_{up}} \text{ or } 1.0\right)$$

c) Fin plate in shear: Gross section

$$V_{Rd3} = \dfrac{h_p t_p}{1.27} \dfrac{f_{yp}}{\sqrt{3}\,\gamma_{M0}}$$

d) Fin plate in shear: Net section

$$V_{Rd4} = A_{v,net} \dfrac{f_{up}}{\sqrt{3}\,\gamma_{M2}}$$

with $A_{v,net} = t_p \left(h_p - n_1 d_0\right)$

e) Fin plate in shear: Shear block

$$V_{Rd5} = F_{eff,2,Rd}$$

where:

$$F_{eff,2,Rd} = 0.5 \frac{f_{up} A_{nt}}{\gamma_{M2}} + \frac{1}{\sqrt{3}} f_{yp} \frac{A_{nv}}{\gamma_{M0}}$$

A_{nt} = net area subjected to tension
- for one vertical bolt row ($n_2 = 1$):

$$A_{nt} = t_p \left(e_2 - d_0/2 \right)$$

- for two vertical bolt rows ($n_2 = 2$):

$$A_{nt} = t_p \left(p_2 + e_2 - 3d_0/2 \right)$$

A_{nv} = net area subjected to shear

$$A_{nv} = t_p \left(h_p - e_1 - (n_1 - 0.5) d_0 \right)$$

(see clause 3.10.2 in EN 1993 Part 1-8)

f) Fin plate in bending
- If $h_p \geq 2.73 z$: $V_{Rd6} = \infty$
- else: $V_{Rd6} = \dfrac{W_{el}}{z} \dfrac{f_{yp}}{\gamma_{M0}}$

with:

$$W_{el} = \frac{t_p h_p^2}{6}$$

g) Buckling of the fin plate

Derived from BS 5950-1 (BSI, 2000; BCSA/SCI Connections Group *et al*, 2002):

$$V_{Rd7} = \frac{W_{el}}{z_p} \frac{f_{pLT}}{0.6 \gamma_{M1}} \leq \frac{W_{el}}{z_p} \frac{f_{yp}}{\gamma_{M0}} \qquad \text{if } z_p > t_p/0.15$$

$$V_{Rd7} = V_{Rd6} \qquad \text{if } z_p \leq t_p/0.15$$

5. SIMPLE JOINTS

where:

$$W_{el} = \frac{t_p h_p^2}{6}$$

f_{pLT} is the lateral torsional buckling strength of the plate, see Annex B

h) Beam web in bearing

$$V_{Rd8} = \frac{1}{\sqrt{\left(\dfrac{\dfrac{1}{n}+\alpha}{F_{b,ver,Rd}}\right)^2 + \left(\dfrac{\beta}{F_{b,hor,Rd}}\right)^2}}$$

– for $n_2 = 1$:

$$\alpha = 0$$

$$\beta = \frac{6z}{p_1 n(n+1)}$$

– for $n_2 = 2$:

$$\alpha = \frac{z}{I}\frac{p_2}{2}$$

$$\beta = \frac{z}{I}\frac{n_1-1}{2}p_1$$

with $I = \dfrac{n_1}{2}p_2^2 + \dfrac{1}{6}n_1\left(n_1^2-1\right)p_1^2$

According Table 3.4 in EN 1993 Part 1-8:
– bearing resistance in vertical direction:

$$F_{b,ver,Rd} = \frac{k_1 \alpha_b f_{ubw} d t_{bw}}{\gamma_{M2}}$$

where:

$$k_1 = \min\left(2.8\frac{e_{2b}}{d_0}-1.7\,;\,1.4\frac{p_2}{d_0}-1.7\,;\,2.5\right)$$

$$\alpha_b = \min\left(\frac{p_1}{3d_0} - \frac{1}{4}; \frac{f_{ub}}{f_{ubw}} \text{ or } 1.0\right)$$

– bearing resistance in horizontal direction:

$$F_{b,hor,Rd} = \frac{k_1 \alpha_b f_{ubw} d t_{bw}}{\gamma_{M2}}$$

where:

$$k_1 = \min\left(1.4\frac{p_1}{d_0} - 1.7; 2.5\right)$$

$$\alpha_b = \min\left(\frac{e_{2b}}{3d_0}; \frac{p_2}{3d_0} - \frac{1}{4}; \frac{f_{ub}}{f_{ubw}} \text{ or } 1.0\right)$$

i) Beam web in shear: Gross section

$$V_{Rd\,9} = A_{b,v} \frac{f_{ybw}}{\sqrt{3}\,\gamma_{M0}} \qquad \text{(clause 5.4.6 in Eurocode 3)}$$

j) Beam web in shear: Net section

$$V_{Rd\,10} = A_{b,v,net} \frac{f_{ubw}}{\sqrt{3}\,\gamma_{M2}}$$

with $A_{b,v,net} = A_{b,v} - n_1 d_0 t_{bw}$

k) Beam web in shear: Shear block

$$V_{Rd\,11} = F_{eff,2,Rd}$$

where

$$F_{eff,2,Rd} = 0.5\frac{f_{ubw} A_{nt}}{\gamma_{M2}} + \frac{1}{\sqrt{3}} f_{ybw} \frac{A_{nv}}{\gamma_{M0}}$$

A_{nt} = net area subjected to tension

- for one vertical bolt row ($n_2 = 1$):

$$A_{nt} = t_{bw}\left(e_{2b} - d_0/2\right)$$

5. SIMPLE JOINTS

- for two vertical bolt rows ($n_2 = 2$):
$$A_{nt} = t_{bw}\left(p_2 + e_{2b} - 3d_0/2\right)$$

A_{nv} = net area subjected to shear
$$A_{nv} = t_{bw}\left(e_{1b} + (n_1 - 1)p_1 - (n_1 - 0.5)d_0\right)$$

(see clause 3.10.2 in EN 1993 Part 1-8)

l) Shear resistance of the joint

$$V_{Rd} = \min_{i=1}^{11}\{V_{Rd\,i}\}$$

Note: The design shear resistance of the joint can only be considered if all the requirements given in sections 5.2.5.2.2, 5.2.5.2.3 and 5.2.5.2.5 are satisfied.

5.2.5.2.5 Requirements to permit a plastic redistribution of internal forces

All the following inequalities have to be satisfied.

1. $V_{Rd} < \min(V_{Rd\,1}; V_{Rd\,7})$

2. For $n_2 = 1$:

   - for the beam web $\quad F_{b,hor,Rd} \leq \min\left(F_{v,Rd}; V_{Rd\,7}\,\beta\right)$

   OR

   - for the fin plate: $\quad F_{b,hor,Rd} \leq \min\left(F_{v,Rd}; V_{Rd\,7}\,\beta\right)$

   For $n_2 = 2$:

   - for the beam web

   $$\max\left(\frac{1}{F_{v,Rd}^2}\left(\alpha^2 + \beta^2\right); \frac{1}{V_{Rd\,7}^2}\right) \leq \left(\frac{\alpha}{F_{b,ver,Rd}}\right)^2 + \left(\frac{\beta}{F_{b,hor,Rd}}\right)^2$$

   OR

   - for the fin plate

   $$\max\left(\frac{1}{F_{v,Rd}^2}\left(\alpha^2 + \beta^2\right); \frac{1}{V_{Rd\,7}^2}\right) \leq \left(\frac{\alpha}{F_{b,ver,Rd}}\right)^2 + \left(\frac{\beta}{F_{b,hor,Rd}}\right)^2$$

OR

$$- V_{Rd\,6} \leq \min\left(\frac{2}{3\sqrt{\alpha^2+\beta^2}} F_{v,Rd}; \frac{2}{3} V_{Rd\,7}\right)$$

3. Moreover, if $V_{Rd} = V_{Rd\,3}, V_{Rd\,4}, V_{Rd\,5}, V_{Rd\,6}, V_{Rd\,9}, V_{Rd\,10}$ or $V_{Rd\,11}$, the following inequality has to be checked:

$$V_{Rd\,1} > \min\left(V_{Rd\,2}; V_{Rd\,8}\right)$$

5.3 COMPOSITE JOINTS

5.3.1 Composite joints for simple framing

The joint configurations considered in the present section are basically similar to those investigated in sub-chapter 5.2. But their particularity is that at least one of the connected members is a steel-concrete composite one.

But this condition is however not sufficient to declare the joint as "composite", as illustrated in Figure 5.20 where:

- all beams are composite ones
- but only the three first configurations are simple ones

Indeed, in these beam-to-perimeter column joint configurations, the forces have to be transferred to the column only through the steel connections. To realise a composite joint, the slab should be extended beyond the column and reinforcement bars should "loop" around the column to ensure their anchorage. The joint would then become a moment resisting joint.

On the other hands, for the two last configurations presented in Figure 5.20, the steel slab reinforcements are continuous along the beams what allows to transfer significant bending moments. The response of such configurations with moment resisting joints will be discussed in chapter 6.

5. SIMPLE JOINTS

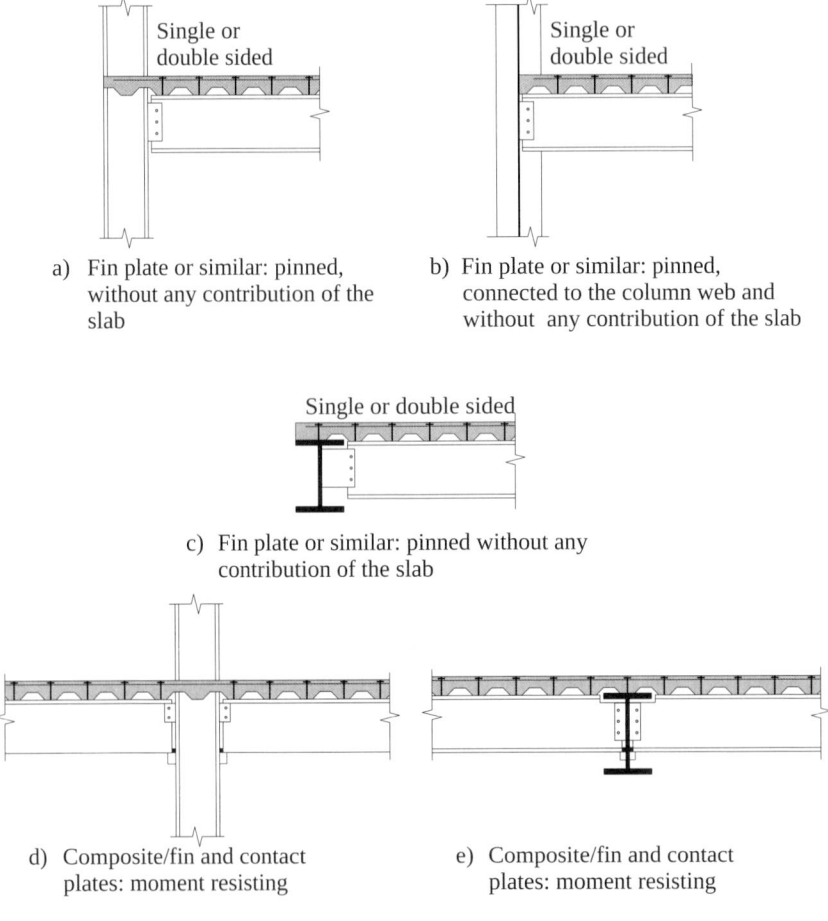

Figure 5.20 – Various composite joints

As a result of the lack of continuity of the reinforcement bars, a nominally pinned composite joint is therefore designed in the same way as a simple steel bare connection.

If, in the configurations illustrated in Figure 5.20c and d, no continuity of the rebars would be achieved (between the left and right composite beams), the left and right joints would be simple ones, but uncontrolled cracking would appear. In such a case, reference has to be made to EN 1992-1-1 (CEN, 2004a) to check whether, according to the exposure class of the structure, specific measures are to be taken.

5.4 COLUMN BASES

5.4.1 Introduction

Simple column bases are assumed to transfer axial (compression or tension) and shear forces. Tension forces may result from an uplift of the structure.

In Figure 5.21 two classical types of column bases, respectively with two and four anchor bolts, are illustrated. The first type, with two anchor bolts, meets more closely the concept of "simple joint" than the second one which may develop significant moments, as a result of the lever arm provided by the two lines of anchor bolts. Obviously, the use of rather thin and unstiffened base plates in the "four anchors" configuration is likely to limit the transfer of bending to the concrete block.

But again here, as stated in section 5.2.3.2, this "actual semi-rigidity" of the column bases will never prevent the designer from considering them as "simple" as long as rotational capacity and ductility will be exhibited by the joint. To achieve it, measures will have to be taken in such a way that anchor bolt failure and weld failure are not relevant failure modes. Reference is made here to section 5.2.4.1.2 where related design requirements have been expressed.

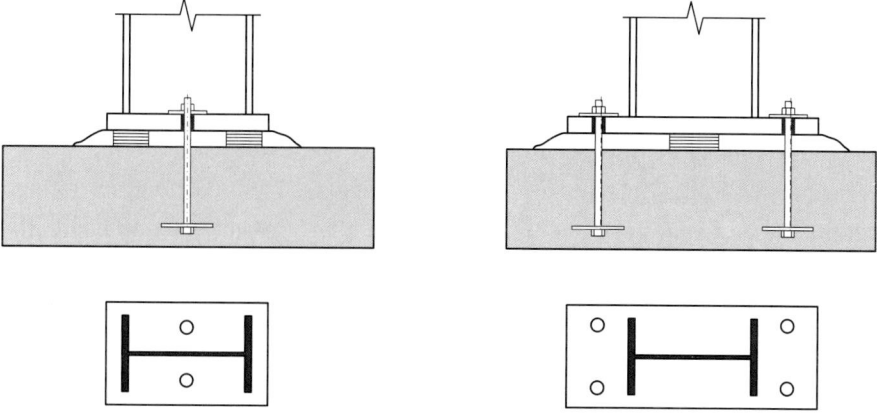

Figure 5.21 – Classical column base detailing, configured with two and four anchor bolts

5. SIMPLE JOINTS

Typical anchor bolt types are illustrated in Figure 5.22; amongst them: hooked bars for light anchoring, cast-in-place headed anchors and anchors bonded to drilled holes.

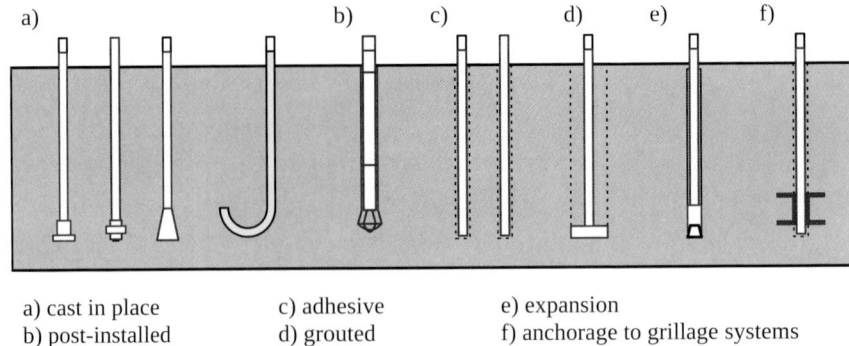

a) cast in place
b) post-installed
c) adhesive
d) grouted
e) expansion
f) anchorage to grillage systems

Figure 5.22 – Various types of anchoring systems

When it is necessary to transfer a big force, more expensive anchoring systems such as grillage beams embedded in concrete are to be selected.

In the next sections, the application of the component method to simple column bases according to EN 1993-1-8 is presented. The interested reader may refer to (Wald *et al*, 2008b; Wald *et al*, 2008a; Steenhuis *et al*, 2008; Jaspart *et al*, 2008b; Gresnigt *et al*, 2008) for detailed information and validations.

5.4.2 Basis for the evaluation of the design resistance

According to the component method, the evaluation of the design resistance of joints, in the present case under axial compression/tension and shear forces, requires the identification, the characterisation and the assembly of the constitutive joint components.

The active components in simple column base joints are illustrated in Figure 5.23. Two of them, namely the "base plate and concrete block in compression" and the "base plate in bending and anchor bolts in tension" are contributing to the transfer of the axial force through the joint, respectively under compression and tension action. The third one, the "anchors bolt in shear" relates to the transfer of the shear force.

5.4 COLUMN BASES

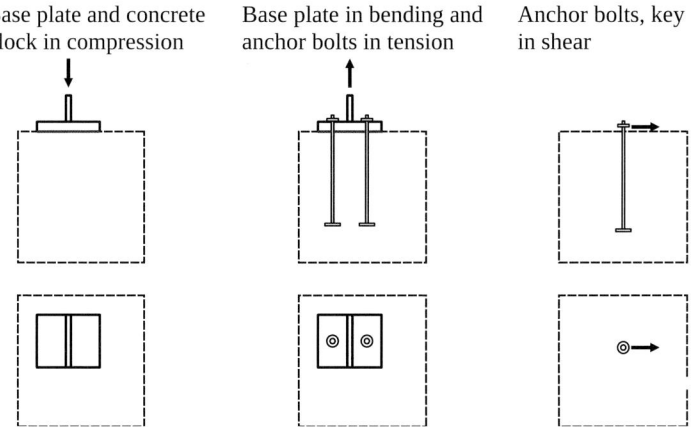

Figure 5.23 – Components in a simple column base

In the following pages, the aspects of characterisation and assembly are addressed and the resulting design procedures for simple column bases under axial and shear forces are respectively presented.

5.4.3 Resistance to axial forces

5.4.3.1 Component "base plate and concrete block in compression"

When considering the "base plate and concrete block in compression" component, the grout layer located between the base plate and the concrete is considered, as it influences the resistance of the component.

To model the response of the base plate in compression, two approaches are found in the literature:

– The base plate is assumed to be rigid in bending.
– The base plate is assumed to be flexible in bending.

The difference between rigid and flexible plates can be explained using a base plate connection loaded by an axial force only. In case of rigid plates it is assumed that the stresses under the plate are uniformly distributed. In case of flexible plates, the stresses are concentrated around the footprint of the column section under the plate. In EN 1993-1-8, the flexible model is considered as (i) it better corresponds to the reality and as (ii) it has been explained before that the need for rotational ductility in simple column bases implies the use of unstiffened and not too thick plates (especially for configurations with four anchor bolts).

5. SIMPLE JOINTS

Various researchers (Shelson, 1957; Hawkins, 1968b, 1968a; DeWolf, 1978) experimentally investigated the resistance of the "base plate and concrete block in compression" component. Factors influencing this resistance are the concrete strength, the plate area, the plate thickness, the grout, the location of the plate on the concrete foundation, the size of the concrete foundation and reinforcement. Concerning modelling, Stockwell (1975) introduced the concept of replacing a flexible plate with a non-uniform stress distribution by an equivalent rigid plate with a uniform stress distribution. Steenhuis and Bijlaard (1999) and Murray (1983) verified this simple practical method with experiments and suggested improvements. EN 1993-1-8 (CEN, 2005c) has adopted this method in a form suitable for standardisation. This method is described below.

The resistance is so determined through an equivalent rigid plate concept. Figure 5.24 shows how an equivalent rigid plate can replace a flexible plate when the base plate connection is loaded by axial force only. The symbol A is the area of top surface of the concrete block, A_p is the area of the plate, A_{eq} is the area of the equivalent rigid plate and c is the equivalent width of footprint. The resistance is determined by two parameters: the bearing strength of the concrete and the dimensions of the equivalent rigid plate.

a) Bearing strength of the concrete

The bearing strength of the concrete underneath the plate is dependent on the size of the concrete block. The edge effect is taken into account by the following definition of the concentration factor

$$k_j = \sqrt{\frac{a_1 \, b_1}{a \, b}} \qquad (5.20)$$

where the geometrical edge conditions, see Figure 5.25, are introduced by

$$a_1 = \min \begin{Bmatrix} a + 2 a_r \\ 5 a \\ a + h \\ 5 b_1 \end{Bmatrix}, \; a_1 \geq a \qquad (5.21)$$

$$b_1 = \min \begin{Bmatrix} b+2b_r \\ 5b \\ b+h \\ 5a_1 \end{Bmatrix}, \quad b_1 \geq b \qquad (5.22)$$

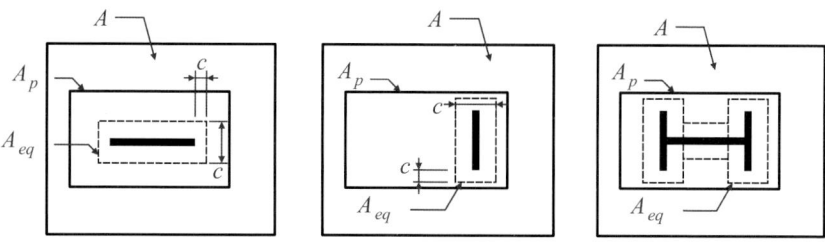

Figure 5.24 – Flexible base plate modelled as a rigid plate of equivalent area

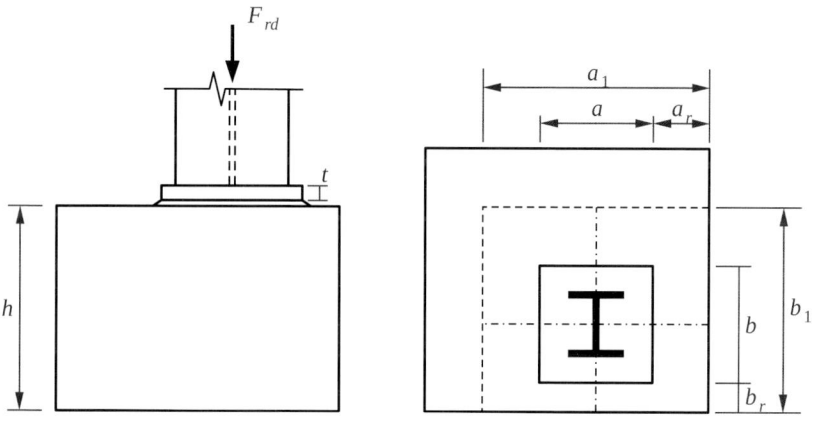

Figure 5.25 – Concrete block geometrical dimensions

If there is no edge effect, it means that the geometrical position of the column base is sufficiently far away from the edges of the concrete block and the value for k_j according to EN 1993-1-8 (CEN, 2005c) is 5. This concentration factor is used for the evaluation of the design value of the bearing strength as follows

$$f_{j,d} = \frac{\beta_j k_j f_{c,k}}{\gamma_c} \qquad (5.23)$$

5. SIMPLE JOINTS

where γ_c is a partial safety factor for concrete. A reduction factor β_j is used for taking into account that the resistance under the plate might be smaller due to the quality of the grout layer. The value $\beta_j = 2/3$ may be used if the grout characteristic strength is more than 0.2 times the characteristic strength of the concrete foundation, $f_{c,g} \geq 0.2 f_c$, and the thickness of the grout is smaller than 0.2 times the minimum base plate width, $t_g \leq 0.2 \min(a; b)$. In case where the thickness of the grout is more than 50 mm, the characteristic strength of the grout should be at least the same as that of the concrete foundation. These conditions are usually fulfilled. If not, the grout should be checked separately, see Steenhuis *et al* (1999).

b) Dimensions of the equivalent rigid plate

The flexible base plate, with area A_p, can be replaced by an equivalent rigid plate with area A_{eq}, see Figure 5.26. This rigid plate area A_{eq} is built up from one T-stub under the column web and two T-stubs under the column flanges.

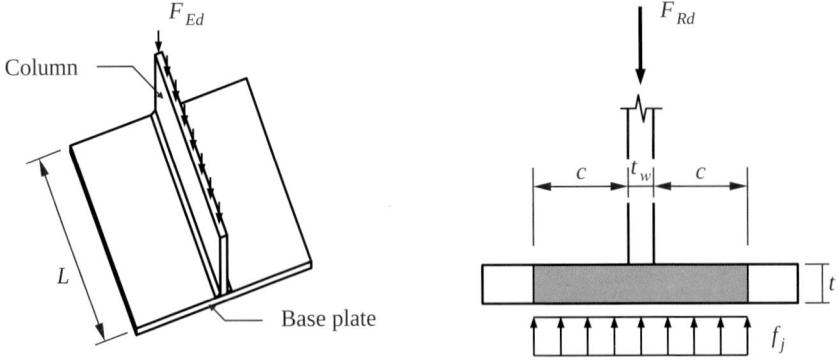

Figure 5.26 – T-stub under compression

The equivalent width c of the T-stub, see Figure 5.26, can be determined assuming that no plastic deformations will occur in the flange of the T-stub. Therefore, the resistance per unit length of the T-stub flange is taken as the elastic resistance:

$$M' = \frac{1}{6} t^2 f_y \qquad (5.24)$$

5.4 COLUMN BASES

It is assumed that the T-stub is loaded by a uniform stress distribution. The bending moment per unit length in the base plate acting as a cantilever with span c is:

$$M' = \frac{1}{2} f_{j,d} \, c^2 \qquad (5.25)$$

where $f_{j,d}$ is concrete bearing strength. The equivalent width c can be resolved by combining Eqs. (5.24) and (5.25):

$$c = t \sqrt{\frac{f_y}{3 f_{j,d} \, \gamma_{M0}}} \qquad (5.26)$$

The width of the T-stub is now:

$$a_{eq,R} = t_w + 2c = t_w + 2t \sqrt{\frac{f_y}{3 f_{j,d} \, \gamma_{M0}}} \qquad (5.27)$$

The resistance F_{Rd} of the T-stub, see Figure 5.26, should be higher than the loading F_{Ed}:

$$F_{Ed} \leq F_{Rd} = A_{eq} f_{j,d} = a_{eq,R} \cdot L \cdot f_{j,d} \qquad (5.28)$$

The base plate is stiffer in bending near the intersections of web and flanges. This stiffening effect is not taken into consideration in the equivalent area A_{eq}. Studies (Wald, 1993) show that this stiffening effect may yield to a 3% higher resistance for open sections and a 10% higher resistance for tubular sections in comparison to the method of determination of A_{eq}. The calculation of the concentration factor k_j based on Eq. (5.20) leads to conservative results. This can be improved by modification of the procedure of Eqs. (5.20) - (5.22). In that case, the equivalent area instead of the full area of the plate should be considered. However, this iterative procedure is not recommended for practical purposes. In case of high quality grout, a less conservative procedure with a distribution of stresses under 45° may be adopted, see Figure 5.27.

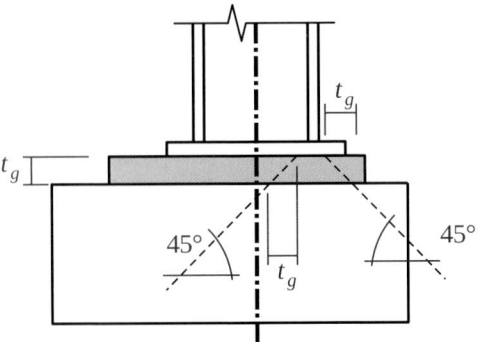

Figure 5.27 – Stress distribution in the grout

The bearing stress under the plate increases with a larger eccentricity of the axial force (DeWolf, Ricker, 1990; Penserini, Colson, 1989). In this case, the base plate is in larger contact with the concrete block due to its bending and the stress in the edge under the plate increases. However, the effect of these phenomena is limited.

The influence of packing under the steel plate may be neglected for practical design (Wald et al, 1993). The influence of the washer under plate used for construction can be also neglected in case of grout quality $f_{c,g} \geq 0.2 f_c$. The anchor bolts and base plate resistance should be taken into account explicitly in case of grout quality $f_{c,g} < 0.2 f_c$.

5.4.3.2 Component "base plate in bending and anchor bolts in tension"

Anchor bolts are longer compared to the bolts used in end plates due to presence of washer plates and grout, thick base plate and the part embedded in the concrete block. As a result, the elongation of the anchor bolts could lead to the separation of the base plate from the concrete block when the anchor bolts are loaded in tension. Different responses are therefore to be considered when the strength of the base plate loaded by bending moment and/or tension axial force has to be predicted.

The failure of the "base plate in bending and anchor bolts in tension" component may result from the yielding of the plate, the failure of the anchor bolts or a combination of both phenomena.

5.4 COLUMN BASES

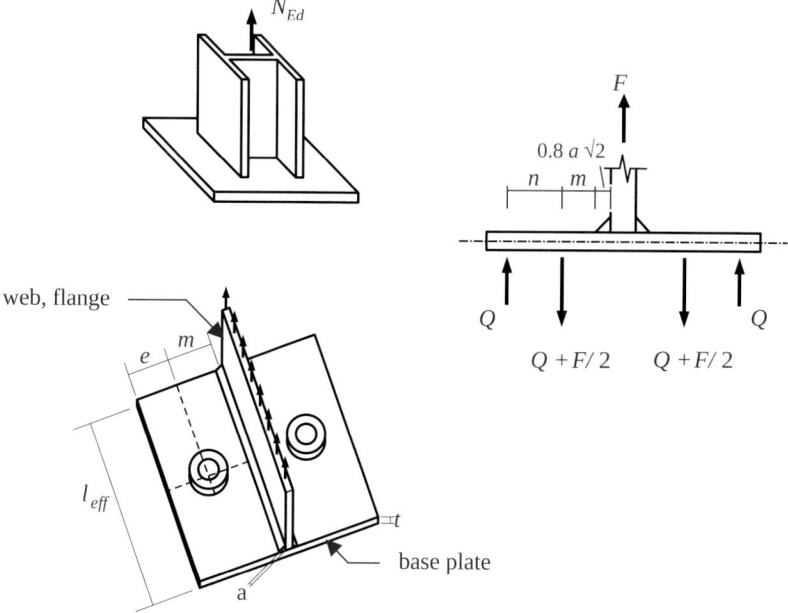

Figure 5.28 – T-stub idealisation (case with prying effects)

In order to model the base plate in bending and the bolts in tension, reference is again made to the so-called "equivalent T-stub" presented in section 3.5.2. This is illustrated in Figure 5.28 for one of the extended parts of the base plate (and the corresponding anchor bolts).

In the case the bolts elongate significantly and the plate is stiffened, the plate separates from the concrete foundation. In the other case, the edge of the plate is in contact with the concrete resulting, what results in the development of prying forces Q (see Figure 5.28).

In section 3.5.2, the design formulae for T-stubs in tension are provided, as well as the criterion which allows defining whether prying effects have to be or not considered. The boundary criterion is:

$$L_{b,\lim} = \frac{8.82 \, m^3 A_s}{l_{eff} \, t^3} \quad (5.29)$$

Where L_b ($L_{b,\lim}$) and A_s designate respectively the effective length and the cross section design area of the anchor bolts. l_{eff} is the length of the equivalent T-stub (see section 3.5.2).

5. SIMPLE JOINTS

The use of the T-stub concept requires the preliminary evaluation of the design resistance of the anchors in tension and of the length of the anchors.

As far as resistance is concerned, the lack of anchoring (pull-out of the anchor, failure of the concrete …) should be avoided and collapse of anchor bolts be preferred. This avoids brittle failure of the anchoring. For seismic areas, the failure of the column base should occur in the base plate rather than in the anchor bolts (Astaneh et al, 1992). The plastic mechanism in the plate ensures ductile behaviour and dissipation on energy.

For a single anchor, the following failure modes have to be considered:

- The pull-out failure $N_{p,Rd}$,
- The concrete cone failure $N_{c,Rd}$,
- The splitting failure of the concrete $N_{sp,Rd}$.

Similar verification is required for a group of anchor bolts.

The detailed description of the related design formulae is included in the CEB Guide (CEB, 1994) various types of fastening.

In the case of embedded anchor bolts, the effective length of the bolt L_b consists of the free length L_{bf} and an effective embedded length L_{be}, $L_b = L_{bf} + L_{be}$, see Figure 5.29.

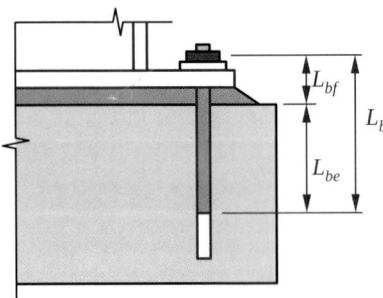

Figure 5.29 – Effective length of an embedded anchor bolt

According to EN 1993-1-8, the effective embedded length of bolts with regular surface is assumed to be equal to $L_{be} \cong 8d$. The minimum actual embedded length of the anchor allowing to justify the use $L_{be} \cong 8d$ is $24d$. This value has therefore to be considered as the minimum actual embedded length of the anchor bolts. This length may be reduced by the use of a headed plate (as illustrated in Figure 5.21).

5.4.3.3 Assembly of components for resistance evaluation

The simple joints under consideration in the present section are illustrated in Figure 5.21: base plates with two or four anchor bolts. Except in cases where the uplift of the structure would induce high tension forces in the column, only two anchors bolts will be used so as to limit the actual transfer of bending moments mentioned in section 5.4.1; this one results from the "non perfectly simple" response of the actual joints.

Whatever the number of anchor bolts, the configuration of the joint is hereunder assumed to be symmetrical, with reference to the axis of the column. In fact, should a lake of symmetry in the joint be introduced, the application of an axial force in the column would result in the development of an axial force and a bending moment in the joint. These aspects of combined loading in the joint will be addressed later in chapter 6.

This being, the axial resistance of a simple column base joint, respectively under axial compression and tension forces may be computed as follows:

a) Design resistance in compression

$$N_{Rd} = A_{eq} f_{j,d} \quad (5.30)$$

where:
- A_{eq} is the "rigid plate" equivalent area of the base plate;
- f_j is the design value of the concrete block bearing strength.

Expressions of these two quantities are provided in section 5.4.3.1.

b) Design resistance in tension

$$N_{Rd} = \sum_{i}^{n} F_{T,Rd,i} \quad (5.31)$$

where:
- $F_{T,Rd,i}$ is the design resistance of the individual T-stub i; the summation extends to all the n individual T-stubs which may be identified in the base plate (one per bolt-row).

Reference is made to 5.4.3.2 for the evaluation of the $F_{T,Rd,i}$ values.

5. SIMPLE JOINTS

5.4.4 Resistance to shear forces

Horizontal shear force in column bases may be resisted by (see Figure 5.30):

- friction between the base plate, grout and concrete footing;
- shear and bending of the anchor bolts;
- a special shear key, for example a block of I-stub or T-section or steel pad welded onto the bottom of the base plate;
- direct contact, e.g. achieved by recessing the base plate into the concrete footing.

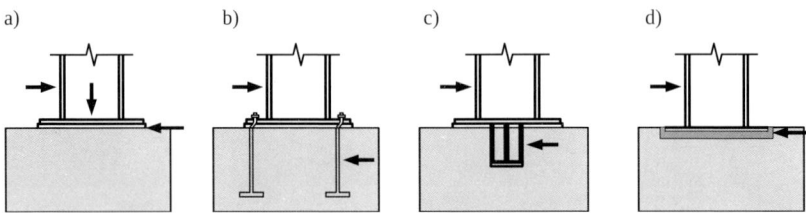

Figure 5.30 – Column bases in shear

In most cases, the shear force can be resisted through friction between the base plate and the grout. The friction depends on the compressive load and on the coefficient of friction. Pre-stressing the anchor bolts will increase the shear force transfer by friction.

Sometimes, for instance in slender buildings, it may happen that due to horizontal forces (wind loading) columns are loaded in tension. In such cases, the horizontal shear force usually cannot be transmitted through friction. If no other provisions are installed (e.g. shear studs), the anchor bolts will have to transmit these shear forces.

Because the grout does not have sufficient strength to resist bearing stresses between the bolt and the grout, considerable bending of the anchor bolts may occur, as is indicated in Figure 5.31. The main failure modes are rupture of the anchor bolts (local curvature of the bolt exceeds the ductility of the bolt material), crumbling of the grout, failure (splitting) of the concrete footing and pull-out of the anchor bolt.

Due to the horizontal displacement, not only shear and bending in the bolts will occur, but also the tensile force in the bolts will be increased due to second order effects. The horizontal component of the increasing tensile force gives an extra contribution to the shear resistance and stiffness. The increasing vertical component gives an extra contribution to the transfer of load by friction and increases resistance and stiffness as well. These factors explain the shape of the load deformation diagram as given in Figure 5.31. The increase of the load continues till fracture occurs in one of the components of the connection. If the connection is well designed and executed, such fracture will occur at very large deformation (much larger than acceptable in serviceability and ultimate limit states).

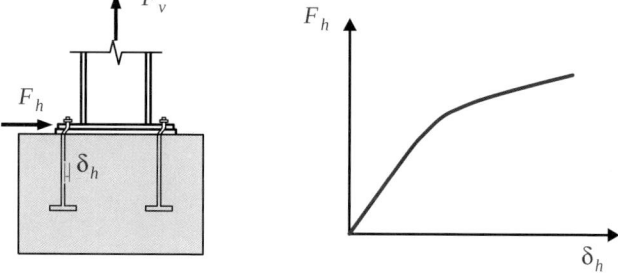

Figure 5.31 – Column base loaded by shear and tension force

The thickness of the grout layer has an important influence on the horizontal deformations. In the tests carried out by the Stevin Laboratory (Bouwman et al, 1989), the deformations at rupture of the anchor bolts were between about 15 and 30 mm, whilst grout layers had a thickness of 15, 30 and 60 mm. The deformations have to be taken into account in the check of the serviceability limit state. Because of the rather large deformations that may occur, this check may govern the design. The size of the holes may have a considerable influence on the horizontal deformations, especially when oversized holes are applied. In such cases it may be useful to apply larger washers under the nuts, to be welded onto the base plate after erection, or to fill the hole by a two component resin. For the application of such resin, reference is made to the ECCS recommendations (ECCS, 1994) and EN 1993-1-8.

In Bouwman et al (1989), a model for the load deformation behaviour of base plates loaded by combinations of normal force and shear has been developed. For the design of fasteners, the CEB has published a Design Guide (CEB, 1996). The CEB model has been compared with the Stevin

5. SIMPLE JOINTS

Laboratory model and the tests. It appears that the CEB model gives very conservative results, especially when large tensile forces are present and / or the thickness of the grout layer is large. The main reason is that the CEB model does not take account of the positive influence of the grout layer. Here below, a summary of the design rules which were proposed in Gresnigt et al (2008) for inclusion in Part 1-8 of EN 1993 and based on the Stevin Laboratory model is presented.

It is noted that in steel construction, usually only the steel part of the base plate connection is considered. In the CEB Design Guide much attention is paid to the concrete part. It is recommended to the steel designer to acquire knowledge about the requirements to the reinforced concrete. For detailed design guidance and the various failure modes that may occur in the concrete, reference is made to the CEB Design Guide (CEB, 1996).

This being the design shear resistance $F_{v,Rd}$ of column base plate joints may be derived as follows:

$$F_{v,Rd} = F_{f,Rd} + nF_{vb,Rd} \qquad (5.32)$$

where:

$F_{f,Rd}$ is the design friction resistance between base plate and grout layer: $F_{f,Rd} = C_{f,d} \cdot N_{c,Sd}$

$C_{f,d}$ is the coefficient of friction between base plate and grout layer. The following values may be used:
- for sand-cement mortar: $C_{f,d} = 0.20$
- for special grout: $C_{f,d} = 0.30$

$N_{c,Ed}$ is the design value of the normal compressive force in the column. If the normal force in the column is a tensile force $F_{f,Rd} = 0$

Note: Also, the preload in the anchor bolts contributes to the friction resistance. However, because of its uncertainty (e.g. relaxation and interaction with the column normal force), it is suggested to neglect this action.

n is the number of anchor bolts in the base plate

$F_{vb,Rd}$ is the smallest of $F_{1,vb,Rd}$ and $F_{2,vb,Rd}$

$F_{1,vb,Rd}$ is the bearing resistance for the anchor bolt - base plate; the bearing resistance of the concrete is provided by EN 1992

5.4 COLUMN BASES

$F_{2,vb,Rd}$ is the shear resistance of the anchor bolt

$$F_{2,vb,Rd} = \frac{\alpha_b f_{ub} A_s}{\gamma_{Mb}}$$

A_s is the tensile stress area of the bolt or of the anchor bolt

α_b is a coefficient depending on the yield strength f_{yb} the anchor bolt: $\alpha_b = 0.44 - 0.0003 f_{yb}$

f_{yb} is the nominal yield strength the anchor bolt, where: $235 \text{ N/mm}^2 \leq f_{yb} \leq 640 \text{ N/mm}^2$

f_{ub} is the nominal ultimate strength the anchor bolt, where: $400 \text{ N/mm}^2 \leq f_{ub} \leq 800 \text{ N/mm}^2$

γ_{Mb} is the partial safety factor

Prerequisites:

In the above equations, it is a prerequisite that:

– the grout layer is of adequate quality (see relevant applicable reference standards),
– the design strength of the anchor – concrete connection is greater than the design rupture strength of the anchor,
– other failure modes, like splitting of the concrete and pull out of the anchor, are prevented by adequate design and execution of the anchor in the concrete block. Reference is made to: EN 1992-1-1 (CEN, 2004a), CEB publications (CEB, 1994) and (CEB, 1996).

Chapter 6

MOMENT RESISTANT JOINTS

6.1 INTRODUCTION

"Moment resistant joints" is a general wording used to cover all joints which transfer significant bending moments between the connected members, but also – in many cases – shear and/or axial forces. These joints may be rigid or semi-rigid, in terms of stiffness, and may exhibit a full or a partial strength resistance level.

The evaluation of their mechanical design properties is a key aspect to which several pages of Eurocode 3 Part 1-8 are devoted. And as already said, it is based on the application of the component approach (see section 1.6.2).

According to the latter, the mechanical properties of the active components are to be computed before the components are assembled so as to derive the global response of the whole joint under the specified considered loading.

In the present chapter (sub-chapter 6.2) the way on how to characterise the mechanical response of the individual components covered by Eurocode 3 Part 1-8 will first be expressed.

In a second step, the assembly procedure problem will be addressed, successively in terms resistance (sub-chapter 6.3), stiffness (sub-chapter 6.4) and deformation capacity (sub-chapter 6.5).

In the following sections, the application of the component approach to specific joints will finally be raised:

- Steel beam-to-column, beam-to-beam and beam splices joint configurations (sub-chapter 6.6)

6. MOMENT RESISTANT JOINTS

- Steel column splices (sub-chapter 6.7)
- Column bases (sub-chapter 6.8)
- Composite joints (sub-chapter 6.9)

6.2 COMPONENT CHARACTERISATION

Joint components are subjected to a compression, a tension of a shear force. In these conditions, they first deform elastically, reach their plastic level of resistance and possibly deform plastically in the absence of brittle failure or development of instability phenomena. In Eurocode 3 Part 1-8 and Eurocode 4 Part 1-1, their characterisation is contemplated through the definition of three specific design properties:

- the elastic stiffness, expressed as equal to Ek_i where k_i is called "stiffness coefficient";
- the plastic resistance $F_{i,Rd}$;
- the plastic deformation capacity (in short, deformation capacity).

In the present section, the determination of these three properties is briefly commented, component by component. For all details about the application of the characterisation formulae, the reader is asked to refer directly to the normative documents.

6.2.1 Column web panel in shear in steel or composite joints

a) Resistance

The shear force $V_{wp,Ed}$ applied to an unstiffened column web panel (see sub-chapter 2.2) has to be compared to the following expression of the design resistance:

$$V_{wp,Rd} = \frac{0.9 f_{y,wc} A_{vc}}{\sqrt{3} \gamma_{M0}} \qquad (6.1)$$

in which:
- A_{vc} is the shear area of the column defined in Eurocode 3 Part 1-1
- $f_{y,wc}$ is the design yield strength of steel
- γ_{M0} is the partial safety factor

In view of the simplified modelling in which reference is made to "joints" (see sub-chapter 2.2), the shear resistance has to be "turned" into a "beam-force" resistance as follows:

$$F_{wp,Rd} = V_{wp,Rd}/\beta \qquad (6.2)$$

In this formula, the 0.9 factor shows the detrimental effect of the longitudinal stresses acting in the column on the shear resistance of the web panel. In practical cases, this influence amounts 5 to 12%. In Eurocode 3 Part 1-8, a fixed reduction of 10% of the resistance has been selected for sake of simplicity in application.

The validity of this formula, and of all those provided in the present section, is limited to panels with a web slenderness $d_c/t_w \leq 69\varepsilon$ (d is the clear depth of the column web, t_w, the web thickness and $\varepsilon = \sqrt{235/f_{y,wc}}$ with $f_{y,wc}$ expressed in N/mm²). Another limitation applies to double-sided beam-to-column joint configurations without diagonal stiffeners on the column web; in such cases, the two beams should have equal depths.

When the resistance of the panel is insufficient, various reinforcement solutions may be contemplated. In EN 1993-1-8, guidelines are provided for supplementary web plates or diagonal/transverse column web stiffeners.

For slender panels exceeding the "$d_c/t_w \leq 69\varepsilon$" rule (for instance, for built-up profiles), shear instability of the column web panel appears which transforms the shear panel into a "truss-like" system providing extra resistance. Appropriate models exist which allows to determine the design resistance of the panel. Such a model is proposed in Jaspart (1997). Similarly, for double-sided beam-to-column joint configurations without diagonal stiffeners on the column web where the two beams would not be of equal depths, information may be found in Jordão (2008).

For a column web panel being part of a concrete encased steel column, the resistance of the component is increased by the shear resistance of the concrete, see EN 1994-1-1 section 8.4.4.1.

b) Stiffness

The stiffness coefficient of an unstiffened column web panel is provided by the expression:

6. MOMENT RESISTANT JOINTS

$$k_1 = \frac{0.38 A_{vc}}{\beta \cdot z} \qquad (6.3)$$

For a column web panel with appropriate diagonal stiffeners:

$$k_1 = \infty \qquad (6.4)$$

For an unstiffened panel (without concrete encasement) in a joint with a steel contact plate connection (see section 6.2.22 below), the stiffness coefficient k_1 may be taken as 0.87 times the value given by Eq. (6.3).

For a column web panel being part of a concrete encased steel column, see Eurocode 4 Part 1-1 section A.2.3.1.

c) Deformation capacity

The deformation capacity of steel column web panels is recognised as being rather high. Plastic redistribution between the components requiring ductility form the shear panel may therefore be accommodated without any risk. Moreover joints in which the web panel in shear is the relevant failure mode will possess enough ductility to perform a plastic structural frame analysis.

No such high deformation capacity may probably be expected from the concrete encasement in shear.

6.2.2 Column web in transverse compression in steel or composite joints

a) Resistance

The design resistance of a column web panel in transverse compression is given by:

$$F_{c,wc,Rd} = \frac{\omega k_{wc} b_{eff,c,wc} t_{wc} f_{y,wc}}{\gamma_{M0}} \qquad (6.5)$$

but with:

$$F_{c,wc,Rd} \leq \frac{\omega k_{wc} \rho b_{eff,c,wc} t_{wc} f_{y,wc}}{\gamma_{M1}} \qquad (6.6)$$

6.2 COMPONENT CHARACTERISATION

This formula is fitted to cover any failure mode in transverse compression, including so crushing, crippling or buckling of the web. Only the "column-sway" buckling mode illustrated in Figure 6.1 is not considered. This one has in fact to be prevented by constructional restraints.

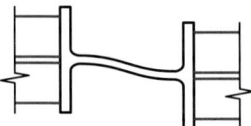

Figure 6.1 – "Column sway" buckling mode of an unstiffened web

In its simplest format, Eq. (6.5) expresses the design resistance of the web as the one of an equivalent web zone area (effective width $b_{eff,c,wc}$ times the web thickness t_{wc}) subjected to a stresses equal to the design strength of the steel material. The value of the effective width $b_{eff,c,wc}$, see Figure 6.2, depends on the connection type (see Eurocode 3 Part 1-8 section 6.2.6.2 and Eurocode 4 Part 1-1 section 8.4.3). Two reduction factors are then added:

- the ω coefficient which allows for the possible stress interaction effects with shear in the column web panel;
- the k_{wc} coefficient which, at first sight, appears to be there to account for the possible stress interaction with the maximum longitudinal compressive stress $\sigma_{com,Ed}$ due to axial force and bending moment in the column; at least as far as $\sigma_{com,Ed} > 0.7 f_{y,wc}$, see Figure 6.3. But in fact, this factor covers the increase risk of column web instability resulting from the application of a high $\sigma_{com,Ed}$ stress in the column. This explains why this coefficient has not to be applied to column webs in transverse tension (see section 6.2.3 below).

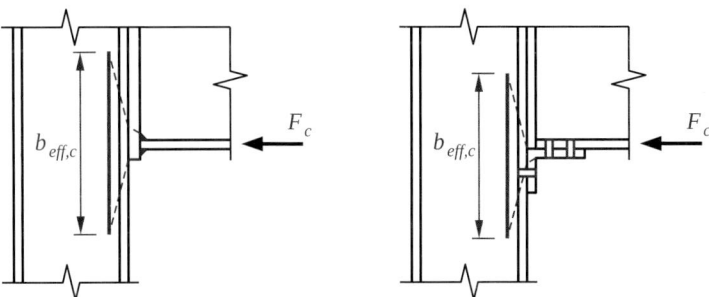

Figure 6.2 – Spread of compression stresses to the column web

6. MOMENT RESISTANT JOINTS

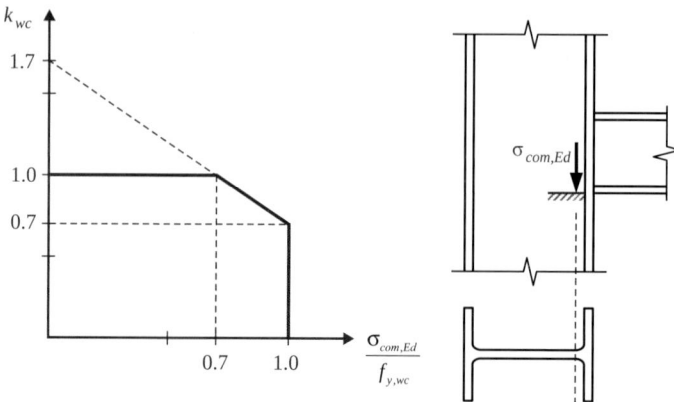

Figure 6.3 – Reduction factor k_{wc}

From a practical point of view, the question has to be raised on how to deal with such interaction factors in the design process. For the ω coefficient, value are provided in Table 6.1. For the k_{wc} coefficient, Eurocode 3 Part 1-8 suggests to consider it as equal to 1.0 as no reduction is necessary in most of the cases. It can therefore be omitted in preliminary calculations when the longitudinal stress is unknown and checked later on.

Table 6.1 – Reduction factor ω for interaction with shear

Transformation parameter β	Reduction factor ω
$0 \leq \beta \leq 0.5$	$\omega = 1$
$0.5 < \beta < 1.0$	$\omega = \omega_1 + 2(1-\beta)(1-\omega_1)$
$\beta = 1$	$\omega = \omega_1$
$1.0 < \beta < 2.0$	$\omega = \omega_1 + (\beta-1)(\omega_2 - \omega_1)$
$\beta = 2.0$	$\omega = \omega_2$
$\omega_1 = \dfrac{1}{\sqrt{1+1.3(b_{eff,c,wc} t_{wc}/A_{vc})^2}}$ and $\omega_2 = \dfrac{1}{\sqrt{1+5.2(b_{eff,c,wc} t_{wc}/A_{vc})^2}}$	
where: A_{vc} is the shear area of the column β is the transformation parameter, see section 2.2.4	

In Eq. (6.6), an additional resistance limitation is addressed through the ρ coefficient. This one is covering the possible risk of instability in the web in the case of slender column sections.

6.2 COMPONENT CHARACTERISATION

Specific rules to apply to stiffened column webs (transverse stiffeners or supplementary web plates) are provided in Part 1-8.

In the case of concrete-encased columns, an increased level of resistance is reached. Eurocode 4 Part 1-1 section 8.4.4.2 provides design recommendations to take full profit of this reinforcement.

b) Stiffness

The stiffness coefficient for the "unstiffened column web in transverse compression" component is expressed as:

$$k_2 = \frac{0.7 b_{eff,c,wc} t_{wc}}{d_c} \quad (6.7)$$

For a column web with an appropriate transverse stiffener:

$$k_2 = \infty \quad (6.8)$$

For an unstiffened column web in a steel contact plate connection, the stiffness coefficient k_2 may be evaluated through a particular expression provided in Eurocode 4 Part 1-1 section A.2.2.2.

For a concrete-encased steel column, see Eurocode 4 Part 1-8 section A.2.3.2.

c) Deformation capacity

Little practical information is unfortunately available nowadays on the plastic deformation capacity of the column web in transverse compression. In Jaspart (1991), it is specified that a significant deformation capacity may be expected from this component as long as instability is not governing the failure mode, but to ensure that this component alone would provide enough ductility to the joint to which it belongs so as to allow a plastic hinge to form and rotate in the joint is quite questionable. If a plastic global analysis of the structure being studied has to be performed, it is seems so preferable to design the joint by privileging another failure mode which exhibits a clear ductile behaviour (as, for instance, the column web panel in shear, see 6.2.1).

No information is provided in Eurocode 4 for concrete-encased steel columns.

6. MOMENT RESISTANT JOINTS

6.2.3 Column web in transverse tension

a) Resistance

The design resistance of the "column web in transverse tension" component is expressed through a quite similar expression as the one presented in section 6.2.2:

$$F_{t,wc,Rd} = \frac{\omega b_{eff,t,wc} t_{wc} f_{y,wc}}{\gamma_{M0}} \qquad (6.9)$$

In the absence of potential instability phenomena, no k_{wc} or ρ coefficient is to be considered. The factor ω, on the other hand, is maintained so as to allow for the possible stress interaction effects with shear in the column web panel.

The main difference between the compression and tension responses is the definition of the effective width, at least there where tension forces are transferred through bolt-rows. In compression (or in the tension zone of a welded beam-to-column joint), the effective width results from the direct transfer of the forces through the endplate, the column flange, the weld radii of fillet, etc. while, in the tension zone of a bolted connection, the diffusion of the forces is less direct. In this context, the concept of "equivalent T-stub" described in section 3.5.2 is of great help as it allows defining the effective width $b_{eff,t,wc}$ as equal to the equivalent length of the T-stub (ℓ_{eff}) which may be substituted to the column flange (see section 6.2.4). According to this, $b_{eff,t,wc}$ is clearly seen as the part of the column web which may be devoted to the transfer of the tension forces by the bolts through the column flange.

Here again, specific rules to apply to stiffened column webs (transverse stiffeners or supplementary web plates) are provided in Part 1-8.

b) Stiffness

The stiffness coefficient for the "unstiffened column web in transverse tension" component is expressed as:

$$k_3 = \frac{0.7 b_{eff,t,wc} t_{wc}}{d_c} \qquad (6.10)$$

6.2 COMPONENT CHARACTERISATION

In a bolted connection, this expression has to be applied at the level of individual bolt-rows and of groups of bolt-rows in tension and $b_{eff,t,wc}$ should be taken equal, for each bolt-row (whether it is considered individually or as part of a group), to the smallest of the possible effective lengths (individually or as part of a group of bolt-rows).

For a column web with an appropriate transverse stiffener:

$$k_3 = \infty \qquad (6.11)$$

c) Deformation capacity

Significant ductility in tension may be expected from normalised steels and so no limitation in plastic deformation capacity may be assumed from this component.

6.2.4 Column flange in transverse bending

a) Resistance

Two different situations can be distinguished to determine the design resistance of a column flange in transverse bending: (i) the connected members is welded to the column flange or (ii) the connected member is bolted to the column flange. In both cases, the column flange can be unstiffened or stiffened. The loading of the component *column flange in transverse bending* is a force in the connected member, perpendicular to the column flange, resulting in bending of the column flange itself.

– Column flange in a welded connection

In a welded joint, the resistance $F_{fc,Rd}$ of an *unstiffened* column flange is determined as follows:

$$F_{fc,Rd} = b_{eff,b,fc} \, t_{fb} \, f_{y,fb} / \gamma_{M0} \qquad (6.12)$$

with:

    $b_{eff,b,fc}$ is the effective width for bending of the unstiffened column flange

    $b_{eff,b,fc} = t_{wc} + 2s + 7kt_{wc}$

    where:

        $s = r_c$ for a rolled I or H section

6. MOMENT RESISTANT JOINTS

$$s = \sqrt{2}\, a_c \quad \text{for a welded I or H section}$$

$$k = \left(t_{fc}/t_{fb}\right) \cdot \left(f_{y,fc}/f_{y,fb}\right) \text{ but } k \leq 1.0$$

From Eq. (6.12) it can be seen that, in reality, the resistance $F_{fc,Rd}$ of an *unstiffened* column flange is the (plastic) resistance of the beam flange, reduced to account for the unequal stress distribution due to the weaker *unstiffened* column flange. Further explanations and details can be found in section 4.4.2.2

Even if $b_{eff,b,fc} \leq b_{fb}$, where b_{fb} is the width of the beam flange, the welds connecting the beam flange to the column flange should be designed to resist a force equal to the plastic resistance of the beam flange $b_{fb} t_{fb} f_{y,fb} / \gamma_{M0}$, i.e. full strength welds are required.

Note that the design resistance of an unstiffened column flange in a welded connection is the same for both an applied tension force and a compression force.

If the effective width $b_{eff,b,fc} \leq \left(f_{y,b}/f_{u,b}\right) b_{fb}$, the column flange should be stiffened. If the following conditions for transverse stiffeners in the column web are satisfied, it may be assumed that the resistance of component *column flange in transverse bending* need not to be checked:

- the stiffeners should be aligned with the beam flange;
- the steel grade of the stiffeners is not lower than that of the beam flange;
- the thickness of the stiffeners is not smaller than the thickness of the beam flange;
- the outstand of the stiffeners is not less than $\left(b_b - t_{wc}\right)/2$.

- Unstiffened column flange in a bolted connection

In section 3.5.2, design methods for the so-called equivalent T-stub are presented. Note that T-stub in section 3.5.2 is a T-stub with two bolts per row. However, in practice, also configurations with four bolts per row are used. An extension of the design methods for a T-stub with four bolts per row can be found in Demonceau *et al* (2011). The resistance and failure mode of an *unstiffened column flange in bending*, together with the associated *bolts in tension*, should be taken as similar to those of an equivalent T-stub flange, see Figure 6.4, for both:

- each individual bolt-row required to resist tension;
- each group of bolt-rows required to resist tension.

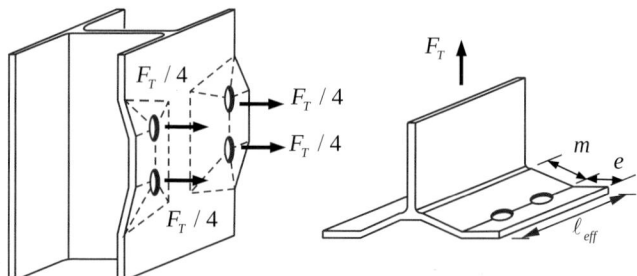

Figure 6.4 – Equivalent T-stub flange representing a column flange in bending

The dimensions e_{min} and m for use in 3.5.2 should be determined from Figure 3.30.

The effective length of the equivalent T-stub flange ℓ_{eff} should be determined for the individual bolt-rows and the bolt-group as explained in section 3.5.2 using the values given in Table 6.2. Note that the figures in Table 6.2 illustrating the yield line patterns for the "end bolt row" only show cases not yet shown for the "inner bolt row".

– Stiffened column flange in a bolted connection

To increase the resistance of the column flange in bending in a bolted connection, transverse stiffeners and/or appropriate arrangements of diagonal stiffeners may be used. The design resistance is determined in the same way as described in the section before for an *unstiffened column flange*. However, a stiffener welded to the column flange will split possible groups of bolt rows, i.e. the groups of bolt rows on either side of a stiffener should be modelled as separate equivalent T-stubs, see Figure 6.5. The design resistance and the failure modes should be determined separately for each equivalent T-stub.

The effective length of equivalent T-stub flanges ℓ_{eff} should be determined for the individual bolt-rows and the bolt-group as explained in section 3.5.2 from the values given for each bolt-row in Table 6.2. For those bolt-rows adjacent to a stiffener effective length values are given in Table 6.3. Note that the figures in Table 6.3 illustrating the yield line patterns only show cases not yet shown in Table 6.2.

6. MOMENT RESISTANT JOINTS

Table 6.2 – Effective lengths for an unstiffened column flange in a bolted connection

Bolt-row location	Bolt-row considered individually		Bolt-row considered as part of a group of bolt-rows	
	Circular patterns $\ell_{eff,cp}$	Non-circular patterns $\ell_{eff,nc}$	Circular patterns $\ell_{eff,cp}$	Non-circular patterns $\ell_{eff,nc}$
Inner bolt-row	$2\pi m$	$4m + 1.25e$	$2p$	p
End bolt-row	The smaller of: $2\pi m$ $\pi m + 2e_1$	The smaller of: $4m + 1.25e$ $2m + 0.625e + e_1$	The smaller of: $\pi m + p$ $2e_1 + p$	The smaller of: $2m + 0.625e + 0.5p$ $e_1 + 0.5p$
Mode 1	$\ell_{eff,1} = \min\left(\ell_{eff,cp}; \ell_{eff,nc}\right)$		$\Sigma\ell_{eff,1} = \min\left(\Sigma\ell_{eff,cp}; \Sigma\ell_{eff,nc}\right)$	
Mode 2	$\ell_{eff,2} = \ell_{eff,nc}$		$\Sigma\ell_{eff,2} = \Sigma\ell_{eff,nc}$	

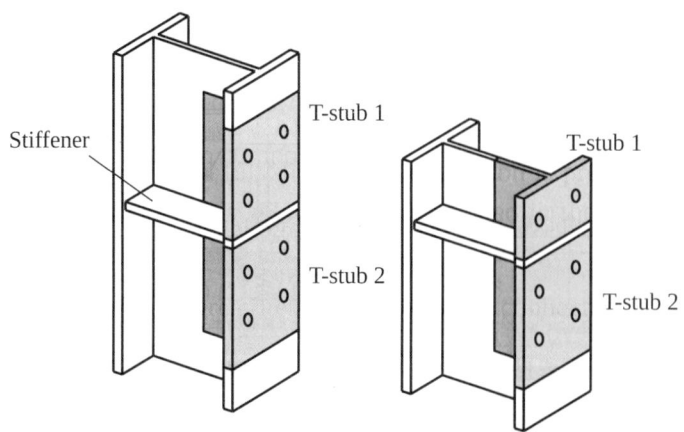

Figure 6.5 – Modelling a stiffened column flange as separate T-stubs

6.2 COMPONENT CHARACTERISATION

Table 6.3 – Effective lengths for a stiffened column flange in a bolted connection

Bolt-row location	Bolt-row considered individually		Bolt-row considered as part of a group of bolt rows	
	Circular patterns $\ell_{eff,cp}$	Non-circular patterns $\ell_{eff,nc}$	Circular patterns $\ell_{eff,cp}$	Non-circular patterns $\ell_{eff,nc}$
Bolt-row adjacent to a stiffener	$2\pi m$	αm	$\pi m + p$	$0.5p + \alpha m - (2m + 0.625e)$
End bolt-row adjacent to a stiffener	The smaller of: $2\pi m$ $\pi m + 2e_1$	The smaller of: $e_1 + \alpha m - (2m + 0.625e)$	not relevant	not relevant
Values for α should be obtained from Figure 6.5 For *inner bolt-rows* and *end bolt-row* not adjacent to a stiffener, see Table 6.2				

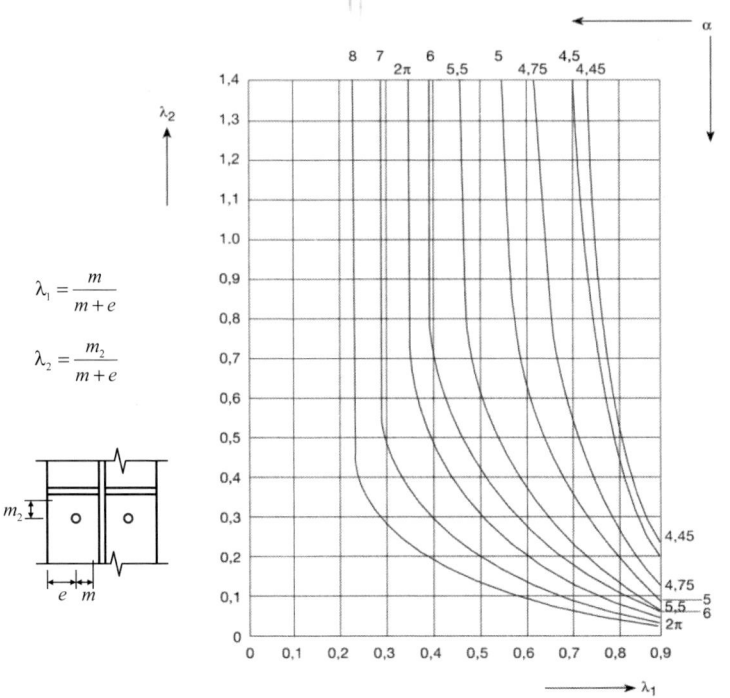

$$\lambda_1 = \frac{m}{m+e}$$

$$\lambda_2 = \frac{m_2}{m+e}$$

Figure 6.6 – Values for α for effective length of bolt-rows adjacent to a stiffener

6. MOMENT RESISTANT JOINTS

b) Stiffness

In bolted connections or in unstiffened welded connection, the stiffness coefficient for the component *column flange in transverse bending* is expressed as:

$$k_4 = \frac{0.9 \, \ell_{eff} \, t_{fc}^3}{m^3} \tag{6.13}$$

In a bolted connection, this expression has to be applied at the level of individual bolt-rows and of groups of bolt-rows in tension and ℓ_{eff} should be taken as equal, for each bolt-row (whether it is considered individually or as part of a group), to the smallest of the possible effective lengths (individually or as part of a group of bolt-rows) given in Table 6.2 and Table 6.3 as appropriate. m is defined in Figure 3.29.

For a welded connection, unstiffened or stiffened with appropriate transverse stiffeners:

$$k_4 = \infty \tag{6.14}$$

c) Deformation capacity

In bolted connections, significant ductility may be assumed for the *column flange in transverse bending* if yielding of the flange will be the governing failure. It may be assumed that this component has even sufficient rotation capacity for a plastic analysis, if the design resistance of the joint is governed by the resistance of the *column flange in transverse bending* and the thickness t_{fc} of the column flange satisfy the following condition:

$$t_{fc} \leq 0.36 \, d \sqrt{f_{ub}/f_{y,fc}} \tag{6.15}$$

where:
- d is the bolt diameter;
- f_{ub} is the ultimate tensile strength of the bolts;
- $f_{y,fc}$ is the yield strength of the column flange.

6.2.5 End-plate in bending

a) Resistance

Similar to a column flange in bending in a bolted connection, see section 6.2.4a), the design resistance of the *end-plate in bending* component is calculated using the equivalent T-stub model introduced in section 3.5.2, i.e. the resistance and failure mode of an end-plate in bending, together with the associated bolts in tension, should be taken as similar to those of an equivalent T-stub flange for both:

- each individual bolt-row required to resist tension;
- each group of bolt-rows required to resist tension.

The groups of bolt-rows at each side of any stiffener connected to the end-plate should be treated as separate equivalent T-stubs. In extended end-plates, the bolt-row in the extended part should also be treated as a separate equivalent T-stub, see Figure 6.7. The resistance and failure mode should be determined separately for each equivalent T-stub.

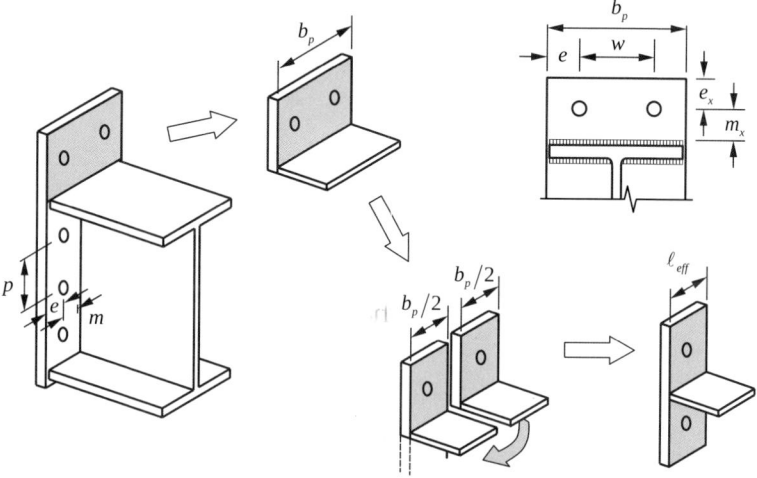

Figure 6.7 – Modelling an extended end-plate as separate T-stub

The dimensions e_{min} and m for use in 3.5.2 should be determined from Figure 3.30 for the part of the end-plate located between the beam flanges. For the end-plate extension e_{min} should be taken as equal to e_x, see Figure 6.7.

The effective lengths of equivalent T-stub flange ℓ_{eff} should be determined in accordance with section 3.5.2. Values are given for

- the bolt-row outside tension flange of beam in Table 6.4;
- the first bolt-row below tension flange of beam in Table 6.3 (bolt-row adjacent to a stiffener);
- other inner bolt-rows or other end bolt-rows in Table 6.2.

6. MOMENT RESISTANT JOINTS

The values of m_x, w, e and e_x for use in Table 6.4 should be obtained from Figure 6.7.

Table 6.4 – Effective lengths for an end-plate

Bolt-row location[*)]	Bolt-row considered individually		Bolt-row considered as part of a group of bolt-rows	
	Circular patterns $\ell_{eff,cp}$	Non-circular patterns $\ell_{eff,nc}$	Circular patterns $\ell_{eff,cp}$	Non-circular patterns $\ell_{eff,nc}$
Bolt-row outside tension flange of beam	Smallest of: $2\pi m_x$ $\pi m_x + w$ $\pi m_x + 2e$	Smallest of: $4m_x + 1.25e_x$ $2m_x + 0.625e_x + e$ $0.5b_p$ $2m_x + 0.625e_x + 0.5w$	not relevant	not relevant
Mode 1	$\ell_{eff,1} = \min(\ell_{eff,cp}; \ell_{eff,nc})$		$\Sigma\ell_{eff,1} = \min(\Sigma\ell_{eff,cp}; \Sigma\ell_{eff,nc})$	
Mode 2	$\ell_{eff,2} = \ell_{eff,nc}$		$\Sigma\ell_{eff,2} = \Sigma\ell_{eff,nc}$	

[*)] For the first bolt-row below tension flange of beam, see *Bolt-row adjacent to a stiffener* in Table 6.3.
[*)] For other inner bolt-rows or other end bolt-rows, see Table 6.2

b) Stiffness

The stiffness coefficient for the component *end-plate in bending* is expressed for a single bolt-row in tension as:

$$k_5 = \frac{0.9\ell_{eff} t_p^3}{m^3} \qquad (6.16)$$

This expression has to be applied at the level of individual bolt-rows and of groups of bolt-rows in tension and ℓ_{eff} should be taken as equal, for each bolt-row (whether it is considered individually or as part of a group), to the smallest of the possible effective lengths (individually or as part of a group of bolt-rows) given in Table 6.2, Table 6.3 and Table 6.4 as appropriate. *m* is generally defined in Figure 3.29, but for a bolt-row located in the extended part of an extended end-plate, $m = m_x$ where m_x is as defined in Figure 6.7.

c) Deformation capacity

For the component *end-plate in bending*, significant ductility may be assumed if yielding of the end-plate will be the governing failure. It may be assumed that an *end-plate in bending* has even sufficient rotation capacity for a plastic analysis, if the design resistance of the joint is governed by the resistance of the *end-plate in bending* and the thickness t_p of end-plate satisfy the following condition:

$$t_p \leq 0.36 \, d\sqrt{f_{ub}/f_{y,p}} \tag{6.17}$$

where:

    d is the bolt diameter;
    f_{ub} is the ultimate tensile strength of the bolts;
    $f_{y,p}$ is the yield strength of the end-plate.

6.2.6 Flange cleat in bending

a) Resistance

The design resistance of a bolted flange cleat in bending is determined in a very similar way than that of the extended part of an extended end-plate, 6.2.5a). The resistance and failure mode of an angle cleat in bending, together with the associated bolts in tension, should be taken as similar to those of an equivalent T-stub flange, see section 3.5.2. The effective length ℓ_{eff} of the equivalent T-stub flange should be taken as $0.5 b_a$ where b_a is the length of the angle cleat. The dimensions e_{min} and m for use in 3.5.2 should be determined from Figure 6.8. Note that, dependent on the gap g between the beam end and the column flange, the plastic hinge in the angle cleat may form either in the leg connected to the column flange or in the leg connected to the beam flange, see Figure 6.8.

6. MOMENT RESISTANT JOINTS

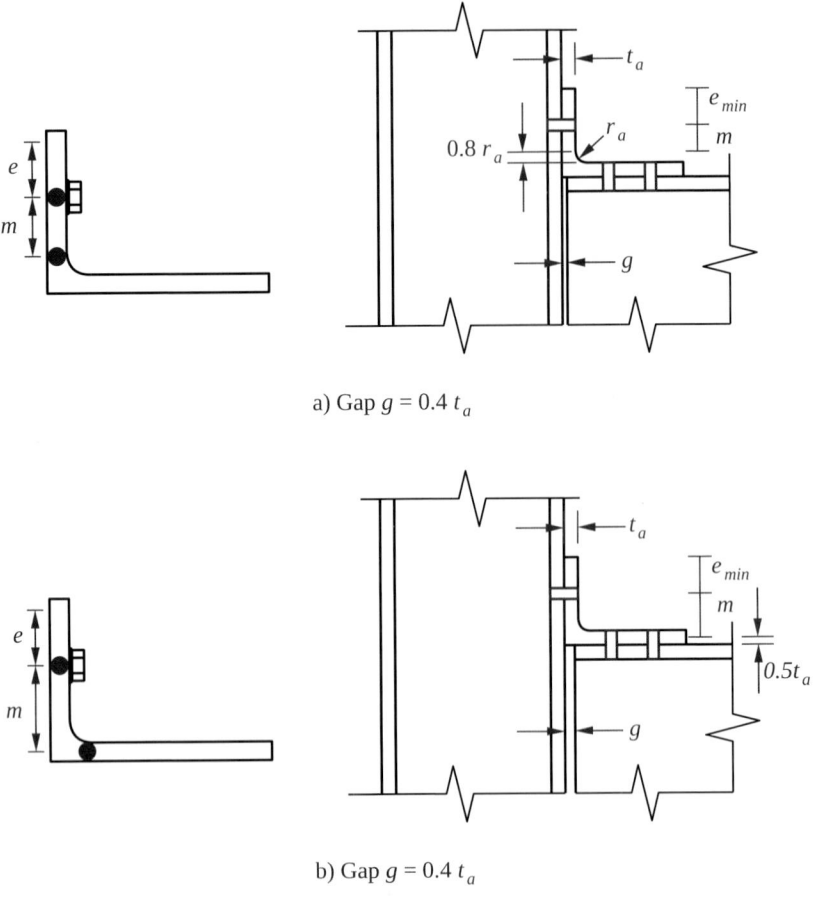

a) Gap $g = 0.4\,t_a$

b) Gap $g = 0.4\,t_a$

Figure 6.8 – Influence of the gap between the beam and the column

b) Stiffness

The stiffness coefficient is determined as follows:

$$k_6 = \frac{0.9\,\ell_{eff}\,t_a^3}{m^3} \qquad (6.18)$$

where:
$\ell_{eff} = 0.5\,b_a$;
b_a is the length of the angle cleat;
t_a is the thickness of the angle cleat;
m is defined in Figure 6.8.

6.2 COMPONENT CHARACTERISATION

c) Deformation capacity

For the component *flange cleat in bending*, significant ductility may be assumed if yielding of the angle cleat will be the governing failure. It may be assumed that this component has even sufficient rotation capacity for a plastic analysis, if the design resistance of the joint is governed by the resistance of the *flange cleat in bending* and the thickness of the angle cleat t_a satisfy the following condition:

$$t_a \leq 0.36\, d\sqrt{f_{ub}/f_{y,a}} \tag{6.19}$$

where:
- d is the bolt diameter;
- f_{ub} is the ultimate tensile strength of the bolts;
- $f_{y,a}$ is the yield strength of the angle cleat.

6.2.7 Beam or column flange and web in compression

a) Resistance

As a result, for instance, of a bending moment applied to the beam, compression develops in the vicinity of one of the beam flange and the adjacent part of the web. The possible development of plasticity or buckling of this zone is covered by the component "flange and web in compression". As shown in Figure 6.9, the concentrated force F_c may be significantly higher than the compression force F induced by the same beam loading, at a certain distance from the connection. In Figure 6.9, forces F_c and F are applied at the centre of gravity of the beam flange.

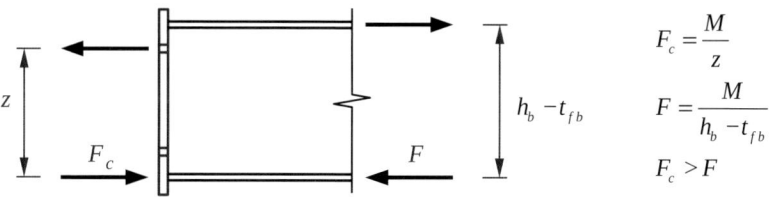

Figure 6.9 – Concentrated force F_c and compression force F

6. MOMENT RESISTANT JOINTS

This reference is conventionally adopted irrespectively of the fact that it does not correspond to the reality, as the compression zone is not only limited to the flange. The localised force F_c is likely to generate premature failure of the zone by plasticity or instability. The verification of this failure mode is achieved by checking that F_c is smaller or equal to:

$$F_{c,fb,Rd} = \frac{M_{c,Rd}}{h_b - t_{fb}} \qquad (6.20)$$

where:

$M_{c,Rd}$ is the design bending moment resistance of the beam reduced, if necessary, by the presence of a shear force; this moment takes into account the risk of plate buckling in the compressed part of the web or in the flange;

h_b is the total beam height;

t_{fb} is the beam flange thickness.

The resistance provided by Eq. (6.20) integrates naturally the resistance of the flange and that of the part of the web in compression. It has to be clearly stated that the formula applies independently of the connection type (steel bolted, composite …) and of the beam loading (for instance, a beam subjected to compression and bending). In other words, the resistance of the component is "as it is" and the effect of the above-mentioned factors will be covered in the "assembly of components" step of the component method.

b) Stiffness

In EN 1993-1-8, the stiffness coefficient for is component is given as equal to:

$$k_7 = \infty \qquad (6.21)$$

Few reasons may be raised to explain it:

- the deformation of the component contributes to that of the beam section, and not to that of the joint; and the deformation of the beam is directly accounted for through, for instance, the beam stiffness EI_b/L_b;
- even if the premature yielding of the component leads to localised plastic deformations not considered through EI_b/L_b, these ones remain quite limited;

6.2 COMPONENT CHARACTERISATION

- the reduction of the actual dimensions of the joint to one point located at the intersection between the connected members results, in the frame analysis, to the consideration of higher beam lengths than in reality. The extra deformations of the fictitious beam stubs along the length of the end joints compensates the localised plastic deformations addressed in the previous bullet point.

c) Deformation capacity

The ductility of this component may be estimated through the definition of the beam cross section class which has to be achieved to define the value of $M_{c,Rd}$ in Eq. (6.20).

6.2.8 Beam web in tension

a) Resistance

The similarity between the *beam web in tension* component and the *column web in transverse tension* one is so obvious that similar formulae are used to derive their design resistance. For the beam web in tension, the expression is as follows:

$$F_{t,wb,Rd} = \frac{b_{eff,t,wb} t_{wb} f_{y,wb}}{\gamma_{M0}} \quad (6.22)$$

The only difference with Eq. (6.9), besides the replacement of the subscript "c" by "b", is the absence of the reduction factor ω, justified by the far much lower risk of stress interaction with shear stresses in the beam web (see section 6.3.5.2 showing how shear stresses acting in the beam section close to the connection may be distributed, so avoiding interactions).

Here the effective width $b_{eff,t,wb}$ is taken as equal to the equivalent length of the T-stub (ℓ_{eff}) which may be substituted to the end-plate in bending obtained for an individual bolt-row or a bolt-group (see section 6.2.5a).

b) Stiffness

The stiffness coefficient for the *beam web in tension* component is expressed as:

$$k_8 = \infty \quad (6.23)$$

6. MOMENT RESISTANT JOINTS

As for the component *column or beam flange and web in compression*, the deformation of the *beam web in tension* is assumed to be included, in view of the frame analysis, in the deformation of the beam element and has therefore not to be taken into account a second time through the definition of a non-infinite value of the k_8 stiffness coefficient.

c) Deformation capacity

Significant ductility in tension may be here again expected from normalised steels and so no limitation in plastic deformation capacity may be assumed from this component.

6.2.9 Plate in tension or compression

a) Resistance

The resistance of a plate in tension may result from the yielding of its transverse cross section or from the failure of its net section (cross sectional area reduced by bolt holes or any other openings). Related design formulae are provided in section 6.2 of Eurocode 3 Part 1-1.

When compression is applied to a plate, the effect of fastening holes is generally disregarded as long as these ones are filled by fasteners, except for oversize and slotted holes. So the cross section check reduces to the verification of the yielding of the plate, again according to section 6.2 of EN 1993-1-1. Plate buckling is practically prevented by an adequate connection detailing (for instance, maximum distance between connecting bolts).

b) Stiffness

The stiffness coefficient k_9 is assumed to be infinite, for the same reasons than those explained for other components like *beam flange and web in compression, beam web in tension*, etc.

c) Deformation capacity

The ductile behaviour of a plate in tension or compression may only result from the yielding of the cross section. From this point of view, the premature failure in the net cross section of the plate subjected to tension has to be avoided through a proper selection of the layout and dimensions of the plate and bolt holes with the objective to make sure that the plastic resistance of the gross-section is lower than the more brittle failure load in the net-section, as explained in sub-chapter 3.2 of the ECCS Design Manual on "Design of Steel Structures" (Simões da Silva et al, 2010).

6.2.10 Bolts in tension

a) Resistance

Design rules for bolts in tension have been presented in section 3.4.2. In fact, they act in joints as parts of T-stub connections simulating the response of column flanges, cleats or end-plate in bending. Related design formulae are provided in section 3.5.2.

b) Stiffness

A bolt in tension may be idealised as a beam element in tension and its stiffness coefficient can as first side be evaluated as equal to A_s/L_b where A_s and L_b are respectively defined as the tensile stress area of the bolt and the bolt elongation length, taken as equal to the grip length (total thickness of materials and washers) plus half the sum of the height of the bolt head and the height of the nut.

However, in EN 1993-1-8, another expression of the stiffness coefficient k_{10} is given:

$$k_{10} = \frac{1.6 A_s}{L_b} \qquad (6.24)$$

The factor "1.6" may be justified as follows:

- EN 1993-1-8 refers to a "bolt-group", and not to a single bolt, what result in the application of a factor "2.0" (in fact, one should read, in EN 1993-1-8, "a bolt-group with two bolts in tension");
- as a result of the development of prying forces in the T-stub, the bolts are overloaded by a factor, derived from the "T-stub theory" (Jaspart, 1997), equal to 1.25, so decreasing their stiffness by $1/1.25 = 0.8$.

For sure, in the case of a "Mode 3" T-stub failure, or a "Mode 1 without prying" failure mode (see section 3.5.2), the application of the "0.8"

6. MOMENT RESISTANT JOINTS

factor is not relevant and one could count on a value of $k_{10} = 2.0 A_s / L_b$. On the other hand, in the two here-above mentioned situations, globally, the stiffness of the bolts remains quite high in comparison of that of the T-stub flange and the "error" remains quite limited. This is why, for sake of simplicity, the "1.6" factor is always selected.

Finally, it has to be mentioned that EN 1993-1-8 does not make any distinction between the stiffness of a non-preloaded or preloaded bolt, even if the use of preloading may result in an increase of the joint stiffness, as shown in Jaspart (1991) where distinct stiffness coefficients are suggested.

c) Deformation capacity

To rely on the ductility of bolts in tension is definitively not acceptable. Even if not strictly forbidden, the design of joints which would reach their design resistance through bolt failure should never be recommended. Moreover, this failure mode is prohibited as soon as the user wants to profit from plasticity. Part 1-8 limits practically the potential risk of brittle failure through various clauses. For instance:

- Eq. (6.32) in section 6.4.2 which limits the thickness of the connected plates in comparison to the diameter of the bolts; this formula [see background in (Jaspart, 1997)] has to be fulfilled when a plastic global analysis of the frame involving the development of plastic hinges in the joints is targeted;
- EN 1993-1-8 clause 6.2.7.2(9) where the "$1.9 B_{t,Rd}$" limitation (see section 6.3.1, Eq. (6.35)) is just of way to limit the redistribution of internal forces in a joint when a risk of premature failure of a bolt-row is likely to occur.

6.2.11 Bolts in shear

a) Resistance

The shear resistance of a bolt (and also of a bolt group) has been defined in chapter 3, in the case of prestressed and non-prestressed bolts.

b) Stiffness

The stiffness coefficient k_{11} relates in EN 1993-1-8 to a bolt-group with two bolts, sheared along one plane only, see background (Jaspart, 1991):

- for non-preloaded bolts:
$$k_{11} = \frac{16 n_b d^2 f_{ub}}{E d_{M16}}$$ (6.25)

- for preloaded bolts:
$$k_{11} = \infty$$ (6.26)

where n_b, d, f_{ub} and d_{M16} designate respectively the number of bolt-rows, the bolt diameter, the ultimate bolt strength and the nominal diameter of a M16 bolt.

c) Deformation capacity

The ductility of a bolt in shear is quite limited and has therefore not to be relied on. This point has been already addressed in section 3.5.1.2.

6.2.12 Bolts in bearing (on beam flange, column flange, end-plate or cleat)

a) Resistance

The bearing of the bolts against the plates is ruled by specific design formulae which have been presented in chapter 3, and more especially in section 3.4.5.

b) Stiffness

As for bolts in shear, the stiffness coefficient k_{12} corresponds to the deformation in one of the connected plates induced by a bolt-group with two bolts:

- for non-preloaded bolts
$$k_{12} = \frac{24 n_b k_b k_t d f_u}{E}$$ (6.27)

- for preloaded bolts
$$k_{12} = \infty$$ (6.28)

where n_b, d and f_u are respectively the number of bolt-rows, the bolt diameter and the ultimate strength of the plate material. k_b and k_t depends on the connection detailing (see EN 1993-1-8).

c) Deformation capacity

The bearing deformation may be assumed to be quite ductile, as previously mentioned (section 3.5.1.2).

6. MOMENT RESISTANT JOINTS

6.2.13 Concrete in compression including grout

a) Resistance

This component has been described in section 5.4.3.1, in terms of resistance.

b) Stiffness

The stiffness coefficient k_{13} is given as (Steenhuis et al, 2008):

$$k_{13} = \frac{E_c \sqrt{b_{eff} \ell_{eff}}}{1.275 E} \qquad (6.29)$$

where b_{eff} and ℓ_{eff} are the effective width and length of the T-stub flange, see section 5.4.3.1.

c) Deformation capacity

In the absence of precise information concerning the ductility of this component, it seems reasonable not to rely on it.

6.2.14 Base plate in bending under compression

The component which has been presented in section 5.4.3.1 should be named: *concrete and base plate under compression*. And, in fact, the information provided in section 6.2.13, here-above, covers the response of this "couple of components". The design resistance and the stiffness coefficient k_{13} is therefore defined as infinite in EN 1993-1-8.

6.2.15 Base plate in bending under tension

a) Resistance

This component is modelled as a T-stub and therefore the model presented in section 3.5.2 again applies. The only specificity of base plates, in comparison to steel-to-steel connected plates, is the higher probability that prying forces do not develop, as a result of the elongation of the anchor bolts. This aspect has been raised in section 3.5.2.4 where relevant appropriate formulae are presented.

b) Stiffness

Two expressions are recommended here for the stiffness coefficient k_{15}, according to the development or not of prying forces as a result of the elongation of the anchors:

- when prying forces develop $\quad k_{15} = \dfrac{0.85 \ell_{eff} t_p^3}{m^3}$ (6.30)

- when prying forces do not develop $\quad k_{15} = \dfrac{0.425 \ell_{eff} t_p^3}{m^3}$ (6.31)

where ℓ_{eff} and t_p are defined as the effective length of the T-stub and the thickness of base plate thickness. m is a geometrical characteristic of the T-stub (see section 3.5.2). No explanation else than a lack of consistency between Eq. (6.30) and Eqs. (6.13), (6.16) and (6.18) may explain the difference between the coefficients "0.85" and "0.9" in the expressions of k_{15} (with prying) and k_4, k_5 and k_6.

c) Deformation capacity

Reference may her be directly done to what has been said for the deformation capacity of a *column flange in bending* (section 6.2.4) and an *end-plate in bending* (section 6.2.5).

6.2.16 Anchor bolts in tension

Basically, an anchor behaves as a bolt in tension. And so the properties may be derived in a similar way than what has been described in section 6.2.10.
Anyway, some particular features need to be addressed:
- The design tension resistance of the anchor may be taken equal to the tensile resistance of its threaded part as long as no premature failure occurs by pull-out failure, concrete cone failure or splitting failure of the concrete. This has been addressed in section 5.4.3.2.
- The stiffness coefficient k_{17} to be used for an anchor is the same than for a bolt (k_{10}) as long as prying forces develop between the base plate and the concrete foundation (see section 6.2.15). If it is not the case, than the factor "1.6" has to be replaced by "2.0" in

6. MOMENT RESISTANT JOINTS

Eq. (6.24). And, as raised in section 5.4.3.2, an appropriate definition of the "bolt length" has to be adopted for anchors.

6.2.17 Anchor bolts in shear

The topic has been already extensively discussed in section 5.4.4. No further information is required here.

6.2.18 Anchor bolts in bearing

As in 6.2.17, reference is here made to section 5.4.4 where relevant information is provided.

6.2.19 Welds

a) Resistance

The evaluation of the design resistance of welds has been extensively covered in sub-chapter 4.3 "Design of welds". So, no complementary information is provided here.

b) Stiffness

The stiffness coefficient for welds, k_{19}, is assumed to be infinite. Welds are therefore not contributing to the global flexibility of the joints.

c) Deformation capacity

No significant plastic deformation capacity may be expected from welds, as already expressed in section 4.4.1. Weld failure should always be considered as a quite brittle and so as an undesirable failure mode.

6.2.20 Haunched beam

a) Resistance

The design resistance of a beam flange and of the adjacent zone of the web in compression is given by Eq. (6.20) in section 6.2.7. For a haunched beam, a similar formula applies, but in which $M_{c,Rd}$ is now evaluated neglecting the intermediate flange.

If the height of the beam including the haunch exceeds 600 mm, EN 1993-1-8 specifies that the contribution of the beam web to the design compression resistance should be limited to 20%. This point is still subject to discussions in the scientific community, recent tests indicating that this limitation could be relaxed.

If a beam is reinforced with haunches, the latter should be arranged such that:

- the steel grade of the haunch should match that of the member;
- the flange size and the web thickness of the haunch should not be less than that of the member;
- the angle of the haunch flange to the flange of the member should not be greater than 45°.

Moreover, if a beam is reinforced with haunches, the design resistance of beam web in compression should be determined according to section 6.2.2. In this evaluation, the length of stiff bearing to the beam should be taken as equal to the thickness of the haunch flange parallel to the beam.

b) Stiffness

As for the component *beam or column flange and web in compression*, the relevant stiffness coefficient k_{20}, is assumed to be infinite.

c) Deformation capacity

Similarly to what has been explained for the component *beam or column flange and web in compression*, the ductility of this component may be estimated through the definition of the beam cross section class which has to be achieved to define the value of $M_{c,Rd}$ in Eq. (6.20).

6.2.21 Longitudinal steel reinforcement in tension

a) Resistance

The effective width of the concrete slab has first to be determined for the beam cross section just at the location of the connection, according to EN 1994-1-1 clause 5.4.1.2. For each layer of reinforcement over the slab thickness, it is be assumed that the effective area of longitudinal reinforcement in tension (i.e., for the considered layer, the area of the

6. MOMENT RESISTANT JOINTS

reinforcement bars over the effective width) is stressed to its design yield strength f_{sd}. The resistance of the layer is consequently obtained by multiplying the effective area of longitudinal reinforcement by f_{sd}.

Where unbalanced loading occurs in a double-sided joint configuration, a strut-tie model may be used to verify the introduction of the forces in the concrete slab into the column, see Figure 6.10.

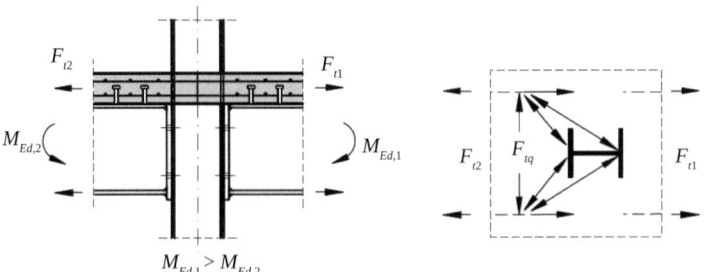

Figure 6.10 – Strut-tie model [Fig 8.2 from EC4]

For a single-sided configuration designed as a composite joint, the effective longitudinal slab reinforcement in tension should be anchored sufficiently well beyond the span of the beam to enable the design tension resistance to be developed.

b) Stiffness

The stiffness coefficient $k_{s,r}$, provided for a single layer r of reinforcement, may be obtained by referring to Table A.1 extracted from EN 1994 Part 1-1, section A.2.1.2 (CEN, 2004b).

c) Deformation capacity

Plastic deformation capacity may be obtained through a proper design of the slab and of its longitudinal and transverse steel reinforcements. Corresponding requirements are expressed in section 6.9.2.2.

6.2.22 Steel contact plate in compression

a) Resistance

Contact plates are sometimes used in composite joints to enable an appropriate transfer of the compression forces form the beam to the column (see example in Figure 6.11).

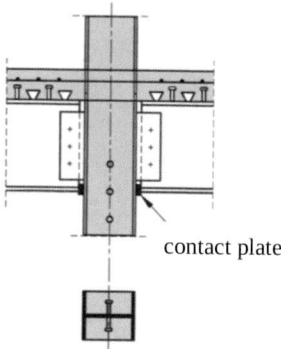

Figure 6.11 – Example of joint with contact plates in compression

In EC4, the resistance of this element is simply defined as the effective area of the contact plate in compression times the design yield strength f_{yd}. Where the height or breadth of the contact plate exceeds the corresponding dimension of the compression flange of the steel section, the corresponding effective dimension should be determined assuming dispersion at 45° through the contact plate.

b) Stiffness

The related stiffness coefficient may be taken as equal to infinity.

c) Deformation capacity

No information is provided by Eurocode 4. However, no risk of brittle failure is expected to occur.

6.3 ASSEMBLY FOR RESISTANCE

6.3.1 Joints under bending moments

The procedure for strength assembly as suggested in EN 1993-1-8 is aimed at deriving the value of the so-called design resistance of the joint. For the sake of clarity, it is not presented at first in a general way but is illustrated in the particular case of beam splices with flush end-plates.

For the connection represented in Figure 6.12, the distribution of internal forces is quite easy to obtain: the compressive force is transferred at

6. MOMENT RESISTANT JOINTS

the centroid of the beam flange and the tension force, at the level of the upper bolt-row. The resistance possibly associated to the lower bolt-row is usually neglected as it contributes in a quite modest way to the transfer of bending moment in the joint (small lever arm).

The design resistance of the joint $M_{j,Rd}$ is associated to the design resistance F_{Rd} of the weakest joint component which can be one of the following: the beam flange and web in compression, the beam web in tension, the plate in bending or the bolts in tension. For the two last components (plate and bolts), reference is made to the concept of "idealized T-stub" introduced in 3.5.2. The bending resistance so equals:

$$M_{j,Rd} = F_{Rd} \cdot z \qquad (6.32)$$

where z is the lever arm ($z = h$).

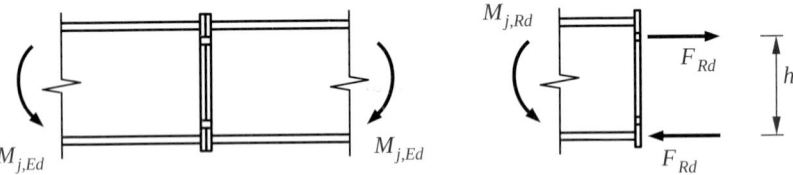

Figure 6.12 – Joint with one bolt-row in tension

When more than one bolt-row has to be considered in the tension zone (Figure 6.13), the distribution of internal forces is more complex.

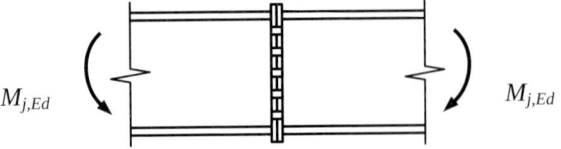

Figure 6.13 – Joint with more than one bolt-row in tension

Let us assume, initially, that the design of the joint leads to adopting a particularly thick end-plate in comparison to the bolt diameter (Figure 6.14). In this case, the distribution of internal forces between the different bolt-rows is linear according to the distance from the centre of compression. The compression force F_{Rd} which equilibrates the tension forces acts at the level of the centroid of the lower beam flange. For the sake of clarity, it is shown only in Figure 6.14, but not in the subsequent figures.

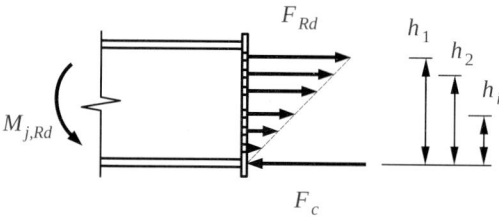

Figure 6.14 – Joint with a thick end-plate

The design resistance $M_{j,Rd}$ of the joint is reached as soon as the bolt-row subjected to the highest stresses - in reality, that which is located the farthest from the centre of compression - reaches its design resistance in tension F_{Rd} ($2B_{t,Rd}$ if a Mode 3 failure is assumed in the "plate in bending" component and there are two bolts per row).

As a matter of fact, the quite limited deformation capacity of the bolts in tension does not allow any redistribution of forces to take place between bolt-rows.

It is assumed here that the design resistance of the beam flange and web in compression is sufficient to transfer the compression force F_c. The tensile resistance of the beam web is also assumed not to limit the design resistance of the joint. So, $M_{j,Rd}$ is expressed as (Figure 6.14):

$$M_{j,Rd} = \frac{F_{Rd}}{h_i} \cdot \sum h_i^2 \qquad (6.33)$$

For thinner end-plates, the distribution of internal forces requires much more attention. When the moment starts to be progressively increased, forces distribute amongst the bolt-rows according to the relative stiffnesses of the latter. This stiffness is namely associated to that of the part of the end-plate adjacent to the considered bolt-row. In the particular case of Figure 6.15, the upper bolt-row is characterized by a higher stiffness because of the presence there of the beam flange and web welded to the end-plate.

Because of the higher stiffness, the upper bolt-row is capable of transferring a higher load than the lower bolt-rows (Figure 6.15b).

6. MOMENT RESISTANT JOINTS

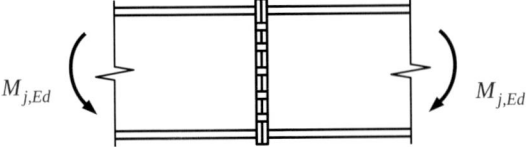

a) Configuration

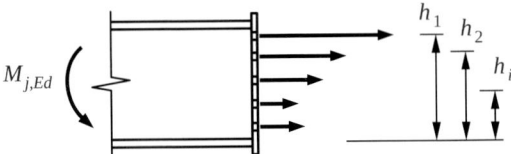

b) Distribution of the internal forces at the beginning of the loading

Figure 6.15 – Joint with a thin end-plate

In EN 1993-1-8, it is assumed that the upper bolt-row will reach its design resistance first. This assumption is quite justified in the particular case being considered but is probably less justified for other connection types such as for end-plate connections with an extended part in the tension zone where it is usual that the second bolt-row - that just below the flange in tension - reaches its design resistance first.

The design resistance of the upper bolt-row may be associated to one of the following components: the bolts only, the end-plate only, the bolts-plate assembly or the beam web in tension. If its failure mode is ductile, a redistribution between the bolt-rows can take place : as soon as the upper bolt-row reaches its design resistance, the supplementary bending moments applied to the joint are carried out by the lower bolt-rows (the second, then the third, etc.) which, each one in their turn, may reach their own design resistance.

The failure may occur in three different ways:

a) The plastic redistribution of the internal forces extends to all bolt-rows when they have sufficient deformation capacity. The redistribution is said to be "complete" and the resulting distribution of internal forces is called "plastic".

The design moment resistance $M_{j,Rd}$ is expressed as (Figure 6.16):

$$M_{j,Rd} = \sum_i F_{Rd,i} \cdot h_i \qquad (6.34)$$

The plastic forces $F_{Rd,i}$ vary from one bolt-row to another according to the failure modes (bolts, plate, bolt-plate assembly, beam web, etc.).

Eurocode 3 considers that a bolt-row possesses sufficient deformation capacity to allow plastic redistribution of internal forces to take place when:

- it is associated to the failure of the beam web in tension or;
- it is associated to the failure of the bolt-plate assembly (including failure of the bolts alone or of the plate alone)

and:

$$F_{Rd,i} \leq 1.9 B_{t,Rd} \qquad (6.35)$$

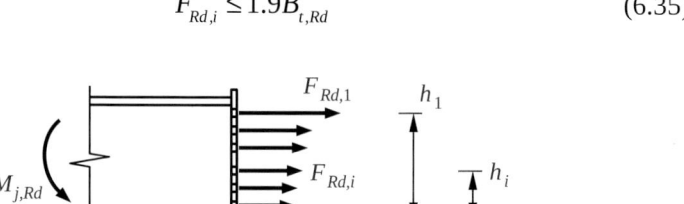

Figure 6.16 – Plastic distribution of forces

b) The plastic redistribution of forces is interrupted because of the lack of deformation capacity in the last bolt-row which has reached its design resistance ($F_{Rd,i} \leq 1.9 B_{t,Rd}$ and linked to the failure of the bolts or of the bolt-plate assembly).

In the bolt-rows located lower than bolt-row k, the forces are then linearly distributed according to their distance to the point of compression (Figure 6.16).

The design moment resistance equals:

$$M_{j,Rd} = \sum_{i=1}^{k} F_{Rd,i} + \frac{F_{Rd,k}}{h_k} \cdot \sum_{j=k+1}^{n} h_j^2 \qquad (6.36)$$

where:

- n is the total number of bolt-rows;
- k is the number of the bolt-row of which the deformation capacity is not sufficient.

6. MOMENT RESISTANT JOINTS

In this case, the distribution is said "elasto-plastic".

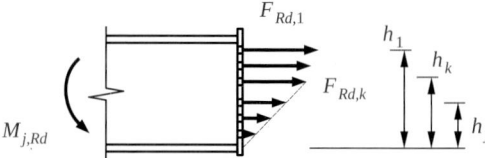

Figure 6.17 – Elasto-plastic distribution of internal forces

c) The plastic or elasto-plastic distribution of internal forces is interrupted because the compression force F_c attains the design resistance of the beam flange and web in compression. The moment resistance $M_{j,Rd}$ is evaluated with similar formulae to Eqs. (6.34) and (6.36) in which, obviously, only a limited number of bolt-rows are taken into consideration. These bolts rows are such that:

$$\sum_{\ell=1}^{m} F_\ell = F_{c,Rd} \qquad (6.37)$$

where:
- m is the number of the last bolt-row transferring a tensile force;
- F_ℓ is the tensile force in bolt-row number ℓ;
- $F_{c,Rd}$ is the design resistance of the beam flange and web in compression.

The application of the above-described principles to beam-to-column joints is quite similar. The design moment resistance $M_{j,Rd}$ is, as for the beam splices, likely to be limited by the resistance of:

- the end-plate in bending,
- the bolts in tension,
- the beam web in tension,
- the beam flange and web in compression,

but also by that of:

- the column web in tension,
- the column flange in bending.
- the column web in compression;
- the column web panel in shear.

In the assembly procedure, the possibility to form individual and group yield mechanisms has to be properly considered too. As explained in section 3.5.2.2, various yield mechanisms are likely to form in the connected plates (end-plate or column flange) when adjacent bolt-rows are subjected to tension forces.

- individual mechanisms (see Figure 6.18a) which develop when the distance between the bolt-rows are sufficiently large;
- group mechanisms (see Figure 6.18b) including more than one adjacent bolt rows.

Each of these mechanisms are associated to specific design resistances.

When distributing the internal forces, Eurocode 3 recommends never to transfer in a given bolt-row:

- a higher load than that which can be carried when it is assumed that the considered bolt-row is the only one able to transfer tensile forces (individual resistance);
- a load such that the resistance of the whole group to which the bolt-row belongs is exceeded.

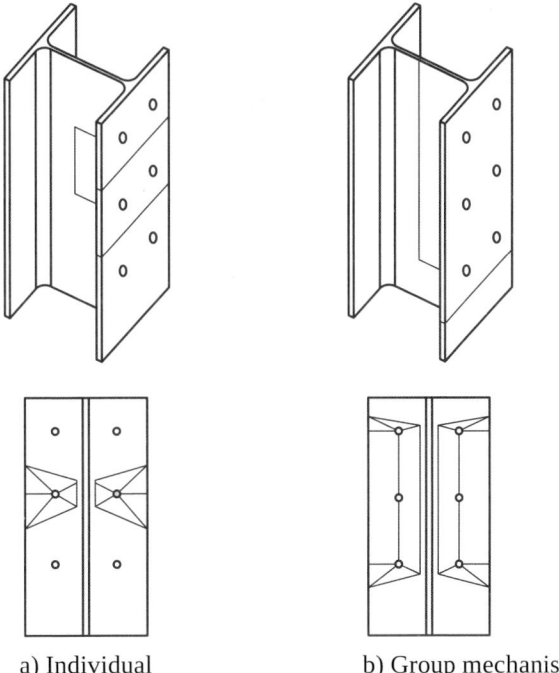

a) Individual   b) Group mechanism

Figure 6.18 – Plastic mechanisms

6. MOMENT RESISTANT JOINTS

Figure 6.19 gives, in more general terms, a complete overview of the assembly procedure proposed by EN 1993-1-8 which has simply to be seen as a practical application of the static theorem depicted in section 1.6.1.

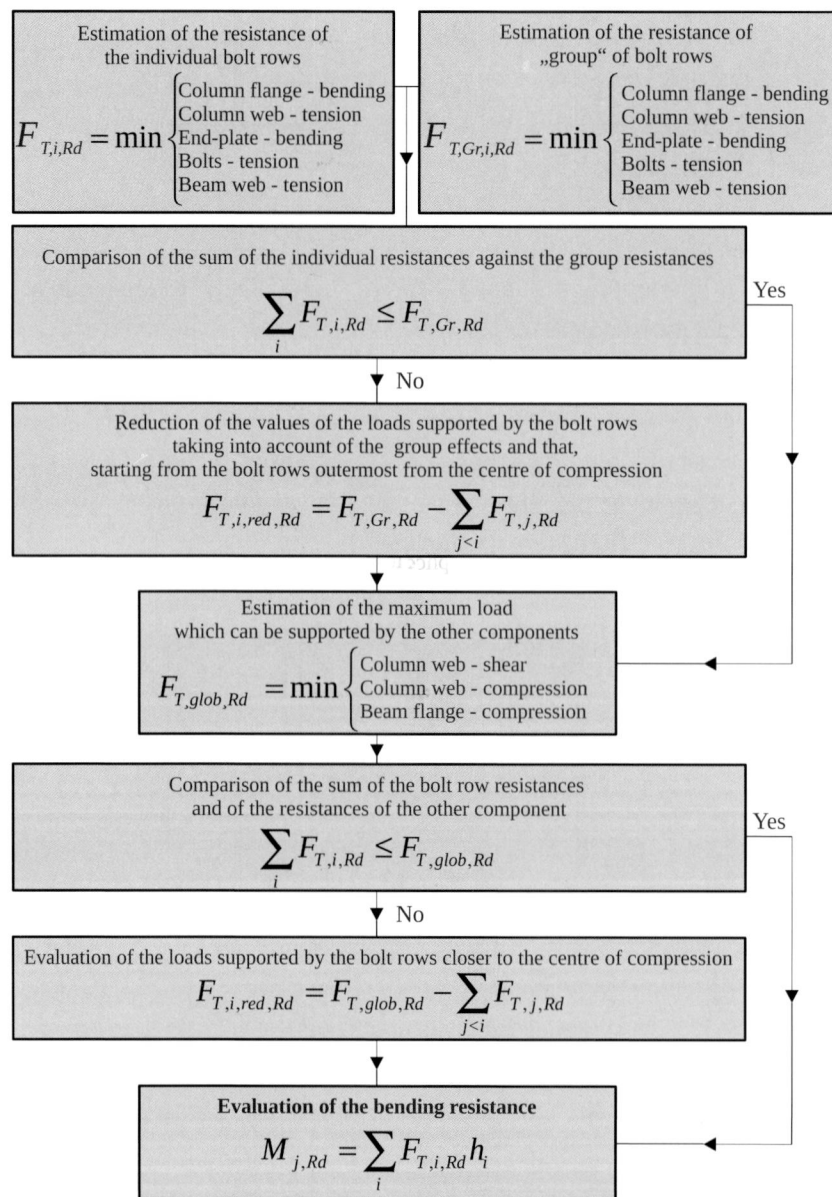

Figure 6.19 – Flow-chart for the assembly procedure

6.3.2 Joints under axial forces

The same principles than those applied in the previous section may be applied here. In Figure 6.20a and b, the specific case of a beam splice in compression and tension is illustrated.

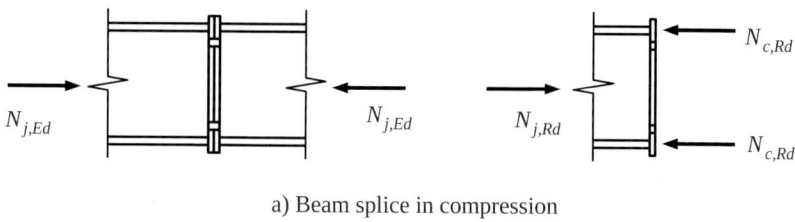

a) Beam splice in compression

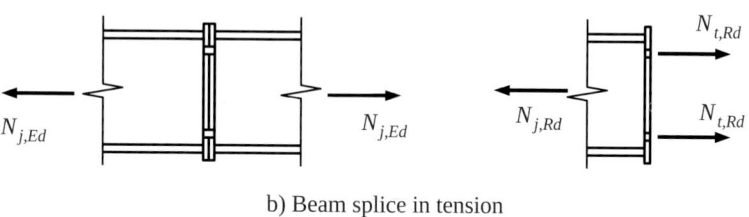

b) Beam splice in tension

Figure 6.20 – Joint with a symmetrical geometry

The design resistance of this joint under compression equals:

$$N_{j,Rd} = 2 \cdot N_{c,Rd} \quad (6.38)$$

where $N_{c,Rd}$ is the design resistance of the "beam flange and web in compression" component. In a more general case, $N_{j,Rd}$ could be also limited by the resistance of the column web in compression (beam-to-column joint) or by the concrete in compression (column bases).

Under tension forces, the design resistance of the joint is expressed as:

$$N_{j,Rd} = 2 \cdot N_{t,Rd} \quad (6.39)$$

where $N_{t,Rd}$ is the smaller of the design resistances of the "endplate in bending" or the "beam web inn tension" components. If more than 2 bolts-rows are present, the design resistance of the joint writes:

6. MOMENT RESISTANT JOINTS

$$N_{j,Rd} = \sum N_{ti,Rd} \qquad (6.40)$$

where $\sum N_{ti,Rd}$ extends to all bolt-rows. Here also the possible limitation of the tension resulting from group effects has to be checked.

When the joint detailing is such that no symmetry is contemplated with respect to the force axis, the joint is subjected to a combination of axial force and bending moment. This situation is explicitly covered in the next section.

6.3.3 Joints under bending moments and axial forces

6.3.3.1 Introduction

In EN 1993-1-8, it is assumed that the design resistance of a joint in bending is not influenced by the application of an additional axial force $N_{j,Ed}$ as long as the latter remains limited, i.e. lower than 5% of the axial design resistance $N_{pl,Rd}$ of the connected beam (and not of the joint, what is quite surprising as far as the influence of the applied axial load on the joint response is of concern):

$$\left| \frac{N_{j,Ed}}{N_{pl,Rd}} \right| \leq 0.05 \qquad (6.41)$$

This value seems to be a fully arbitrary one and is not at all scientifically justified.

However, in many cases, the joints are subjected to significant combined axial loads and bending moments, for instance at the extremities of inclined roof beams or in frames subjected to an exceptional event leading to the loss of a column, situation where significant tying forces can developed in the structural beams above the lost column.

If the criterion given above is not satisfied, Eurocode 3 Part 1-8 recommends checking the resistance by referring to a "M-N" interaction diagram defined by the polygon linking the four points corresponding respectively to the hogging and sagging bending resistances in absence of axial forces and to the tension and compression axial resistances in absence of bending. The interaction formulae for normal forces greater than 5% of the axial design resistance $N_{pl,Rd}$ is expressed as follows:

$$\frac{M_{j,Ed}}{M_{j,Rd}} + \frac{N_{j,Ed}}{N_{j,Rd}} \leq 1.0 \qquad (6.42)$$

where:
- $M_{j,Rd}$ is the design moment resistance of the joint without the presence of normal forces;
- $N_{j,Rd}$ is the design normal resistance of the joint in the absence of bending moments.

How to derive the "tension and compression axial resistances in absence of bending" is however not expressed in EC3.

The combination of both criteria in Eqs. (6.41) and (6.42) is illustrated in Figure 6.21 by a solid line. In (Cerfontaine, 2003), a more precise analytical design procedure, based on the component method concept is presented. It has been (i) developed by Cerfontaine to predict the response of ductile and non-ductile steel joints subjected to combined axial loads and bending moments and (ii) extended to composite joints in (Demonceau, 2008). This method is briefly introduced in the present section.

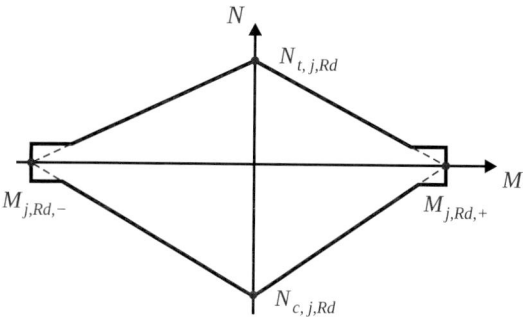

Figure 6.21 – M-N resistant curve for a joint proposed in EN 1993-1-8

6.3.3.2 Brief description of the advanced analytical procedure for steel joints

Within this section, a brief description of the analytical procedure developed by (Cerfontaine, 2003) is given which is in full agreement with the component approach recommended in the Eurocodes.

6. MOMENT RESISTANT JOINTS

The general concept of the method is first presented. Then, the principles on which it is based to predict the resistance of the joints are given.

a) General concept

The advanced design method is based on a component approach. As the behaviour of the components is not dependent on the type of loading applied to the whole joint, the challenge is to develop an appropriate assembly procedure for bending moments and axial forces. Another aspect to be dealt with is the modification of the list of active components within the joints according to the relative importance of the bending moment and axial load, and obviously according to the respective signs of the applied forces.

Finally, two particular features of the component method have also to be carefully considered within the developed procedure:

- Group effects which may affect the response of "T-stub related" components: column flange in bending, endplate in bending, column web in tension and the beam web in tension.
- Component interactions: these ones may occur in "column" components where three types of stresses interact: shear stresses in the web panel, longitudinal stresses due to axial and bending forces in the column and transversal stresses due to the load-introduction in the joint area (column web in tension, column web in compression and column web in shear).

Only the procedure to apply to joints activating ductile components at collapse is presented herein. So, the behaviour of each of the constitutive joint components is assumed to be infinitely ductile. As a result, a full plastic redistribution of the internal forces in the joint under bending moment and axial load may be contemplated.

b) Conventions

Within this section, developments are presented for the case of a bolted end-plate steel connection with N_b bolt rows; these ones may potentially, when activated, transfer tension forces; besides that, two compression zones centred at mid-thickness of the upper and lower beam flanges are identified (respectively noted "upper" or "up" and "lower" or "lo"). The compression zones activate two components: the beam flange and web in compression and the column flange in compression.

So, this leads to a total of $N_b + 2 = n$ rows where internal forces may be transferred from the beam to the column. Conventionally, the tension forces are assumed to be positive (or equal to zero) while a compression force has a negative (or zero) value. All the rows are numbered from 1 to n by starting from the upper row. An example of row numbering is given in Figure 6.22.

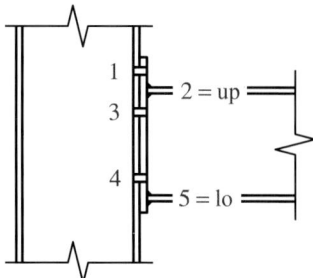

Figure 6.22 – Example of row numbering with an extended end-plate connection

c) Equilibrium equations for the connection and load eccentricity

The evaluation of the resistance of the connection based on the static theorem requires at failure equilibrium between the distribution of internal forces and the external applied loads. For connections subjected to a bending moment M_{Ed} and an axial load N_{Ed}, the equilibrium criteria are:

$$M_{Ed} = \sum_{i=1}^{n} h_i \cdot F_i \text{ and } N_{Ed} = \sum_{i=1}^{n} F_i \qquad (6.43)$$

where F_i is the load in row i and h_i is the corresponding lever arm; the latter is defined as the vertical distance between the reference beam point where M_{Ed} and N_{Ed} are computed and the row itself (h_i values are positive for rows located on the upper side of the reference point).

d) Resistance criteria

According to the static theorem (see 1.6.1), the resistance of each row, which is equal to the resistance of the weakest component in the row, should never be exceeded. At first sight, it looks easy as long as individual resistances of bolt-rows are considered but it is much more questionable when group effects develop in the connections.

6. MOMENT RESISTANT JOINTS

Within the developed procedure, any group of rows noted $[m,p]$, i.e. from row m to row p, in which group effects appear is considered as an equivalent fictitious row with an equivalent lever arm and a group resistance equal to that of the weakest group component. Therefore, the resistance criteria for each of the rows belonging to the $[m,p]$ group may write, for any constitutive component α:

$$\sum_{i=m}^{p} F_i \leq F_{mp}^{Rd,\alpha} \text{ with } m=1,...,p \text{ and } p=m,m+1,...,n \quad (6.44)$$

where $F_{mp}^{Rd,\alpha}$ is the resistance of the component α for the group of rows m to p. When m is equal to p, $F_{mp}^{Rd,\alpha}$ designates the individual resistance of the component α for row m. Such a resistance of the group of rows $[m,p]$, noted F_{mp}^{Rd}, is defined as the smallest of the $F_{mp}^{Rd,\alpha}$ values.

This situation is illustrated in Figure 6.23 from (Cerfontaine, 2003) for a connection with three bolt rows (1, 2 and 3) but more generally covers the case of any connection with n rows in which group effects would develop in three bolt rows.

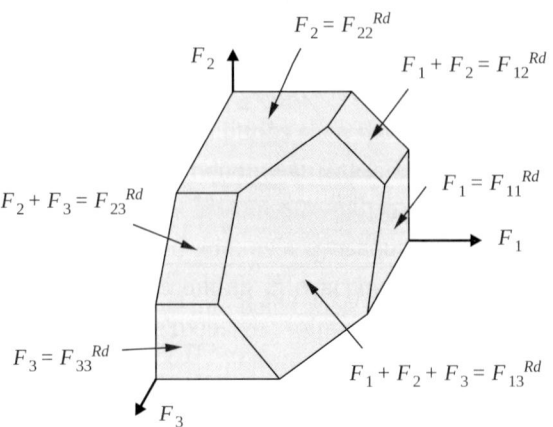

Figure 6.23 – Possible group effects between three bolt rows

e) Definition of the failure criterion for the whole connection

Details about the application of the static theorem to a connection with n rows are given in (Cerfontaine, 2003). This application leads to the following definition and writing of the $M-N$ resistance interaction diagram: the interaction criterion between the bending moment and the axial force at failure is described by a set of $2n$ parallel straight line segments; the slope of each of the $2n$ parallel segments is equal to the value of the lever arm h_k and along these segments, the force F_k in row k varies between 0 at one end and the maximum row resistance at the other end. An example of a $M-N$ resistance interaction diagram is given in Figure 6.24 for a joint with four bolt rows (i.e. $2n = 2 \times (4 + \text{up} + \text{lo}) = 12$ segments); for instance, to pass from point A to point B, the row 6 in compression is progressively activated ($F_6 = 0$ at point A where only the rows in tension are activated and $F_6 = F_{Rd,6}$ at point B).

It can be observed in Figure 6.24 (Cerfontaine, 2003) that different values of bending moments correspond to the maximum tensile load; this phenomenon can be easily justified by the evolution of the loads within the rows involved in a group behaviour.

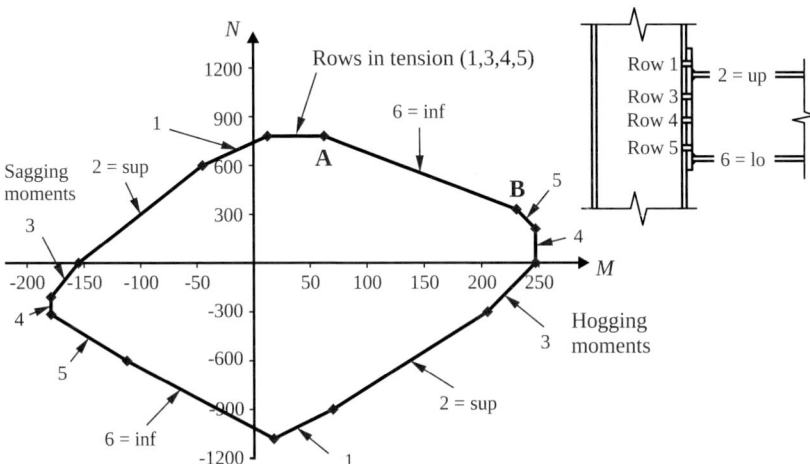

Figure 6.24 – Example of a *M-N* resistance interaction curve obtained for a four bolt row joint

Each point of the curve is obtained by expressing the quite complex failure criteria given here below for k from 1 to n; this criterion is defined so as to get maximum resistance bending moments by optimising the distributions of the loads amongst the bolt rows with account of the group effects.

Within this expression, it can be observed that two different resistances are attributed to the same row i (F_i^{Rd+} and F_i^{Rd-}). The evaluation procedure of the F_i^{Rd+} and F_i^{Rd-} values is illustrated in Figure 6.25 (Cerfontaine, 2003) for a connection with three bolt rows where the black and white dots respectively show the successive steps for the evaluation of

6. MOMENT RESISTANT JOINTS

F_i^{Rd+} and F_i^{Rd-}; the objective of the evaluation procedure is to obtain the maximum resistant bending moment by maximising (or minimising if the sign of the bolt row resistance is negative) the loads in the bolt rows which are the most distant from the bolt row k.

$$M = h_k \cdot N + \sum_{i=1}^{n} (h_i - h_k) \cdot F_i^c$$

either
$$\left. \begin{array}{l} F_i^c = \max(F_i^{Rd+};0) \text{ if } i < k \\ F_i^c = \min(F_i^{Rd+};0) \text{ if } i > k \end{array} \right\} \text{ tension in top rows} \quad (6.45)$$

or
$$\left. \begin{array}{l} F_i^c = \min(F_i^{Rd-};0) \text{ if } i < k \\ F_i^c = \max(F_i^{Rd-};0) \text{ if } i > k \end{array} \right\} \text{ tension in bottom rows}$$

with

$$F_i^{Rd+} = \min\left(F_{mi}^{Rd} - \sum_{\substack{j=m \\ i \neq up, lo}}^{i-1} F_j^{Rd+}, m=1,...,i \right) \quad \text{for } i < k$$

$$F_i^{Rd+} = F_i^{Rd} \quad \text{for } i = up, lo > k$$

$$F_i^{Rd-} = \min\left(F_{im}^{Rd} - \sum_{\substack{j=i+1 \\ i \neq up, lo}}^{m} F_j^{Rd-}, m=i,...,n \right) \quad \text{for } i > k$$

$$F_i^{Rd-} = F_i^{Rd} \quad \text{for } i = up, lo < k$$

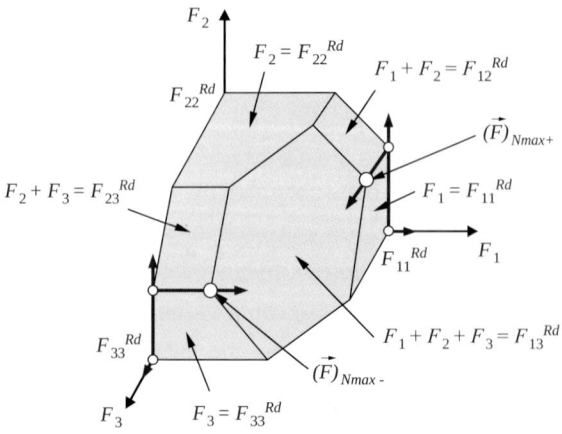

Figure 6.25 – Successive steps for the evaluation of F_i^{Rd+} and F_i^{Rd-} (black and white dots respectively)

The interested reader will find in (Cerfontaine, 2003) information about the way on how to account for non-ductile components response in the here-above described $M - N$ assembly procedure.

6.3.4 M-N-V

In endplate connections, the shear forces applied the joints in addition to bending moments and/or axial forces are transferred, between the connected members, through the bolts and by bearing of the bolts against the endplate or column flange. Shear failure can so result from a lack of bearing resistance of the connected plates or from a lack of bolt resistance.

The verification of the bearing resistance may be achieved as specified in section 3.4.5 while the resistance of a bolt simultaneously subjected to shear and tension forces is expressed as follows (see 3.4.3):

$$\frac{F_{v,Ed}}{F_{v,Rd}} + \frac{F_{t,Ed}}{1.4 F_{t,Rd}} \leq 1.0 \tag{6.46}$$

where:
- $F_{v,Rd}$ is the design resistance of the bolt in shear
- $F_{t,Rd}$ is the design resistance of the bolt in tension

From this formula, it is seen that the full tension resistance of the bolt $F_{t,Rd}$ may be reached as long as the shear force is not exceeding 0.286 times the shear design resistance $F_{v,Rd}$.

Based on this conclusion, the following procedure is suggested in EN 1993-1-8 to check the shear resistance of the joint in presence of bending moments and/or axial forces:

- identify as an outcome of the $M - N$ assembly procedure the bolt-rows which are subjected to tension (n_t) and those bolt rows which are located in the compression zone or which are not activated in tension (n_{nt});
- consider that the bolt-rows which are not subjected to tension may transfer a shear force equal to $2 F_{v,Rd}$ (bolt-rows with two bolts each)
- consider that the bolt-rows which are subjected to tension may at least transfer a shear force equal to $2 \times 0.286\, F_{v,Rd}$ (bolt-rows with two bolts each)

- evaluate the capacity of the joint to transfer shear forces, in addition to bending moments and/or axial forces, as equal to

$$V_{Rd} = n_t \times 2 \times 0.286\, F_{v,Rd} + n_{nt} \times 2 \times F_{v,Rd} \quad (6.47)$$

A similar but less safe alternative is to calculate the actual force acting in the n_t bolts subjected tension and evaluating through Eq. (6.46) the actual available resistance of the bolts shear instead of conservatively lowering this value to $0.286\, F_{v,Rd}$.

6.3.5 Design of welds

The verification of the welds connecting the beam and the end-plate (possibly fitted with haunch or stiffeners) consists in a "global" verification of a particular cross section made of welds only. This section is called here after the "weld section" (Figure 6.26). In the here defined method a part of the weld section is reserved for undertaking the shear force and does not contribute to the bending nor the axial force resistance. Similarly, the weld area that resists the bending moment and the axial force does not contribute to shear resistance.

The verification that is performed for the axial stresses is an elastic verification of the axial stress distribution in the weld section as no ductility leading to potential plastic redistribution may be contemplated in welds. The distribution of the axial stresses is based on the Navier assumption. The stresses in the welds are obtained from the internal forces in the frame at the position of the connection.

The here-below described procedure is appropriate for an "elastic-elastic" or "elastic-plastic" analysis and verification structural design process. For a "plastic-plastic" one, further recommendations are to be considered (see section 6.3.5.3).

6.3.5.1 Definition of the weld section

The weld section is defined as the one aimed at transferring the axial forces and bending moments. As a part of it, a "tension zone" (top of the connection in Figure 6.26) may be first defined; it is located in the tension part of the joint and it comprises the welds which are "covered by the bolt-rows", including welds related to stiffeners and haunch, if any. The term "covered by a bolt-row" means: located within the "effective length" obtained from the study of the representative T-stub of the considered bolt-row, this length being centred on the bolt-row position. As indicated in the Figure 6.26, these lengths should of course not be accounted twice if they overlap.

Moreover, if there would be a gap between successive "effective heights", it would be ignored.

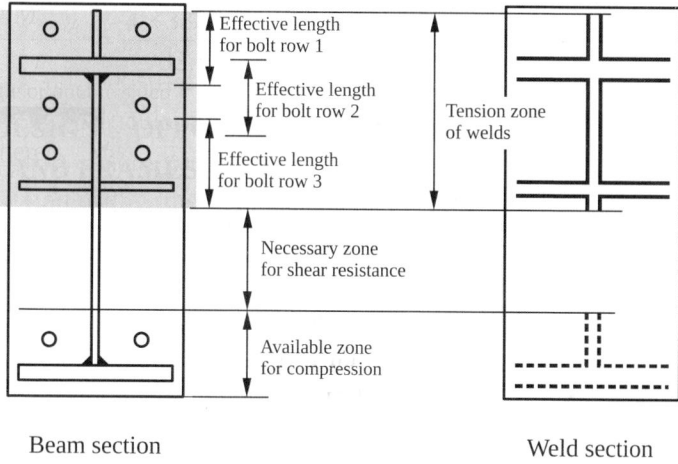

Figure 6.26 – Definition of the weld section

After the tension side of the weld section has been defined, a part of the web welds is defined in order to resist the applied shear force. This part of the web welds is arbitrarily located just below the tension area of the weld section.

This being, the compression area of the weld section can then take place below the necessary shear area in the remaining available area of the welds. The way on how the longitudinal stresses may distribute in the weld section part which is not reserved for shear is defined below.

6.3.5.2 Position of the neutral axis and calculation of the axial stresses

An elastic distribution of axial stresses being assumed, the following two equilibrium equations must be respected:

6. MOMENT RESISTANT JOINTS

$$N_{Ed} = \int_{WS} \sigma \cdot d\Omega$$
$$M_{yEd} = \int_{WS} \sigma \cdot z \cdot d\Omega \qquad (6.48)$$

In these ones, the integration must be achieved on the weld section ("WS"), σ is the axial stress at position z and Ω is the weld section.

Two unknowns are identified: the position of the neutral axis and the value of the maximum axial tension stress in the weld section. Their definition, through the resolution of the two above equations, is an iterative process for which a practical procedure is suggested below.

But practically speaking, three cases can be met as far as the position of the neutral axis is concerned (see Figure 6.27):

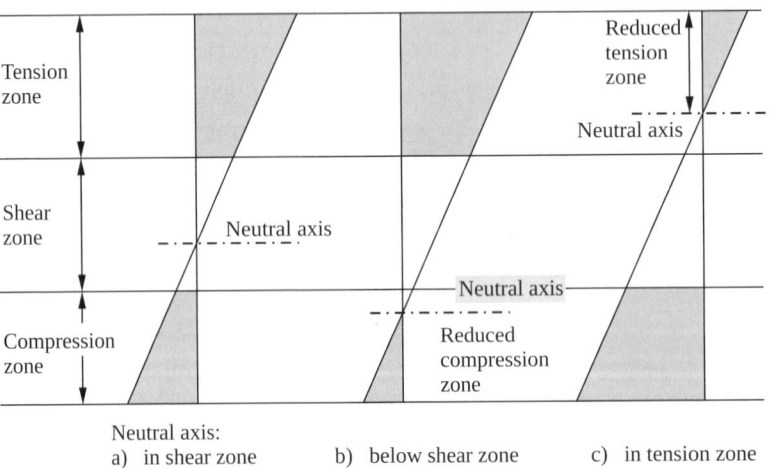

Neutral axis:
a) in shear zone   b) below shear zone   c) in tension zone

Figure 6.27 – Possible stress distributions in the weld section and position of neutral axis

- The neutral axis is located in the shear area (left sketch in Figure 6.27); in this case the part of the shear area that is below the neutral axis cannot be considered for compression.
- The neutral axis is located below the shear area (sketch in the middle); in this case the part of the web that is above the neutral axis but below the shear area cannot be used for tension.
- The neutral axis is located in the tension area (right sketch); in this case the tension zone height must be reduced to account for this specific position of the neutral.

As a starting point of the iterative solving process, a position of the neutral axis is guessed (this one change at each resolution step until convergence is reached). As a result, the height of the compression zone is known and so the weld section is completely defined.

An arbitrary value of the maximum tension stress σ_t is now guessed – it can even be taken as the yield stress; it allows for the calculation of the corresponding maximum compression stress σ_c. The axial force N and the bending moment M_y corresponding to this guessed situation can thus be calculated. The ratio M_y/N is then evaluated and compared with the same ratio corresponding to the actual internal forces $M_{y,Ed}/N_{Ed}$. If the two values are the same, it means that the guessed position of the neutral axis is the correct one. The actual axial stresses can then be obtained by multiplying the guessed values σ_t and σ_c by the factor $M_{y,Ed}/M_y$ or N_{Ed}/N.

If the ratios M_y/N and $M_{y,Ed}/N_{Ed}$ are not the same, the guessed position of the neutral axis is not correct, and it must be further searched.

6.3.5.3 Design requirements according to the analysis and verification structural design process

The design requirements for joints, as for cross sections, depend on the selected global analysis and verification structural design process: elastic/elastic (E-E), elastic/plastic (E-P) or plastic/plastic (P-P).

As an example, for a joint in bending (if $M_{j,Rd}$ is the plastic resistance of the joint as given by EN Part 1-8):

- E-E: $M_{j,Ed} < 2/3 M_{j,Rd}$ no request in terms of ductility
- E-P: $M_{j,Ed} < M_{j,Rd}$ ductility is required to redistribute forces inside the joint
- P-P: $M_{j,Ed} < M_{j,Rd}$ plastic ductility is required to redistribute forces in the structure

In this context, the sizing of the welds has to be addressed accordingly.

a)   E-E design process

The design resistance $M_{j,Rd}$ of the joint is evaluated by assuming that the welds are infinitely resistant.

The elastic resistance of the joint is then derived (it is conventionally assumed to be equal to $2/3 M_{j,Rd}$).

6. MOMENT RESISTANT JOINTS

The resistance of the weld section $M_{w,Rd}$ is evaluated on the basis of the Navier linear distribution.

Finally the resistance of the joint is defined as equal to $\min(2/3 M_{j,Rd}; M_{w,Rd})$.

If one wants to fully profit from the joint resistance (it means to reach $M_{w,Rd}$ equal to $2/3 M_{j,Rd}$), the welds should be designed such that $M_{w,Rd} \geq 2/3 M_{j,Rd}$.

In EN 1993 Part 1-8 clause 6.2.3(4), it is however stated that "in all joints, the sizes of the welds should be such that the design moment resistance $M_{j,Rd}$ of the joint is always limited by the design resistance of its other basic components, and not by the design resistance of the welds". This means that $M_{w,Rd}$ should always be higher than $M_{j,Rd}$. Fundamentally this request appears as unjustified as long as no ductility is requested from the joint in an E-E design process. But for sure, this may be seen as a safe measure to cover the risk of uncertainties on the actual distribution of internal forces in the frame or the risk of unexpected loading situations resulting from exceptional actions.

b) E-P design process

The design resistance $M_{j,Rd}$ of the joint is evaluated by assuming that the welds are infinitely resistant.

The resistance of the weld section $M_{w,Rd}$ is evaluated on the basis of the Navier linear distribution.

Finally the resistance of the joint is defined as equal to $\min(2/3 M_{j,Rd}; M_{w,Rd})$.

If one wants to fully profit from the joint resistance (it means to reach $M_{w,Rd}$ equal to $M_{j,Rd}$), the welds should be designed such that $M_{w,Rd} \geq 2/3 M_{j,Rd}$. Here again (see comment for "E-E design process" just above), Eurocode 3 recommends that "in all joints, the sizes of the welds should be such that the design moment resistance $M_{j,Rd}$ of the joint is always limited by the design resistance of its other basic components, and not by the design resistance of the welds".

c) P-P design process

The design resistance $M_{j,Rd}$ of the joint is evaluated by assuming that the welds are infinitely resistant.

6.4 ASSEMBLY FOR ROTATIONAL STIFFNESS

As the joint has to undergo a plastic rotation and as strain hardening will unavoidably occur at that occasion, resulting in an actual increase of the bending moment in the joint (what is disregarded in a conventional plastic analysis with plastic hinges), the welds – which should be imperatively protected again brittle failure – should be calculated, according to EN 1993 Part 1-8, for a bending moment equal to $1.4M_{j,Rd}$ or $1.7M_{j,Rd}$.

The resistance of the weld section $M_{w,Rd}$ is evaluated on the basis of the Navier linear distribution; but either $M_{w,Rd} \geq 1.4M_{j,Rd}$ or $M_{w,Rd} \geq 1.7M_{j,Rd}$ depending respectively on the non-sway or sway character of the frame. The condition is more severe for non-sway frames than for non-sway frames as higher rotation capacities are required in this case to form a full plastic mechanism in the frame.

Finally the resistance of the joint is defined as equal to $M_{j,Rd}$.

6.4 ASSEMBLY FOR ROTATIONAL STIFFNESS

6.4.1 Joints under bending moments

6.4.1.1 Refined method

According to the component method, the rotational response of a joint is based on the mechanical properties of its different constitutive components. The advantage is that an engineer is able to calculate the mechanical properties of any joint by decomposing the joint into relevant components. Table 6.5 lists the components reported in EN 1993-1-8 and, as an example, identifies the components to be taken into account when calculating the initial stiffness for end-plated, welded and flange cleated beam-to-column joints.

Table 6.5 – Overview of components covered by EN 1993-1-8 and list of active components for different joints

Component	Number	End-plated	Welded	Flange cleated
Column web panel in shear	1	✓	✓	✓
Column web in compression	2	✓	✓	✓
Column web in tension	3	✓	✓	✓
Column flange in bending	4	✓		✓

6. MOMENT RESISTANT JOINTS

Table 6.5 – Overview of components covered by EN 1993-1-8 and list of active components for different joints (continuation)

Component	Number	End-plated	Welded	Flange cleated
End-plate in bending	5	✓		
Flange cleat in bending	6			✓
Bolts in tension	10	✓		✓
Bolts in shear	11			✓
Bolts in bearing	12			✓
Concrete in compression	13			
Base plate in bending under compression	14			
Base plate in bending under compression	15			
Anchor bolts in tension	16			

It is assumed that the deformations of the following components: (i) beam flange and web in compression, (ii) beam web in tension and (iii) plate in tension or compression are included in the deformations of the beam in bending. Consequently they are not assumed to contribute to the flexibility of the joint. Haunches, when present, fall also in this category.

The initial stiffness $S_{j,ini}$ is derived from the elastic stiffnesses of the components. The elastic behaviour of each component is represented by an extensional spring. The force-deformation relationship of this spring is given by:

$$F_i = k_i \cdot E \cdot \Delta_i \qquad (6.49)$$

where:
- F_i is the force in the spring i;
- k_i is the stiffness coefficient of the component i;
- E is the Young modulus;
- Δ_i is the deformation of the spring i.

The spring components in a joint are combined into a spring model. Figure 6.28 shows for example the spring model for an unstiffened welded beam-to-column joint.

6.4 ASSEMBLY FOR ROTATIONAL STIFFNESS

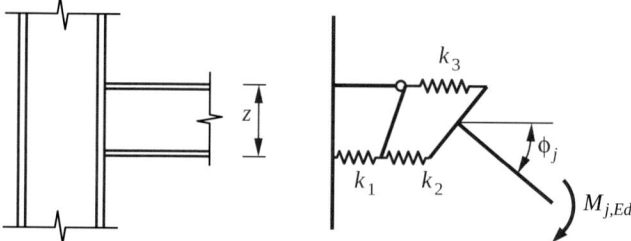

Figure 6.28 – Spring model for an unstiffened welded joint

The force in each spring is equal to F. The moment M_{Ed} acting on the spring model is equal to $F \cdot z$, where z is distance between the centre of tension (for welded joints, located in the centre of the upper beam flange) and the centre of compression (for welded joints, located in the centre of the lower beam flange). The rotation ϕ in the joint is equal to $(\Delta_1 + \Delta_2 + \Delta_4)/z$. In other words:

$$S_{j,ini} = \frac{M_{Ed}}{\phi} = \frac{F \cdot z}{\dfrac{\sum_i \Delta_i}{z}} = \frac{F \cdot z^2}{\dfrac{F}{E} \cdot \sum_i \dfrac{1}{k_i}} = \frac{E \cdot z^2}{\sum_i \dfrac{1}{k_i}} \qquad (6.50)$$

The same formula applies for an end-plated joint with a single bolt-row in tension and for a flange cleated joint. However, components to be taken into account are different, see Table 6.5.

Figure 6.29a shows the spring model adopted for end-plated joints with two or more bolt-rows in tension. It is assumed that the bolt-row deformations for all rows are proportional to the distance to the point of compression (still assumed to be located in the centre of the beam flange), but that the elastic forces in each row are dependent on the stiffness of the components. Figure 6.29b shows how the deformations $k_{i,r}$ of components 3, 4, 5 and 7 are added to an effective spring per bolt-row, with an effective stiffness coefficient $k_{eff,r}$ (r is the index of the row number). It is indicated in Figure 6.29c how these effective springs per bolt-row are replaced by equivalent spring acting at a lever arm z. The stiffness coefficient of this effective spring is k_{eq}. The effective stiffness coefficient k_{eq} can directly be applied in Eq. (6.50). The formulae to determine $k_{eff,r}$, z_{eq} and k_{eq} are as follows:

6. MOMENT RESISTANT JOINTS

$$k_{eff,r} = \frac{1}{\sum_r \frac{1}{k_{i,r}}} \tag{6.51}$$

$$z_{eq} = \frac{\sum_r k_{eff,r} \cdot h_r^2}{\sum_r k_{eff,r} \cdot h_r} \tag{6.52}$$

$$k_{eq} = \frac{\sum_r k_{eff,r} \cdot h_r}{z_{eq}} \tag{6.53}$$

They can be derived from the sketches of Figure 6.29 by expressing that the moment-rotation response is the same for each of the systems represented in Figure 6.29a to Figure 6.29c are equal. An additional condition is that the compressive force in the lower rigid bar is equal in each of these systems.

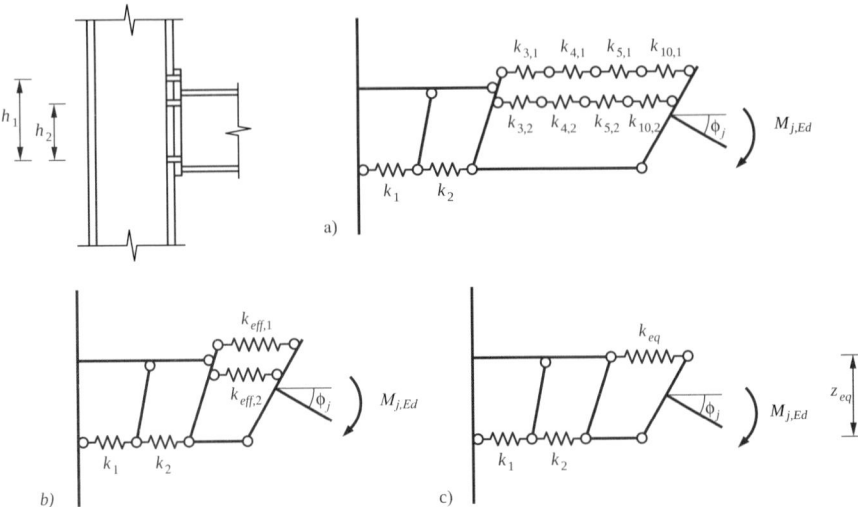

Figure 6.29 – Spring model for a beam-to-column end-plated joint with more than one bolt-row in tension

As a result, in this stiffness model used in Eurocode 3 Part 1-8:

- the internal forces are in equilibrium with the bending moment;

6.4 ASSEMBLY FOR ROTATIONAL STIFFNESS

- the compatibility of the displacements is ensured through the assumption of an infinitely rigid transverse stiffness of the beam cross section;
- the plasticity criterion is fulfilled as long as the elastic resistance of the springs is not reached;
- no ductility requirement is likely to limit the deformation capacity of the springs in the elastic range of behaviour.

The respect of these four main requirements (which any distribution of internal forces should satisfy, from a theoretical point of view) allows therefore be consider the model as an "exact" one.

Beyond the elastic resistance ($M_{j,Ed} > 2/3\, M_{j,Rd}$), the response of the joint becomes non-linear, as depicted in Figure 6.30.

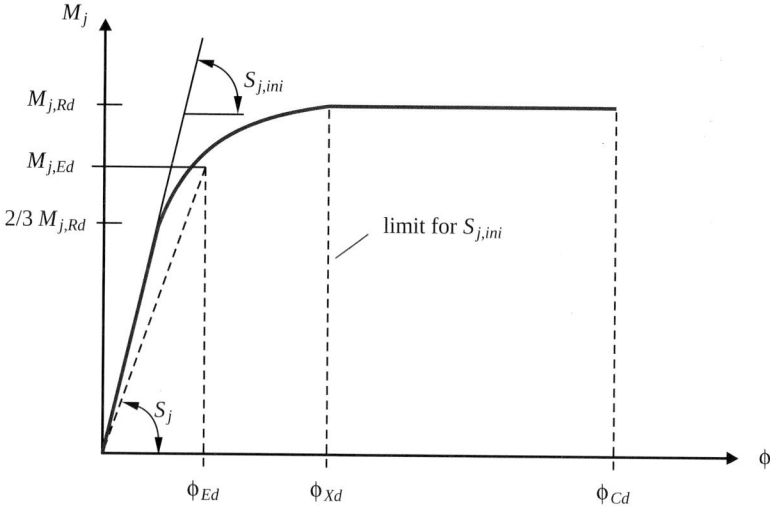

Figure 6.30 – Non-linear response of a joint

In the elastic zone, the joint stiffness is equal to elastic initial stiffness ($S_{j,ini}$). Beyond the elastic resistance limit, the secant stiffness S_j is slowly decreasing until $M_{j,Rd}$ is reached. In Part 1-8, the following expression is provided for S_j (with reference to a factor $\mu = S_{j,ini}/S_j$):

$$S_j = \frac{1}{\mu} \cdot S_{j,ini} = \frac{E \cdot z^2}{\mu \cdot \sum_i \frac{1}{k_i}} \qquad (6.54)$$

6. MOMENT RESISTANT JOINTS

The limit of validity of this formula is indicated in Figure 6.30 (limit 1). Values for μ are derived as follows (see also Table 6.6):

$$\mu = 1.0 \quad \text{for } M_{j,Ed} \leq \frac{2}{3} \cdot M_{j,Rd} \tag{6.55}$$

$$\mu = \left(\frac{1.5 \cdot M_{j,Ed}}{M_{j,Rd}}\right)^{\psi} \quad \text{for } \frac{2}{3} \cdot M_{j,Rd} < M_{j,Ed} \leq M_{j,Rd} \tag{6.56}$$

Table 6.6 – Values of the ψ coefficient

Connection type	μ	ψ
Welded	3.0	2.7
Bolted end-plates	3.0	2.7
Bolted angle flange cleats	3.5	3.1
Base plate connections	3.0	2.7

Beyond $M_{j,Rd}$, a yield plateau is assumed, but obviously as long as some plastic ductility is available (ductility limit ψ_{Cd}).

6.4.1.2 Simple prediction of the initial stiffness

In the preliminary frame design phase, it is difficult to assess the stiffness of the (semi-rigid) joints; indeed the joints have not been designed yet. To overcome this problem, some simplified formulae have been derived by Martin Steenhuis based on EN 1993-1-8 (Maquoi et al, 1998). By means of these formulae, the designer can determine the stiffness of a joint by selecting the joint configuration only.

Of course these formulae are based on some fixed choices regarding the connection detailing. These are:

- For end-plated connections:
  - The connection has two bolt rows only in the tension zone;
  - The bolt diameter is approximately 1.5 times the column flange thickness;

- The location of the bolt is as close as possible to the root radius of the column flange, the beam web and flange (about 1.5 times the thickness of the column flange);
- The end-plate thickness is similar to the column flange thickness.

- For *cleated connections:*
  - The connection has one bolt row in the tension zone;
  - The bolt diameter is approximately 1.5 times the cleat thickness;
  - The location of the bolt is as close as possible to the root radius of the column flange and the cleat (about 1.5 times the thickness of the column flange);
  - The cleat thickness is whenever possible similar to the column flange thickness.

The approximate value $S_{j,app}$ of the initial joint stiffness is expressed as:

$$S_{j,app} = \frac{E\,z^2\,t_{fc}}{C} \qquad (6.57)$$

where the values of the factor C are given in Table 6.7 for different joint configurations and loadings. These values of the joint stiffness involve two parameters only: z and t_{fc}; z is the distance between the compression and tensile resultants and t_{fc} is the thickness of the column flange. In an extended end-plate connection with two bolt rows, the distance z is approximately equal to the beam depth. For the same joint with a haunch, z is equal to the sum of the beam depth and the haunch depth.

Table 6.7 – Good guess of the initial stiffness for typical beam-to-column joint configurations

	Joint configuration		C
1	Extended end-plate, single sided, unstiffened ($\beta=1$)		13

6. MOMENT RESISTANT JOINTS

Table 6.7 – Good guess of the initial stiffness for typical beam-to-column joint configurations (continuation)

	Joint configuration		C
2	Extended end-plates, double sided, unstiffened, symmetrical ($\beta = 0$)		7.5
3	Extended end-plate, single sided, stiffened in tension and compression ($\beta = 1$)		8.5
4	Extended end-plates, double sided stiffened in tension and compression symmetrical ($\beta = 0$)		3
5	Extended end-plate, single sided, Morris stiffener ($\beta = 1$)		3
6	Flush end-plate, single sided ($\beta = 1$)		14

6.4 ASSEMBLY FOR ROTATIONAL STIFFNESS

Table 6.7 – Good guess of the initial stiffness for typical beam-to-column joint configurations (continuation)

	Joint configuration		C
7	Flush end-plates, double sided, symmetrical ($\beta=0$)		9.5
8	Flush end-plate, single sided, cover plate at column top ($\beta=1$)		11.5
9	Flush end-plates, double sided, cover plate at column top, symmetrical ($\beta=0$)		6
10	Welded joint, single sided, unstiffened ($\beta=1$)		11.5
11	Welded joints, double sided, unstiffened, symmetrical ($\beta=0$)		6

6. MOMENT RESISTANT JOINTS

Table 6.7 – Good guess of the initial stiffness for typical beam-to-column joint configurations (continuation)

Joint configuration			C
12	Cleated, single sided ($\beta=1$)		70
13	Cleated, double sided, symmetrical ($\beta=0$)		65

Note:

For the rare cases of double-sided joint configurations where $\beta=2$ (unbalanced moments), the value of the C factor is obtained by adding 11 to the relevant value for symmetrical conditions (balanced moments).

6.4.2 Joints under bending moments and axial forces

For a joint subjected to a combination of axial forces and bending moments, rotational and axial stiffnesses have to be evaluated; these stiffnesses depend of the activated components under the considered combined applied loads.

The flexural and extensional stiffnesses are given by the following formulae (Cerfontaine, 2003):

- Stiffness in bending: $K_M = N/\varphi^{el}$
- Stiffness in tension/compression: $K_N = N/\Delta^{el}$

where φ^{el} is the rotation of the joint section and Δ^{el}, the elongation at the reference axis ($h=0$) (Figure 6.31).

6.4 ASSEMBLY FOR ROTATIONAL STIFFNESS

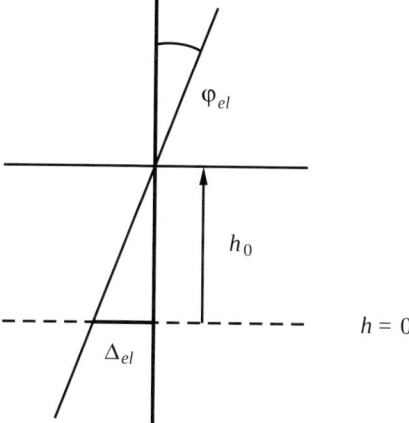

Figure 6.31 – Plane deformation of the joint section

The neutral axis corresponds to the "zero-displacement" level. Its position is given by $h_0 = -\Delta^{el}/\varphi^{el}$, under the assumption of plane cross section.

The elastic displacement of any row i is given by:

$$\Delta_i^{el} = \frac{F_i^{el}}{K_{i,ini}} = \Delta^{el} + h_i . \varphi^{el} \tag{6.58}$$

where $K_{i,ini}$ is the initial stiffness of row i, computed as the geometric mean of the initial rigidities of the components involved in row i.

So, the equilibrium equations become:

$$N^{el} = \sum_i F_i^{el} = \left(\sum K_{i,ini}\right).\Delta^{el} + \left(\sum K_{i,ini}.h_i\right).\varphi^{el} \tag{6.59}$$

$$M^{el} = \sum_i F_i^{el}.h_i = \left(\sum K_{i,ini}.h_i\right).\Delta^{el} + \left(\sum K_{i,ini}.h_i^2\right).\varphi^{el} \tag{6.60}$$

in which only the activated rows are taken into account in the summations.

Further developments lead to the following expressions of the position of the neutral axis and the initial rigidities as functions of $e^{el} = M^{el}/N^{el}$:

6. MOMENT RESISTANT JOINTS

$$h_0^{el} = \frac{\sum K_{i,ini} \cdot h_i \cdot (h_i - e^{el})}{\sum K_{i,ini} \cdot (h_i - e^{el})} \quad (6.61)$$

$$K_N^{el} = \frac{\sum K_{i,ini} \cdot \sum K_{i,ini} \cdot h_i^2 - \left(\sum K_{i,ini} \cdot h_i\right)^2}{\sum K_{i,ini} \cdot h_i \cdot (h_i - e^{el})} \quad (6.62)$$

$$K_M^{el} = \frac{e^{el} \cdot \left[\sum K_{i,ini} \cdot \sum K_{i,ini} \cdot h_i^2 - \left(\sum K_{i,ini} \cdot h_i\right)^2\right]}{\sum K_{i,ini} \cdot (e^{el} - h_i)} = -e^{el} \cdot h_0^{el} \cdot K_N^{el} \quad (6.63)$$

As the summations involve only the activated rows, which are unknown until the position of the neutral axis has been determined, an iterative procedure is required to find h_0^{el}. For a given value of e^{el}, an assumption has first to be made on the value of h_0^{el} and the corresponding activated rows. Then, h_0^{el} can be computed using the previous formula and the assumption can be checked. If it was correct, then K_N^{el} and K_M^{el} can be computed. If it wasn't, another assumption has to be made until the actual activated rows are found.

6.5 ASSEMBLY FOR DUCTILITY

Presently Eurocode 3 is not providing practitioners with a real "component approach for ductility" in which the ductility of the individual components should first be evaluated before an assembly procedure is used to derive the value of the rotational ductility. Few related investigations have been achieved and are still being searched so as to derive such a "component approach for ductility" but it has to be recognised that still nowadays no consistent set of recommendations may be proposed for inclusion in EN 1993-1-8.

So, as already announced in sub-chapter 2.5, the derivation of a maximum rotation value is substituted in Part 1-8 by the proposal of criteria which allow a classification of the joints in three categories:

- *Class 1 joints*: ductile joints exhibiting an adequate rotation capacity for a global plastic analysis with plastic hinges developing in the joints;
- *Class 2 joints*: semi-ductile joints with too limited plastic rotational capacity to achieve a plastic analysis;

6.5 ASSEMBLY FOR DUCTILITY

- *Class 3 joints*: brittle behaviour requesting an elastic frame analysis combined with an elastic verification of the joint resistance.

Before describing these classification criteria, two statements have to be done; they are expressed below.

- If the design resistance of the joint exceeds by at least 20% the design resistance of the weakest of the connected members, it may assumed that the plastic hinge will not form in the joint, even in the case of material over strength. Therefore, no plastic rotational capacity is required from the joint, but ductility should be provided by the connected members.
- Welds are brittle by nature and so their failure should be strictly avoided as soon as ductility is required. This results, in Part 1-8, in specific requirements that have been already addressed in section 6.3.5.3.

This being, the classification criteria for bolted and welded steel joints are presented respectively in sections 6.5.1 and 6.5.2. The ones are valid for S 235, S 275 and S 355 steel grades and for joints in which the design value of the axial force N_{Ed} in the connected member does not exceed 5% of the design plastic resistance $N_{p\ell,Rd}$ of its cross section (see section 6.3.3.1).

In cases not covered by sections 6.5.1 and 6.5.2, the rotation capacity may be determined by testing in accordance with EN 1990, Annex D (CEN, 2002). Alternatively, appropriate calculation models may be used, provided that they are based on the results of tests in accordance with EN 1990.

6.5.1 Steel bolted joints

Figure 6.32 presents the flow chart for the ductility classification of bolted joints.

Sufficient ductile deformations are possible when the following basic components prevail:

- column web in shear;
- column flange, endplate or cleat in bending.

When the column web panel in shear determines the moment design resistance, the available rotation capacity is assumed to be sufficient for a rigid-plastic analysis, provided that the shear buckling of the web does not occur. To respect this condition, the web slenderness has not to exceed a slenderness limit specified in section 6.2.1a).

6. MOMENT RESISTANT JOINTS

When the moment resistance of the joint is limited by the yielding of the column flange, of the endplate or of the connecting cleats, a sufficient rotation capacity for the rigid-plastic analysis is also obtained provided that the failure is due to pronounced plastic bending deformations of the plate components and so is not resulting from a bolt failure. Conditions to respect this requirement are indicated in Figure 6.32. These ones may also be used at preliminary design stages to ensure a ductile response of the joints.

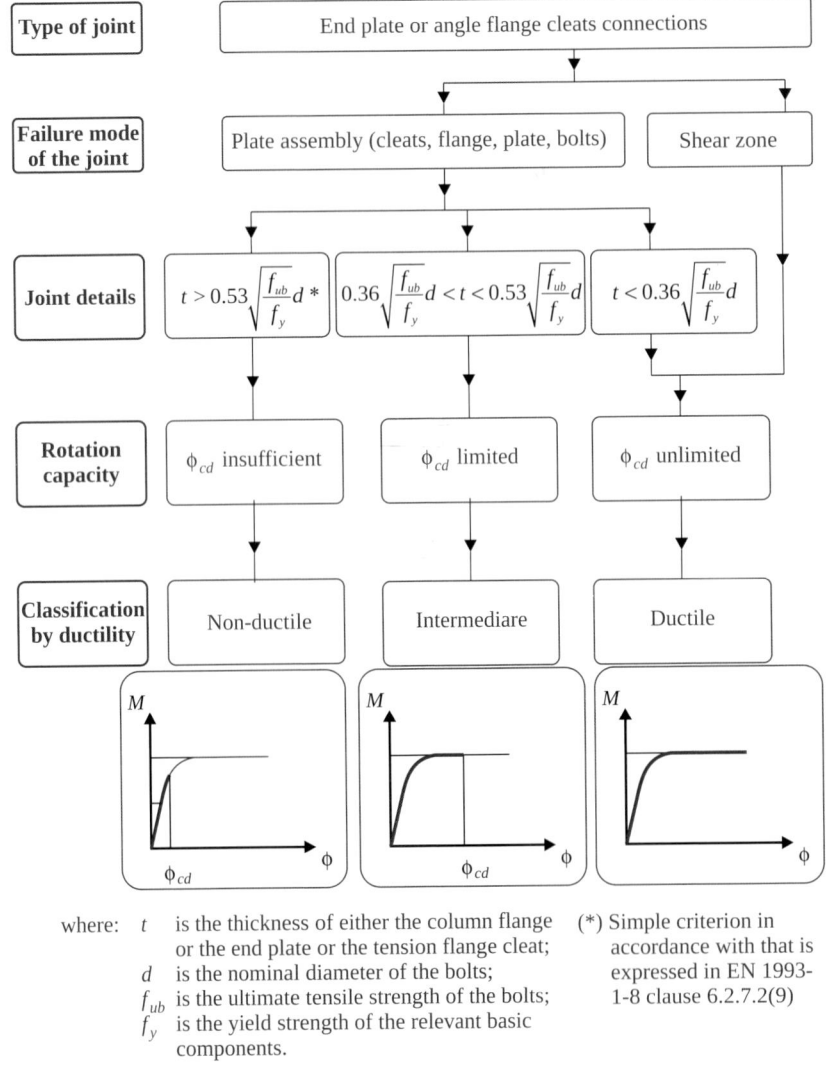

where: t is the thickness of either the column flange or the end plate or the tension flange cleat;
d is the nominal diameter of the bolts;
f_{ub} is the ultimate tensile strength of the bolts;
f_y is the yield strength of the relevant basic components.

(*) Simple criterion in accordance with that is expressed in EN 1993-1-8 clause 6.2.7.2(9)

Figure 6.32 – Classification criteria for the rotational ductility of bolted joints

6.5 ASSEMBLY FOR DUCTILITY

If the moment capacity is limited by the load capacity of the bolts in shear, a sufficient rotation capacity for the rigid-plastic analysis cannot be expected.

The flow chart in Figure 6.32 covers only some of the possible failure modes; for those not covered, the rotation capacity may be determined by testing or through models validated by testing.

6.5.2 Steel welded joints

Figure 6.33 summarises the way on how to classify welded joints for rotational ductility. It is seen that ductile deformations may only be expected from the following basic components:

- column web in shear;
- column web in compression;
- column web in tension

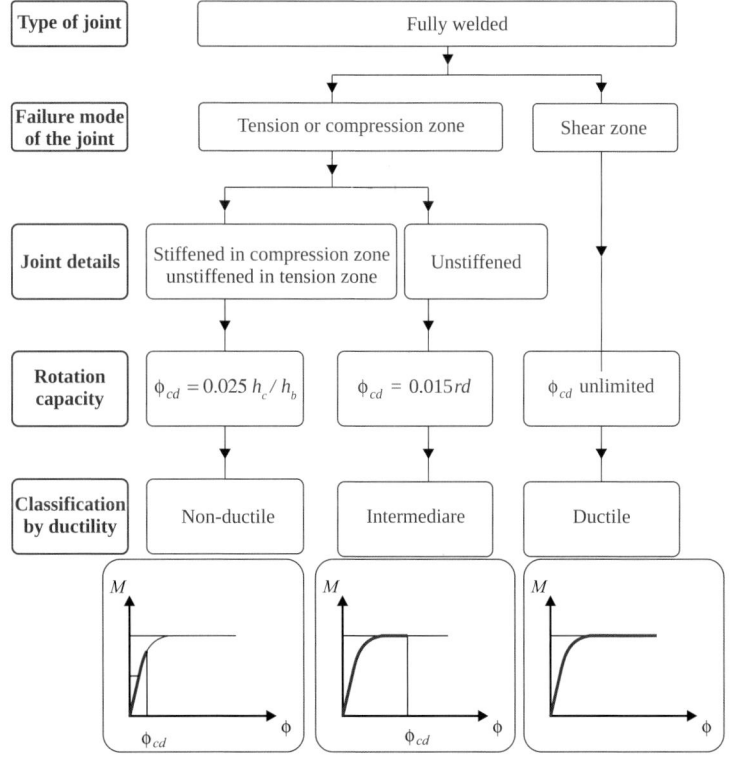

where: h_c is the height of the column
h_b is the height of the beam

Figure 6.33 – Classification criteria for the rotational ductility of welded joints

6. MOMENT RESISTANT JOINTS

When the bending resistance of the joint is limited by the resistance of the column web panel in shear, a rigid-plastic analysis involving the development of plastic hinges in the joints may be contemplated provided that the shear buckling of the web does not occur. As also said for bolted joints, the web slenderness has not to exceed a slenderness limit specified in section 6.2.1a) to respect this last condition. When the two other above-mentioned failure modes prevail, an intermediate (limited) rotational ductility is available. To know whether it is sufficient for a rigid-plastic structural analysis involving plastic hinges in joints, this available rotation capacity should be compared to the required one.

6.6 APPLICATION TO STEEL BEAM-TO-COLUMN JOINT CONFIGURATIONS

6.6.1 Extended scope

According to the component method approach used in EN 1993-1-8, a joint is considered as a set of individual components and the derivation of its design properties relies on a three-step procedure successively involving (section 1.6.2.2):

- the identification of the basic components
- the characterization of the component response
- the assembly of the components

Fundamentally its scope of application is extremely large in terms of covered joint configurations and connection types; but in practice, it is limited by the available knowledge the user has on component characterisation and component assembly. For sure, EN 1993-1-8 is providing information on several commonly met components (EN 1993-1-8, Table 1.2 in section 1.6 and section 6.2) and on assembly procedures (EN 1993-1-8, sections 6.3 to 6.5), but the wide scope of application which is so made available is certainly not likely to satisfy all the designer's needs, in particular when tubular members or innovative/less traditional connection types and connectors are used. Few examples of "not EC3 covered" joints are shown in Figure 6.34 to Figure 6.39. In such cases, one often comes to the conclusion that the lack of knowledge relates to the characterisation of the response of some specific "unknown" individual components.

6.6 APPLICATION TO STEEL BEAM-TO-COLUMN JOINT CONFIGURATIONS

In (Jaspart *et al*, 2005), the authors have gathered most of the scientifically validated component characterisation models that were available from literature in 2005 and which are not yet included in EN 1993-1-8. This report should be progressively reviewed and updated within CIDECT. The combined use of Part 1-8 and (Jaspart *et al*, 2005) is allowed by Eurocode 3 and is therefore an appropriate way for the user to satisfy his actual needs and to extend his possibilities in terms of conceptual design of joints.

Furthermore the component approach is nowadays a reference for many researchers in the world who regularly develop characterisation models for new components, so multiplying the user's design possibilities. As examples, publications (Jaspart, 1997) and (Demonceau *et al*, 2011) provide information for the design of the joints illustrated in Figure 6.34 to Figure 6.36.

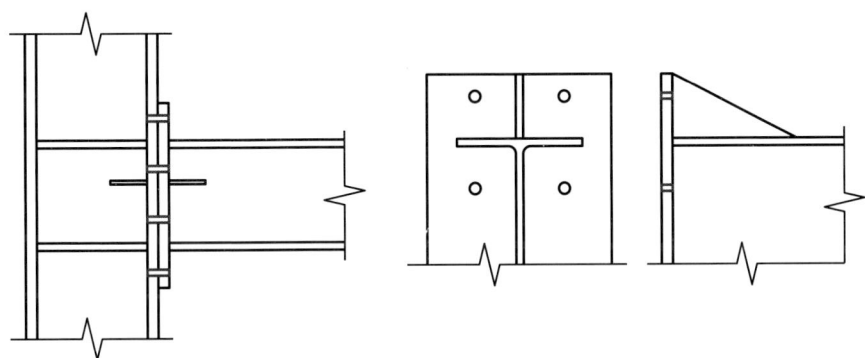

Figure 6.34 – Bolted endplate joints with various reinforcing stiffeners

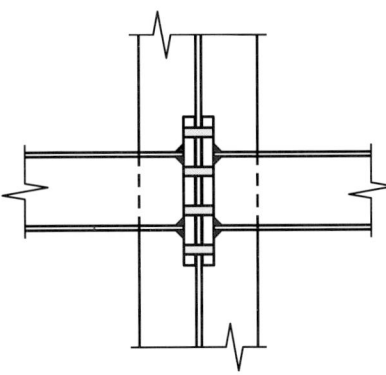

Figure 6.35 – Weak axis beam-to-column joints with H or I members

6. MOMENT RESISTANT JOINTS

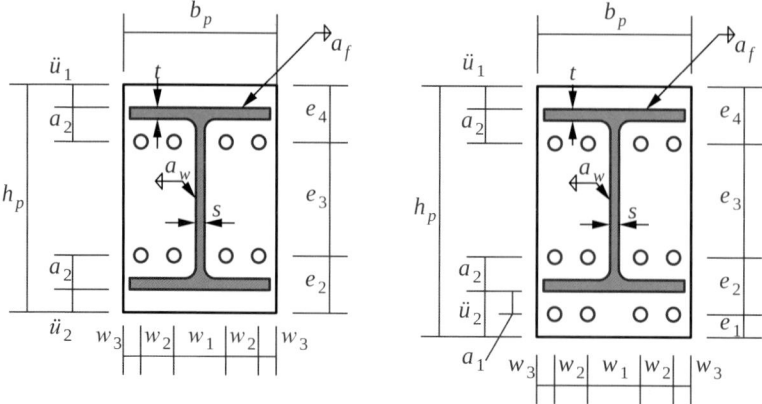

Figure 6.36 – Beam splice with endplate connection exhibiting four bolts per row

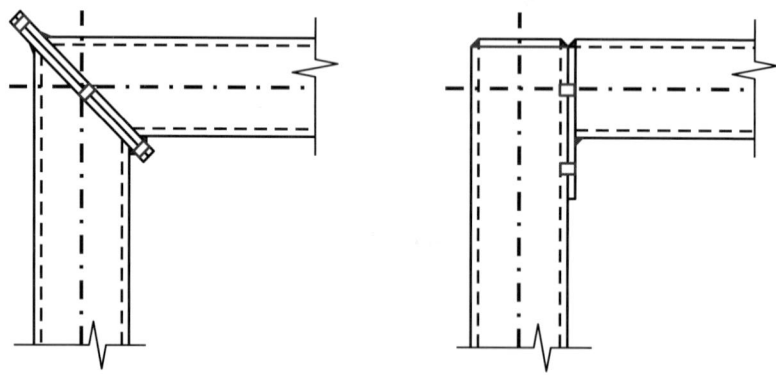

Figure 6.37 – Beam-to-column joints with RHS members

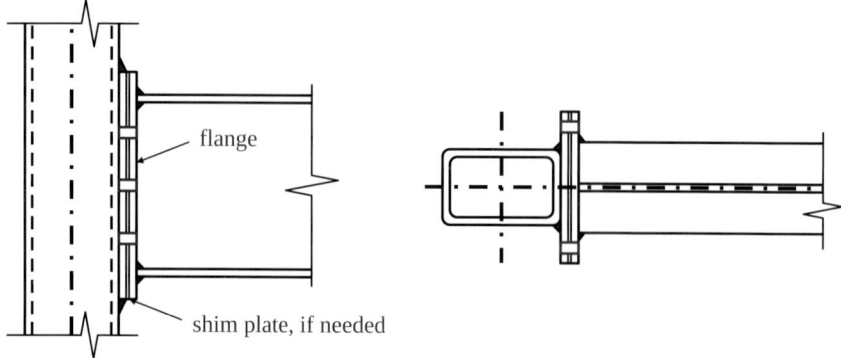

Figure 6.38 – Bolted flange plate connection in a beam-to-column joint with an RHS column

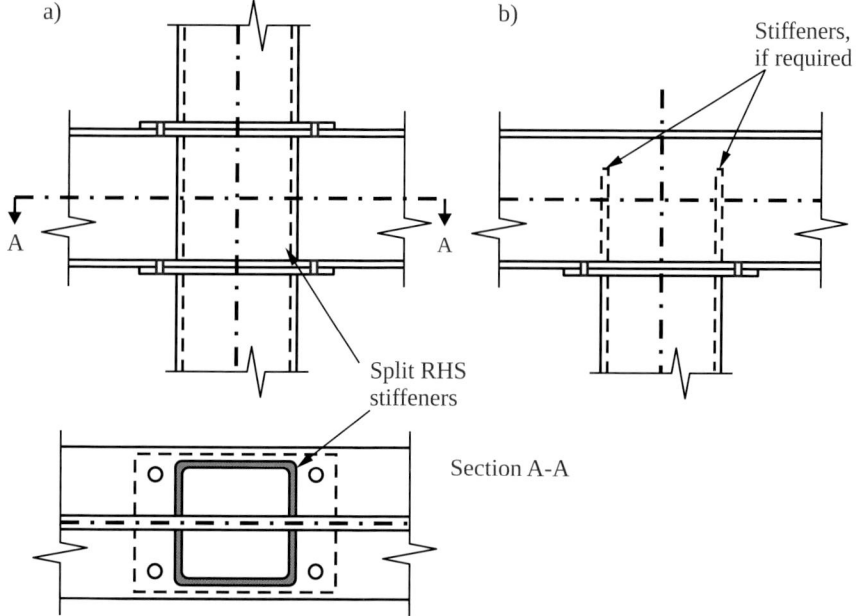

Figure 6.39 – Beam-to-column connections with a continuous RHS beam

6.6.2 Possible design simplifications for endplate connections

EN 1993-1-8 allows using a simplified procedure for the estimation of the resistance and the stiffness of joints with extended end-plates subjected to bending. The following sections detail the nature of this simplification.

6.6.2.1 Design moment resistance

As a conservative simplification, the design moment resistance of an extended end-plate joint with only two rows of bolts in tension may be approximated as indicated in Figure 6.40, provided that the total design resistance F_{Rd} does not exceed $3.8F_{t,Rd}$, where $3.8F_{t,Rd}$ is the design resistance of a bolt in tension. In this case the whole tension region of the end-plate may be treated as a single basic component as long as the two bolt-rows are approximately equidistant either side of the beam flange. $F_{1,Rd}$ is first evaluated and $F_{2,Rd}$ is then assumed to be equal to $F_{1,Rd}$; finally F_{Rd} may be taken as equal to $2 \cdot F_{1,Rd}$. The $3.8F_{t,Rd}$ limitation ensures the safe

6. MOMENT RESISTANT JOINTS

character of the approach by avoiding that the second failure mode exhibits a brittle failure mode before the calculated $M_{j,Rd}$ bending resistance of the whole joint is reached. In (CRIF *et al*), this simplification has been followed to derive practical design tools (simplified design procedure for specific joint configurations and tables of standardized joints).

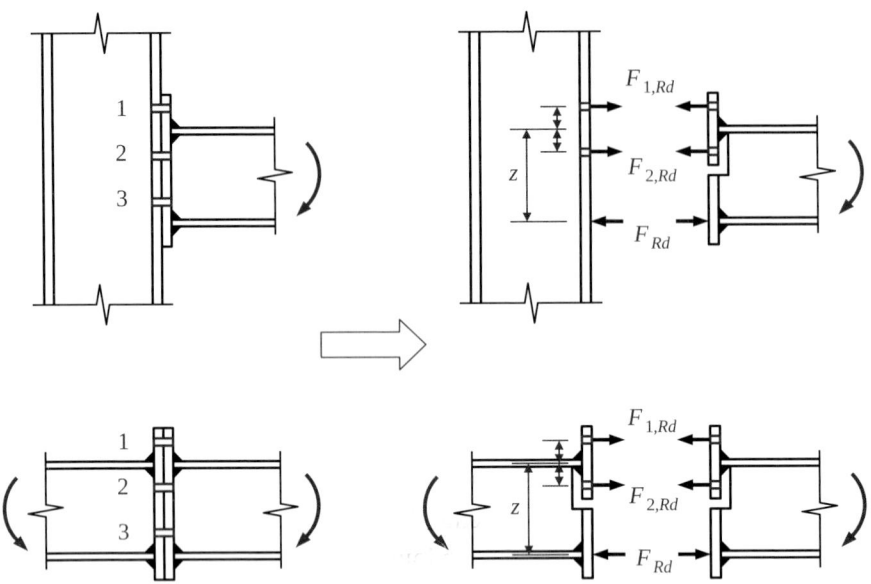

Figure 6.40 – Simplified design model for bending resistance

6.6.2.2 Initial stiffness

A similar simplification is suggested for the computation of the initial stiffness. In this case, a set of modified values may be used for the stiffness coefficients of the related basic components to allow for the combined contribution of both bolt-rows. Each of these modified values should be taken as twice the corresponding value for a single bolt-row in the extended part of the end-plate. This approximation leads to a slightly lower estimate of the rotational stiffness.

When using this simplified method, the lever arm z should be taken as equal to the distance from the centre of compression to a point midway between the two bolt-rows in tension, see Figure 6.41.

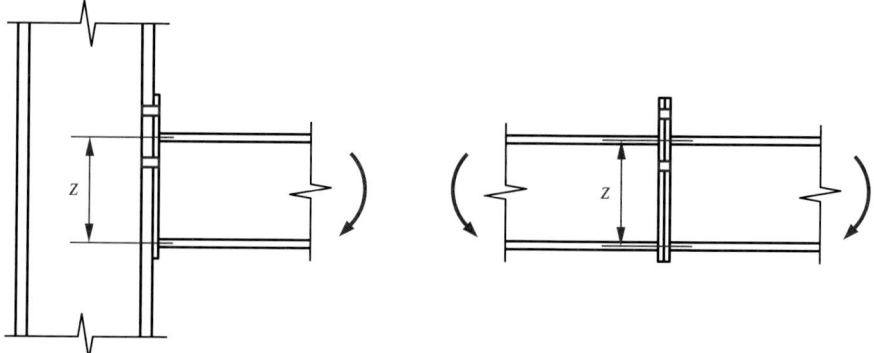

Figure 6.41 – Lever arm for simplified method

6.6.3 Worked example

6.6.3.1 General data

The selected example is the beam-to-column joint in a typical steel frame shown in Figure 6.42. The detailing of the joint are also shown in this Figure. As type of connection an end-plate extended in the tensile zone was chosen. Four bolt rows are met in this joint. A steel grade of S 235 is used.

The objective of this example is to characterise the joint in terms of resistance and stiffness when subjected to hogging moments. The prediction of the resistance in shear will also be addressed.

The partial safety factors (as recommended in the CEN version of Eurocode 3) are as follows:

$$\begin{aligned} \gamma_{M0} &= 1.00 \\ \gamma_{M1} &= 1.00 \\ \gamma_{M2} &= 1.25 \\ \gamma_{Mb} &= 1.25 \\ \gamma_{Mw} &= 1.25 \end{aligned} \qquad (6.64)$$

At the end, the joint will be classified, in view of its modelling in a structural analysis.

6. MOMENT RESISTANT JOINTS

Here below, a detailed computation of the joint is realised according to the recommendations given in the EN 1993-1-8. The main steps described in Figure 6.19 will be followed to compute the resistance in bending.

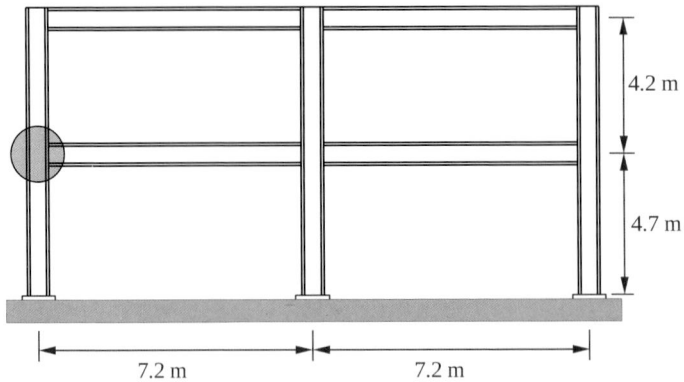

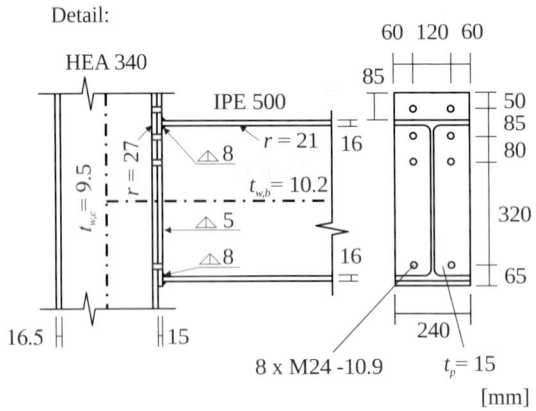

Figure 6.42 – Joint in a steel building frame and joint detailing

6.6.3.2 Determination of the component properties

a) Component 1 – Column web panel in shear

According to the catalogue of profiles, the shear area is equal to 4495 mm². So, the resistance of the column web panel in shear is equal to:

$$V_{wp,Rd} = \frac{0.9 \times 235 \times 4495}{1000\sqrt{3} \times 1.0} = 548.88 \text{ kN} \Rightarrow \frac{548.88}{1.0} = 548.88 \text{ kN} \qquad (6.65)$$

b) Component 2 – Column web in compression

The effective width is equal to:

$$b_{eff,c,wc,2} = t_{fb} + 2\sqrt{2} \times a_{fb} + 5(t_{fc}+s) + s_p$$
$$= 16 + 2\sqrt{2} \times 8 + 5(16.5+27) + 15 + 3.69 \quad (6.66)$$
$$= 274.81 \text{ mm}$$

In most of the cases, the longitudinal stress in the web of the column $\sigma_{com,Ed} \leq 0.7 f_{y,wc}$ and, so $k_{wc} = 1.0$. In the case of a single-sided joint, β can be safely taken as equal to 1.0. Accordingly:

$$\omega = \omega_1 = \frac{1}{\sqrt{1+1.3\left(\dfrac{b_{eff,c,wc} \times t_{wc}}{A_{vc}}\right)^2}}$$

$$= \frac{1}{\sqrt{1+1.3(274.81 \times 9.5/4495)^2}} = 0.834 \quad (6.67)$$

The estimation of the slenderness of the web is used for the definition of the reduction factor ρ.

$$\overline{\lambda}_p = 0.932\sqrt{\frac{b_{eff,c,wc} d_{wc} f_{y,wc}}{E t_{wc}^2}} = 0.932\sqrt{\frac{274.81 \times 243 \times 235}{210000 \times 9.5^2}} = 0.848 \quad (6.68)$$

$$\rho = \frac{(\overline{\lambda}_p - 0.2)}{\overline{\lambda}_p^2} = \frac{(0.848-0.2)}{0.848^2} = 0.901 \quad (6.69)$$

As $\rho < 1.0$, the resistance in compression will be limited by a local buckling phenomenon. At the end:

$$F_{c,wc,Rd} = \frac{0.834 \times 1.0 \times 274.81 \times 9.5 \times 235}{1.0 \times 1000} = 511.67 \text{ kN}$$
$$\overset{!}{\leq} \frac{0.834 \times 1.0 \times 0.901 \times 274.81 \times 9.5 \times 235}{1.0 \times 1000} = 460.9 \text{ kN} \quad (6.70)$$

6. MOMENT RESISTANT JOINTS

c) Component 3 – Column web in tension

The values of effective lengths need to be known for the estimation of the resistance of this component. These values are the same as those to be calculated for the component 4 „column flange in bending" (see section 6.6.3.2d). Four resistant loads have to be determined:

- Individual bolt row resistances
- Resistance of the group "rows 1 and 2"
- Resistance of the group "rows 1, 2 and 3"
- Resistance of the group "rows 2 and 3"

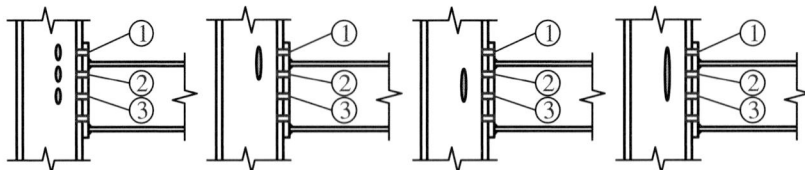

Figure 6.43 – Bolt rows and groups of bolt rows

In a first step, the behaviour of each individual bolt row is considered. At this level, the resistance of the bolt rows 1, 2 and 3 are identical:

$$F_{t,wc,Rd} = \frac{0.86 \times 247.1 \times 9.5 \times 235}{1.0 \times 1000} = 474 \text{ kN} \qquad (6.71)$$

where: $b_{eff,c,wc} = l_{eff,CFB} = 247.1$ mm

$$\omega = \frac{1}{\sqrt{1+1.3(247.1 \times 9.5/4495)^2}} = 0.86$$

The resistance of the group composed of bolt rows 1 and 2 is equal to:

$$F_{t,wc,Rd} = \frac{0.781 \times 332.1 \times 9.5 \times 235}{1.0 \times 1000} = 579.04 \text{ kN} \qquad (6.72)$$

where: $b_{eff,c,wc} = l_{eff,CFB} = (166.05 + 166.05) = 332.1$ mm

$$\omega = \frac{1}{\sqrt{1+1.3(332.1 \times 9.5/4495)^2}} = 0.781$$

6.6 Application to steel beam-to-column joint configurations

For the group composed of bolt rows 1 to 3:

$$F_{t,wc,Rd} = \frac{0.710 \times 412.1 \times 9.5 \times 235}{1.0 \times 1000} = 653.21 \text{ kN} \qquad (6.73)$$

$$\text{where:} \quad b_{eff,c,wc} = l_{eff,CFB} = (166.05 + 82.5 + 163.55) = 412.1 \text{ mm}$$

$$\omega = \frac{1}{\sqrt{1 + 1.3(412.1 \times 9.5/4495)^2}} = 0.710$$

Finally, for the group composed of bolt rows 2 and 3:

$$F_{t,wc,Rd} = \frac{0.785 \times 327.1 \times 9.5 \times 235}{1.0 \times 1000} = 573.25 \text{ kN} \qquad (6.74)$$

$$\text{where:} \quad b_{eff,c,wc} = l_{eff,CFB} = (163.55 + 163.55) = 327.1 \text{ mm}$$

$$\omega = \frac{1}{\sqrt{1 + 1.3(327.1 \times 9.5/4495)^2}} = 0.785$$

d) Component 4 – Column flange in bending

For this component, the T-stub model is used. For the latter, the following parameters need to be determined:

$$m = \frac{w_1}{2} - \frac{t_{wc}}{2} - 0.8 r_c = \frac{120}{2} - \frac{9.5}{2} - 0.8 \times 27 = 33.65 \text{ mm}$$

$$e = \frac{b_c}{2} - \frac{w_1}{2} = \frac{300}{2} - \frac{120}{2} = 90 \text{ mm} \qquad (6.75)$$

$$e_1 = 50 \text{ mm}$$

$$e_{min} = \min(e; w_2) = \min(90; 60) = 60 \text{ mm}$$

Table 6.4 of EN 1993-1-8 allows computing the effective lengths corresponding to different patterns of yielding lines and that, for the following cases:

- Individual bolt rows
- Group involving bolt rows 1 and 2

- Group involving bolt rows 1, 2 and 3
- Group involving bolt rows 2 and 3

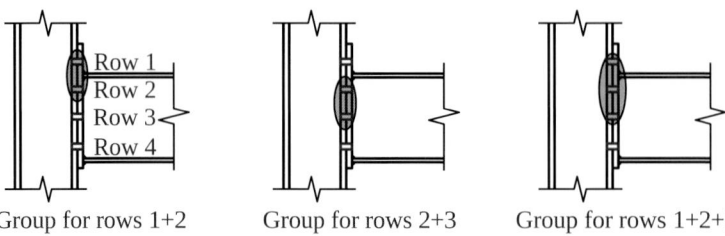

Figure 6.44 – Bolt rows and groups of bolt rows

The contribution of bolt row 4 to the resistance in bending is limited, due to its close position to the centre of compression; accordingly, this contribution is neglected here. In this case bolt row 4 may be assumed to contribute with its full shear resistance to the design shear resistance of the joint. However, if needed, also bolt row 4 may be considered to be in tension zone and hence, its contribution to the moment resistance could be taken into account.

Row 1 – Individual effective lengths:

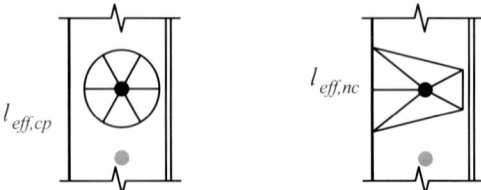

$$l_{eff,cp} = 2\pi m = 2\pi \times 33.65 = 211.43 \text{ mm}$$
$$l_{eff,nc} = 4m + 1.25e = 4 \times 33.65 + 1.25 \times 90 = 247.1 \text{ mm}$$
(6.76)

Row 1 – Effective lengths as first bolt row of a group:

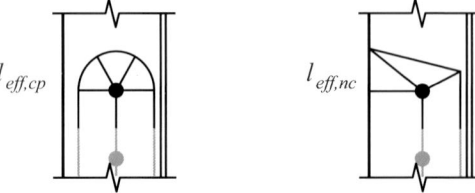

$l_{eff,cp} = \pi m + p = \pi \times 33.65 + 85 = 190.71$ mm

$l_{eff,nc} = 2m + 0.625e + 0.5p$ (6.77)

$= 2 \times 33.65 + 0.625 \times 90 + 0.5 \times 85 = 166.05$ mm

Row 2 – Individual effective lengths:

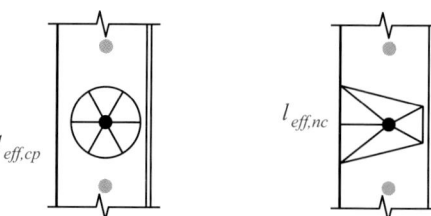

$l_{eff,cp} = 2\pi m = 2\pi \times 33.65 = 211.43$ mm

$l_{eff,nc} = 4m + 1.25e = 4 \times 33.65 + 1.25 \times 90 = 247.1$ mm (6.78)

Row 2 – Effective lengths as first bolt row of a group:

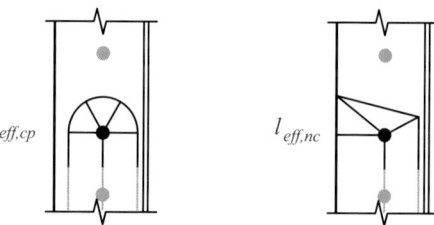

$l_{eff,cp} = \pi m + p = \pi \times 33.65 + 80 = 185.71$ mm

$l_{eff,nc} = 2m + 0.625e + 0.5p$ (6.79)

$= 2 \times 33.65 + 0.625 \times 90 + 0.5 \times 80 = 163.55$ mm

Row 2 – Effective lengths as last bolt row of a group:

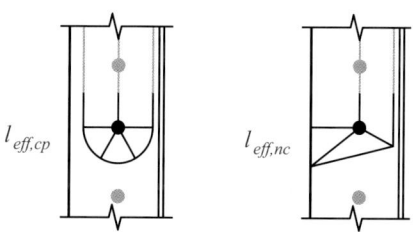

6. MOMENT RESISTANT JOINTS

$$l_{eff,cp} = \pi m + p = \pi \times 33.65 + 85 = 190.71 \text{ mm}$$
$$l_{eff,nc} = 2m + 0.625e + 0.5p$$
$$= 2 \times 33.65 + 0.625 \times 90 + 0.5 \times 85 = 166.05 \text{ mm}$$
(6.80)

Row 2 – Effective lengths as internal bolt row of a group:

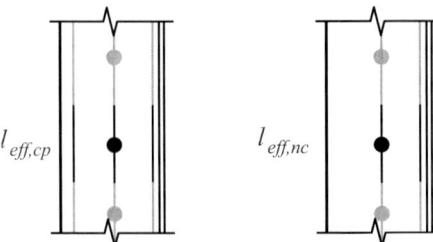

$$l_{eff,cp} = 2p = 2 \times \frac{85+80}{2} = 165.0 \text{ mm}$$
$$l_{eff,nc} = p = \frac{85+80}{2} = 82.5 \text{ mm}$$
(6.81)

Row 3 – Individual effective lengths:

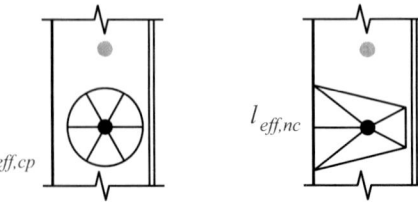

$$l_{eff,cp} = 2\pi m = 2\pi \times 33.65 = 211.43 \text{ mm}$$
$$l_{eff,nc} = 4m + 1.25e = 4 \times 33.65 + 1.25 \times 90 = 247.1 \text{ mm}$$
(6.82)

Row 3 – Effective lengths as last bolt row of a group:

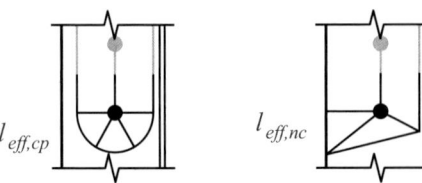

$$l_{eff,cp} = \pi m + p = \pi \times 33.65 + 80 = 185.7 \text{ mm}$$

$$l_{eff,nc} = 2m + 0.625e + 0.5p \qquad (6.83)$$
$$= 2 \times 33.65 + 0.625 \times 90 + 0.5 \times 80 = 163.55 \text{ mm}$$

The resistances are then determined on the basis of the knowledge of the values of the effective lengths.

Individual resistances of bolt rows 1, 2 and 3:
Mode 1

$$F_{T,1,Rd} = \frac{4M_{pl,1,Rd}}{m} = \frac{4 \times 3381.76}{33.65} = 401.99 \text{ kN} \qquad (6.84)$$

$$\text{where:} \quad M_{pl,1,Rd} = \frac{0.25 l_{eff} t_{fc}^2 f_{y,c}}{\gamma_{M0}}$$

$$= \frac{0.25 \times 211.43 \times 16.5^2 \times 235}{1.0 \times 1000} = 3381.76 \text{ kNmm}$$

Mode 2

$$F_{T,2,Rd} = \frac{2 \times 3952.29 + 42.0625 \times 2 \times 254.16}{33.65 + 42.0625} = 386.80 \text{ kN} \qquad (6.85)$$

$$\text{where:} \quad M_{pl,2,Rd} = \frac{0.25 l_{eff,nc} t_{fc}^2 f_{y,c}}{\gamma_{M0}}$$

$$= \frac{0.25 \times 247.1 \times 16.5^2 \times 235}{1.0 \times 1000} = 3952.29 \text{ kNmm}$$

$$F_{t,Rd} = \frac{0.9 \times 1000 \times 353}{1.25 \times 1000} = 254.16 \text{ kN}$$

$$n = \min(e_{min}; 1.25m = 1.25 \times 33.65 = 42.0625)$$
$$= 42.0625 \text{ mm}$$

Mode 3

$$F_{T,3,Rd} = 2 \times 254.16 = 508.32 \text{ kN} \qquad (6.86)$$

Resistance of the group with bolt rows 1 and 2:
In this example, the non-circular yield pattern are decisive.

6. MOMENT RESISTANT JOINTS

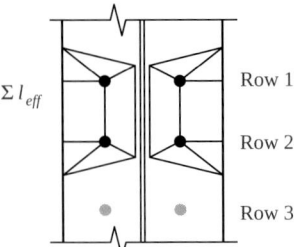

Mode 1

$$F_{T,1,Rd} = \frac{4 \times 5311.84}{33.65} = 631.42 \text{ kN} \qquad (6.87)$$

$$\text{where: } M_{pl,1,Rd} = \frac{0.25(166.05+166.05)\times 16.5^2 \times 235}{1.0\times 1000}$$
$$= 5311.84 \text{ kNmm}$$

Mode 2

$$F_{T,2,Rd} = \frac{2\times 5311.84 + 42.0625 \times 4 \times 254.16}{33.65+42.0625} = 705.12 \text{ kN} \qquad (6.88)$$

$$\text{where: } M_{pl,2,Rd} = \frac{0.25(166.05+166.05)\times 16.5^2 \times 235}{1.0\times 1000}$$
$$= 5311.84 \text{ kNmm}$$

Mode 3

$$F_{T,3,Rd} = 4 \times 254.16 = 1016.64 \text{ kN} \qquad (6.89)$$

Resistance of the group with bolt rows 1 to 3:

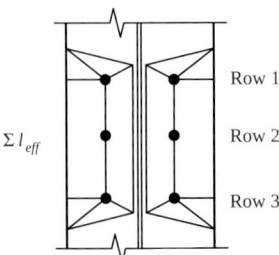

6.6 Application to steel beam-to-column joint configurations

$$F_{T,1,Rd} = \frac{4 \times 6591.41}{33.65} = 783.53 \text{ kN} \qquad (6.90)$$

where: $M_{pl,1,Rd} = \dfrac{0.25(166.05 + 82.5 + 163.55) \times 16.5^2 \times 235}{1.0 \times 1000}$

$= 6591.41 \text{ kNmm}$

$$F_{T,2,Rd} = \frac{2 \times 6591.41 + 42.0625 \times 6 \times 254.16}{33.65 + 42.0625} = 1021.32 \text{ kN} \qquad (6.91)$$

where: $M_{pl,2,Rd} = \dfrac{0.25(166.05 + 82.5 + 163.55) \times 16.5^2 \times 235}{1.0 \times 1000}$

$= 6591.41 \text{ kNmm}$

$$F_{T,3,Rd} = 6 \times 254.16 = 1524.96 \text{ kN} \qquad (6.92)$$

Resistance of the group with bolt rows 2 and 3:

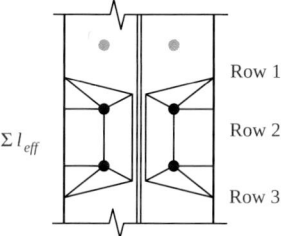

$$F_{T,1,Rd} = \frac{4 \times 5231.86}{33.65} = 621.92 \text{ kN} \qquad (6.93)$$

where: $M_{pl,1,Rd} = \dfrac{0.25(163.55 + 163.55) \times 16.5^2 \times 235}{1.0 \times 1000}$

$= 5231.86 \text{ kNmm}$

$$F_{T,2,Rd} = \frac{2 \times 5231.86 + 42.0625 \times 4 \times 254.16}{33.65 + 42.0625} = 703.00 \text{ kN} \qquad (6.94)$$

where: $M_{pl,2,Rd} = \dfrac{0.25(163.55 + 163.55) + 16.5^2 \times 235}{1.0 \times 1000}$

$= 5231.86 \text{ kNmm}$

6. Moment Resistant Joints

$$F_{T,3,Rd} = 4 \times 254.16 = 1016.64 \text{ kN} \tag{6.95}$$

e) Component 5 – End-plate in bending

Reference is again made to the equivalent T-stub model. Accordingly, the following parameters need to be determined:

$e_x = 50$ mm

$m_x = u_2 - 0.8 a_f \sqrt{2} - e_x = 85 - 0.8 \times 8\sqrt{2} - 50 = 25.95$ mm

$w = 120$ mm

$e = 60$ mm

$m = \dfrac{w}{2} - \dfrac{t_{wb}}{2} - 0.8 a_{fb} \sqrt{2} = 60 - 5.1 - 0.8 \times 5\sqrt{2} = 49.29$ mm

$m_2 = e_x + e_{1-2} - u_2 - t_{fb} - 0.8 a_{fb} \sqrt{2}$
$= 50 + 85 - 85 - 16 - 0.8 \times 8\sqrt{2} = 24.95$ mm

$n_x = \min(e_x;\ 1.25 m_x = 1.25 \times 25.95 = 32.44) = 32.44$ mm

As previously done, the reduced effect of bolt row 4 is neglected for the estimation of the resistance in bending. The values of the effective lengths are extracted from Table 6.6 of Part 1-8; only group 2-3 is considered here as the presence of the beam flange does not allow developing yield lines between bolt row 1 and bolt row 2. Bolt row 2 needs the estimation of a α coefficient as this bolt row is close to the beam flange; the two following parameters are needed for the estimation of this parameter (see figure 6.11 of EN 1993-1-8):

$$\left. \begin{array}{l} \lambda_1 = \dfrac{m}{m+e} = \dfrac{49.24}{49.24+60} = 0.45 \\[2mm] \lambda_2 = \dfrac{m_2}{m+e} = \dfrac{24.95}{49.24+60} = 0.23 \end{array} \right\} \rightarrow \alpha = 7.21 \tag{6.96}$$

Row 1 – Individual effective lengths:

$l_{eff,cp,1} = 2\pi m_x = 2\pi \times 25.95 = 163.05$ mm

$l_{eff,cp,2} = \pi m_x + w = \pi \times 25.95 + 120 = 201.52$ mm

$l_{eff,cp,3} = \pi m_x + 2e = \pi \times 25.95 + 2 \times 60 = 201.52$ mm

$l_{eff,nc,1} = 4m_x + 1.25e_x = 4 \times 25.95 + 1.25 \times 50 = 166.3$ mm

$l_{eff,nc,2} = e + 2m_x + 0.625e_x = 60 + 2 \times 25.95 + 0.625 \times 50 = 143.15$ mm

$l_{eff,nc,3} = 0.5b_p = 0.5 \times 240 = 120$ mm

$l_{eff,nc,4} = 0.5w + 2m_x + 0.625e_x$

$\phantom{l_{eff,nc,4}} = 0.5 \times 120 + 2 \times 25.95 + 0.625 \times 50 = 143.15$ mm

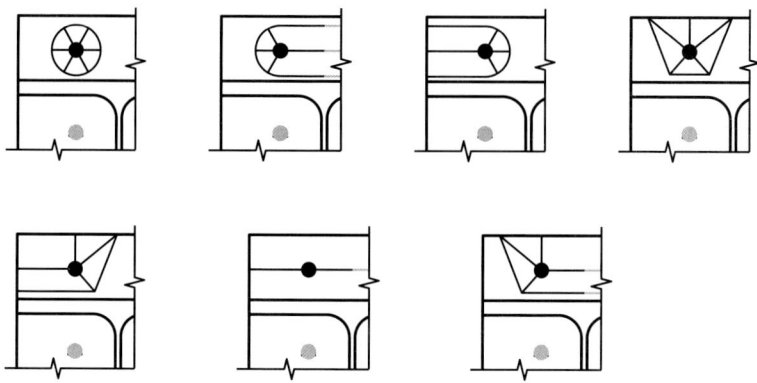

Row 2 – individual effective lengths:

$$l_{eff,cp} = 2\pi m = 2\pi \times 49.24 = 309.40 \text{ mm}$$
$$l_{eff,nc} = \alpha m = 7.21 \times 49.24 = 355.02 \text{ mm} \tag{6.97}$$

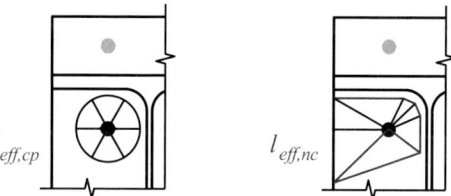

Row 2 – effective lengths as first row of a group:

$$l_{eff,cp} = \pi m + p = \pi \times 49.24 + 80 = 234.69 \text{ mm}$$
$$l_{eff,nc} = 0.5p + \alpha m - (2m + 0.625e) \tag{6.98}$$
$$\phantom{l_{eff,nc}} = 0.5 \times 80 + 7.21 \times 49.24 - (2 \times 49.24 + 0.625 \times 60) = 259.04 \text{ mm}$$

6. MOMENT RESISTANT JOINTS

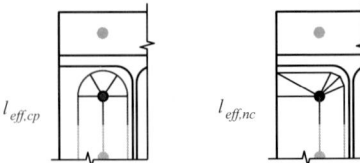

Row 3 – individual effective lengths:

$$l_{eff,cp} = 2\pi m = 2\pi \times 49.24 = 309.40 \text{ mm}$$
$$l_{eff,nc} = 4m + 1.25e = 4 \times 49.24 + 1.25 \times 60 = 271.96 \text{ mm}$$
(6.99)

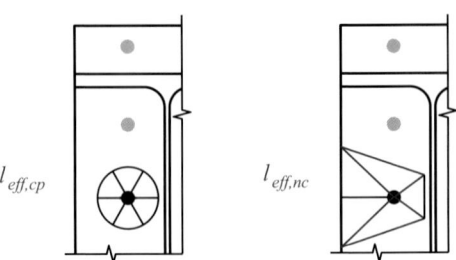

Row 3 – effective lengths as last row of a group:

$$l_{eff,cp} = \pi m + p = \pi \times 49.24 + 80 = 234.69 \text{ mm}$$
$$l_{eff,nc} = 2m + 0.625e + 0.5p$$
$$= 2 \times 49.24 + 0.625 \times 60 + 0.5 \times 80 = 175.98 \text{ mm}$$
(6.100)

The corresponding resistances are then estimated.

Individual resistances of row 1:

$$F_{T,1,Rd} = \frac{4M_{pl,1,Rd}}{m_x} = \frac{4 \times 1586.25}{25.95} = 244.51 \text{ kN}$$
(6.101)

where: $M_{pl,1,Rd} = \dfrac{0.25 \times 120 \times 15^2 \times 235}{1.0 \times 1000} = 1586.25$ kNmm

$$F_{T,2,Rd} = \dfrac{2M_{pl,2,Rd} + n_x \sum F_{t,Rd}}{m_x + n_x} \qquad (6.102)$$

$$= \dfrac{2 \times 1586.25 + 32.44 \times 2 \times 254.16}{25.95 + 32.44} = 336.74 \text{ kN}$$

where: $M_{pl,2,Rd} = \dfrac{0.25 \times 120 \times 15^2 \times 235}{1.0 \times 1000} = 1586.25$ kNmm

$$F_{T,3,Rd} = 2 \times 254.16 = 508.32 \text{ kN} \qquad (6.103)$$

Individual resistances of row 2:

$$F_{T,1,Rd} = \dfrac{4 \times 4089.88}{49.24} = 332.24 \text{ kN} \qquad (6.104)$$

where: $M_{pl,1,Rd} = \dfrac{0.25 \times 309.40 \times 15^2 \times 235}{1.0 \times 1000} = 4089.88$ kNmm

$$F_{T,2,Rd} = \dfrac{2 \times 4692.92 + 60 \times 2 \times 254.16}{49.24 + 60} = 365.11 \text{ kN} \qquad (6.105)$$

where: $M_{pl,2,Rd} = \dfrac{0.25 \times 355.02 \times 15^2 \times 235}{1.0 \times 1000} = 4692.92$ kNmm

$$F_{T,3,Rd} = 2 \times 254.16 = 508.32 \text{ kN} \qquad (6.106)$$

Individual resistances of row 3:

$$F_{T,1,Rd} = \dfrac{4 \times 3594.97}{49.24} = 292.04 \text{ kN} \qquad (6.107)$$

where: $M_{pl,1,Rd} = \dfrac{0.25 \times 271.96 \times 15^2 \times 235}{1.0 \times 1000} = 3594.97$ kNmm

6. Moment Resistant Joints

$$F_{T,2,Rd} = \frac{2 \times 3594.97 + 60 \times 2 \times 254.16}{49.24 + 60} = 345.01 \text{ kN} \quad (6.108)$$

where: $M_{pl,2,Rd} = \dfrac{0.25 \times 271.96 \times 15^2 \times 235}{1.0 \times 1000} = 3594.97 \text{ kNmm}$

$$F_{T,3,Rd} = 2 \times 254.16 = 508.32 \text{ kN} \quad (6.109)$$

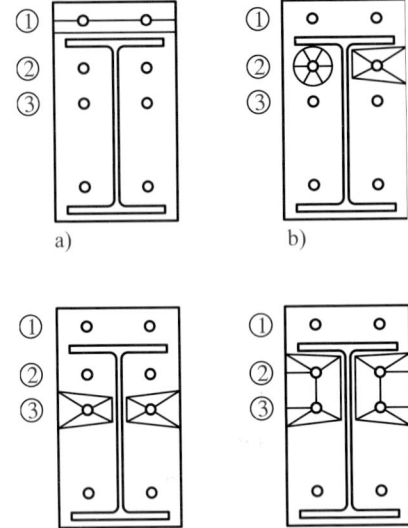

Group resistance - rows 2 and 3:

$$F_{T,1,Rd} = \frac{4 \times 5750.42}{49.24} = 467.13 \text{ kN} \quad (6.110)$$

where: $M_{pl,1,Rd} = \dfrac{0.25(259.04 + 175.98)15^2 \times 235}{1.0 \times 1000} = 5750.42 \text{ kNmm}$

$$F_{T,2,Rd} = \frac{2 \times 5750.42 + 60 \times 4 \times 254.16}{49.24 + 60} = 663.67 \text{ kN} \quad (6.111)$$

where: $M_{pl,2,Rd} = \dfrac{0.25(259.04 + 175.98)15^2 \times 235}{1.0 \times 1000} = 5750.42 \text{ kNmm}$

$$F_{T,3,Rd} = 4 \times 254.16 = 1016.64 \text{ kN} \tag{6.112}$$

f) Component 6 – Flange and web of the beam in compression

This resistance is estimated as follows, on the basis of properties extracted from the catalogue of profiles:

$$F_{c,fb,Rd} = \frac{515.59}{0.5 - 0.016} = 1065.3 \text{ kN} \tag{6.113}$$

g) Component 7 – Beam web in tension

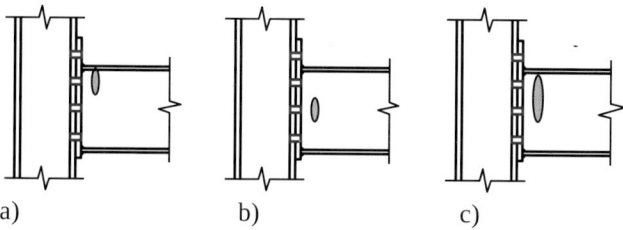

a)   b)   c)

This component has to be considered for bolt rows 2 and 3 only. The individual resistances of rows 2 and 3 (case a and b of the figure) and the group resistance (case c of the figure) are given here after:

Individual resistance of row 2:

$$F_{t,wb,Rd} = \frac{309.40 \times 10.2 \times 235}{1.0 \times 1000} = 741.63 \text{ kN} \tag{6.114}$$

Individual resistance of row 3:

$$F_{t,wb,Rd} = \frac{271.96 \times 10.2 \times 235}{1.0 \times 1000} = 651.89 \text{ kN} \tag{6.115}$$

Group resistance - rows 2 and 3:

$$F_{t,wb,Rd} = \frac{(259.04 + 175.98) \times 10.2 \times 235}{1.0 \times 1000} = 1042.74 \text{ kN} \tag{6.116}$$

6. MOMENT RESISTANT JOINTS

h) Assembling of the components

Load which can be supported by row 1 considered individually

Column flange in bending:	401.99 kN
Column web in tension:	474.00 kN
End-plate in bending:	244.51 kN
$F_{1,min} =$	244.51 kN

Load which can be supported by row 2 considered individually

Column flange in bending:	401.99 kN
Column web in tension:	474.00 kN
End-plate in bending:	332.24 kN
Beam web in tension:	741.63 kN
$F_{2,min} =$	332.24 kN

Load which can be supported by row 3 considered individually

Column flange in bending:	401.99 kN
Column web in tension:	474.00 kN
End-plate in bending:	292.04 kN
Beam web in tension:	651.89 kN
$F_{3,min} =$	292.04 kN

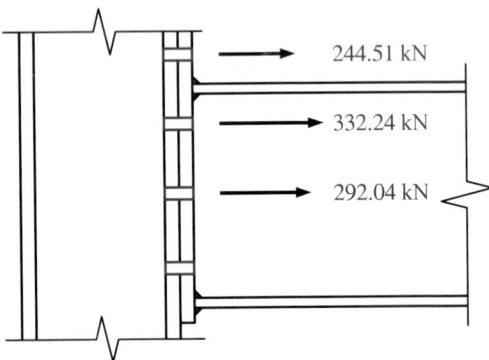

Figure 6.45 – Loads which can be supported individually by the rows

Account for the group effects (group between rows 1-2 only)
The sum of the individual resistances of rows 1 and 2 is equal to:

$$F_{1,min} + F_{2,min} = 244.51 + 332.24 = 576.75 \text{ kN} \tag{6.117}$$

6.6 APPLICATION TO STEEL BEAM-TO-COLUMN JOINT CONFIGURATIONS

This value is compared to the load which can be supported by bolt rows 1 and 2 acting as a group:

Column flange in bending:	631.42 kN
Column web in tension:	579.04 kN
$F_{1+2,min} =$	579.04 kN

As the group resistance is higher than the sum of the individual resistances, the individual resistances of rows 1 and 2 will be reached (i.e. the maximum load which can be supported by these rows is not affected by the considered group effect).

Account for group effects (group between rows 1-2-3 only)

The sum of the individual resistances of rows 1-2-3 is equal to:

$$F_{1,min} + F_{2,min} + F_{3,min} = 244.51 + 332.24 + 292.04 = 868.79 \text{ kN} \tag{6.118}$$

This value is compared to the load which can be supported by rows 1-2-3 acting as a group:

Column flange in bending:	783.53 kN
Column web in tension:	653.21 kN
$F_{1+2+3,min} =$	653.21 kN

The group resistance is smaller than the sum of the individual resistances. Accordingly, the individual resistances in each row cannot be reached and a reduction of the loads which can be supported as to be taken into account. The load which can be supported by bolt row 3 is estimated as follows:

$$\begin{aligned} F_{3,min,red} &= F_{1+2+3,min} - F_{1,min} - F_{2,min} \\ &= 653.21 - 244.51 - 332.24 = 76.46 \text{ kN} \end{aligned} \tag{6.119}$$

Account for the group effects (group between rows 2-3 only)

The sum of the individual resistances of rows 2-3, taking into account of the reduction estimated in the previous section, is equal to:

$$F_{2,min} + F_{3,min,red} = 332.24 + 76.46 = 408.7 \text{ kN} \tag{6.120}$$

6. MOMENT RESISTANT JOINTS

This value is compared to the load which can be supported by the group formed between rows 2 and 3 estimated as follows:

Column flange in bending:	621.92 kN
Column web in tension:	573.25 kN
End-plate in bending:	467.13 kN
Beam web in tension:	1042.7 kN
$F_{2+3,min} =$	467.13 kN

As the resistance of the group is higher than the sum of the individual resistances, it is assumed that each individual resistance can be reached.

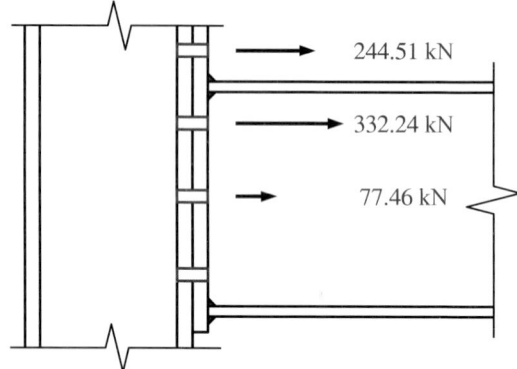

Figure 6.46 – Maximum loads which can be supported by the rows, taking into account of the group effects

In order to ensure equilibrium of the internal forces in the joint, one has to compare the resistance values of the bolt rows in tension with the other components. The sum of the maximal tensile resistances of rows 1, 2 and 3, taking into account of the group effects, is equal to:

$$F_{1,min} + F_{2,min} + F_{3,min,red} = 244.51 + 332.24 + 76.46 = 653.17 \text{ kN} \qquad (6.121)$$

This value has now to be compared to the following loads:

Column web panel in shear:	548.88 kN
Column web in compression:	460.90 kN
Beam flange and beam web in compression:	1065.3 kN
$F_{glob,min} =$	460.90 kN

6.6 APPLICATION TO STEEL BEAM-TO-COLUMN JOINT CONFIGURATIONS

Accordingly, the resistance of the column web in compression limits the maximal loads which can be supported by the rows. The loads in the latter have to be reduced, starting from bolt row 3:

$$F_{1,min} + F_{2,min} = 244.51 + 332.24 = 576.75 \text{ kN} > F_{glob,min} = 460.90 \text{ kN}$$
$$\Rightarrow F_{3,min,red} = 0 \text{ kN} \quad (6.122)$$
$$F_{2,min,red} = F_{glob,min} - F_{1,min} = 460.9 - 244.51 = 216.39 \text{ kN}$$

Because the resistance of the bolt row 2 is limited by a "global" component (here: column web in compression), bolt row 3 cannot be considered to contribute to the design moment resistance of the joint. The final summary of the loads which can be supported by the rows is shown in Figure 6.47

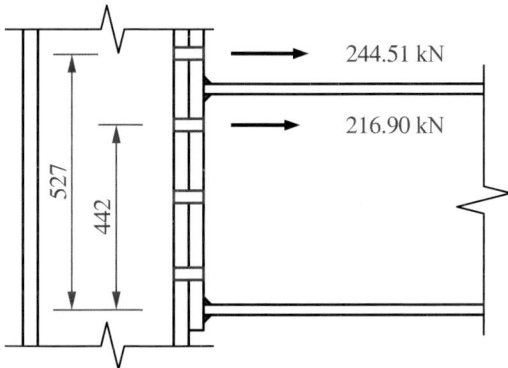

Figure 6.47 – Final loads which can be supported by the rows

6.6.3.3 Determination of the design moment resistance

The design bending resistance is estimated as follows:

$$M_{j,Rd} = 244.5 \times 0.527 + 216.4 \times 0.442 = 224.5 \text{ kNm} \quad (6.123)$$

6.6.3.4 Determination of the rotational stiffness

The first step consists in estimating stiffness coefficients for the different components of the joint, according to Table 6.11 of EN 1993-1-8. These

6. MOMENT RESISTANT JOINTS

coefficients are given in Table 6.8. The formulae as given in Table 6.11 of EN 1993-1-8 are not repeated in the table.

Table 6.8 – Stiffness coefficients

Components	Stiffness coefficients
Column web panel in shear	$k_1 = \dfrac{0.38 \times 4495}{1.0\left(500 - \dfrac{16}{3} + 85 - 50 - \dfrac{85}{2}\right)10^{-3}} = 3.53$
Column web in compression	$k_2 = \dfrac{0.7 \times 274.81 \times 9.5}{243} = 7.52$
Column web in tension	$k_{3,1} = \dfrac{0.7 \times 166.05 \times 9.5}{243} = 4.54$ $k_{3,2} = \dfrac{0.7 \times 82.5 \times 9.5}{243} = 2.26$ $k_{3,3} = \dfrac{0.7 \times 163.55 \times 9.5}{243} = 4.48$ Contribution from row 4 neglected
Column flange in bending	$k_{4,1} = \dfrac{0.9 \times 166.05 \times 16.5^3}{33.65^3} = 17.62$ $k_{4,2} = \dfrac{0.9 \times 82.5 \times 16.5^3}{33.65^3} = 8.75$ $k_{4,3} = \dfrac{0.9 \times 163.55 \times 16.5^3}{33.65^3} = 17.35$
End-plate in bending	$k_{5,1} = \dfrac{0.9 \times 120 \times 15^3}{25.95^3} = 20.86$ $k_{5,2} = \dfrac{0.9 \times 234.69 \times 15^3}{49.24^3} = 5.97$ $k_{5,3} = \dfrac{0.9 \times 175.98 \times 15^3}{49.24^3} = 4.48$
Beam web and beam flange in compression	$k_6 = \infty$
Bolts in tension	$k_{10} = \dfrac{1.6 \times 353}{15 + 16.5 + (15 + 19)/2} = 11.65$

In a second step, the component assembling is realised in agreement with the recommended rules given in Section 6.3.3.1 of EN 1993-1-8. The effective stiffness coefficient of each bolt row is obtained as follows:

6.6 APPLICATION TO STEEL BEAM-TO-COLUMN JOINT CONFIGURATIONS

$$k_{eff,1} = \frac{1}{\frac{1}{4.54}+\frac{1}{17.62}+\frac{1}{11.65}+\frac{1}{20.86}} = 2.43 \quad \text{and} \quad h_1 = 527 \text{ mm}$$

$$k_{eff,2} = \frac{1}{\frac{1}{2.26}+\frac{1}{8.75}+\frac{1}{11.65}+\frac{1}{5.97}} = 1.23 \quad \text{and} \quad h_2 = 442 \text{ mm} \quad (6.124)$$

$$k_{eff,3} = \frac{1}{\frac{1}{4.48}+\frac{1}{17.35}+\frac{1}{11.65}+\frac{1}{4.48}} = 1.70 \quad \text{and} \quad h_3 = 362 \text{ mm}$$

The equivalent stiffness coefficient related to the tensile part of the joint is determined as follows:

$$z_{eq} = \frac{2.43 \times 527^2 + 1.23 \times 442^2 + 1.70 \times 362^2}{2.43 \times 527 + 1.23 \times 442 + 1.70 \times 362} = 466.44 \text{ mm}$$

$$\Rightarrow k_{eq} = \frac{2.43 \times 527 + 1.23 \times 442 + 1.7 \times 362}{466.44} = 5.23 \text{ mm} \quad (6.125)$$

The joint stiffness is then computed through the use of formula 6.27 of EN 1993-1-8:

$$S_{j,ini} = \frac{210000 \times 466.44^2}{\frac{1}{5.23}+\frac{1}{7.52} \times \frac{1}{3.53}} 10^{-6} = 75.214 \text{ MNm/rad} \quad (6.126)$$

6.6.3.5 Computation of the resistance in shear

The shear load is supported by the four bolt rows. The latter being assumed as non-prestressed, the resistance in shear will be limited by the resistance of the bolt shanks in shear or by the bearing resistances of the connected plates (i.e. column flange or end-plate). In the previous case, the use of the design rules for these failure modes as recommended in table 3.4 of EN 1993-1-8 was described; through these rules, it is possible to demonstrate that the failure mode to be considered is the mode "bolts in shear":

$$F_{v,Rd} = \frac{\alpha_v f_{ub} A_s}{\gamma_{M2}} = \frac{0.5 \times 1000 \times 353}{1.25} 10^{-3} = 141.2 \text{ kN} \quad (6.127)$$

6. MOMENT RESISTANT JOINTS

Eight bolts contributes to the shear resistance. However, as indicated here above, half of them (i.e. the upper two bolt rows) support tensile loads to contribute to the bending resistance. Accordingly, their contribution to the shear resistance has to be limited to take into account of these tensile loads; it is recommended in EN 1993-1-8 to multiply the shear resistance of the bolts by the factor $0.4/1.4$. This factor is derived from the tension-shear interaction check assuming the applied tensile force is equal to the tensile resistance of the bolt, see section 3.4.3.

The resistance of the joint to shear force is then obtained through the following formula:

$$V_{Rd} = 4F_{v,Rd} + 4F_{v,Rd}\frac{0.4}{1.4} = 4F_{v,Rd}\left(1+\frac{0.4}{1.4}\right) = 726.2 \text{ kN} \tag{6.128}$$

6.7 APPLICATION TO STEEL COLUMN SPLICES

6.7.1 Common splice configurations

Column splices are required every two or three storeys in a regular building so as to comply with the commercial lengths of the profiles and with the transportation requirements. The splices should hold the connected members in line and transfer the internal forces along the column height.

Splices are predominantly subject to axial forces, and more particularly to compression. According to the position of the splices along the storey height, the amount of bending moment including second order effects may however vary quite significantly. The designer will usually select the location of the splice so as to decrease the value of the bending moments to be locally transferred.

The detailing of the splice should preferably aim at a perfect continuity of the centroidal axes of the upper column stub, the lower column stub and of the splice material. When this is not the case, the bending moment resulting from the axis eccentricities will have to be explicitly considered in the design. In Figure 6.48, few common "perfectly aligned" splice configurations, taken from (Moreno *et al*, 2011), are illustrated. They can be divided into two categories: (i) bolted cover plate splices and (ii) end-plate splices.

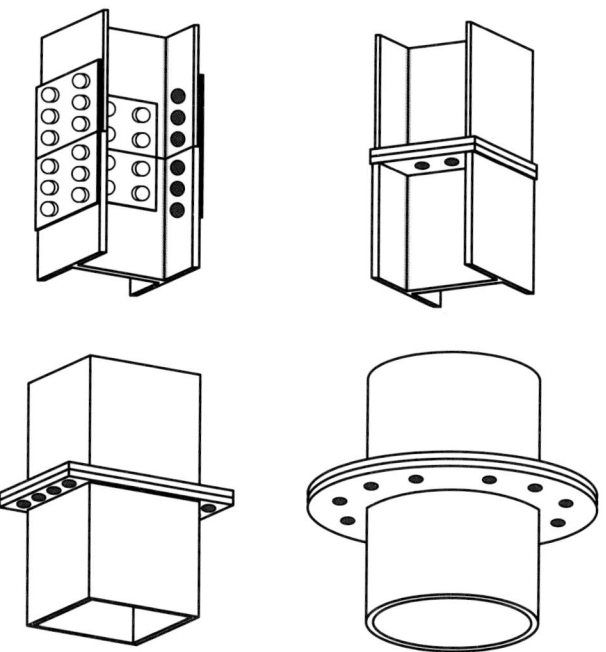

Figure 6.48 – Common splice configurations (Moreno *et al*, 2011)

For the "bolted cover plate" splices, two situations are contemplated:

– Bearing type splices (Figure 6.49)

   The forces are transferred between the column stubs in direct contact or alternatively through a division plate. Direct bearing does not require the machining or end milling of the columns or even a full contact over the whole column area. A normal preparation is usually acceptable, as long as tolerances specified in EN 1090-2 are respected.

– Non-bearing type splices (Figure 6.50)

   The forces are transferred between the column stubs through bolts and cover plates. The forces possibly passing through direct contact are ignored; in most cases, a physical gap between is even often provided between the column stubs.

6. MOMENT RESISTANT JOINTS

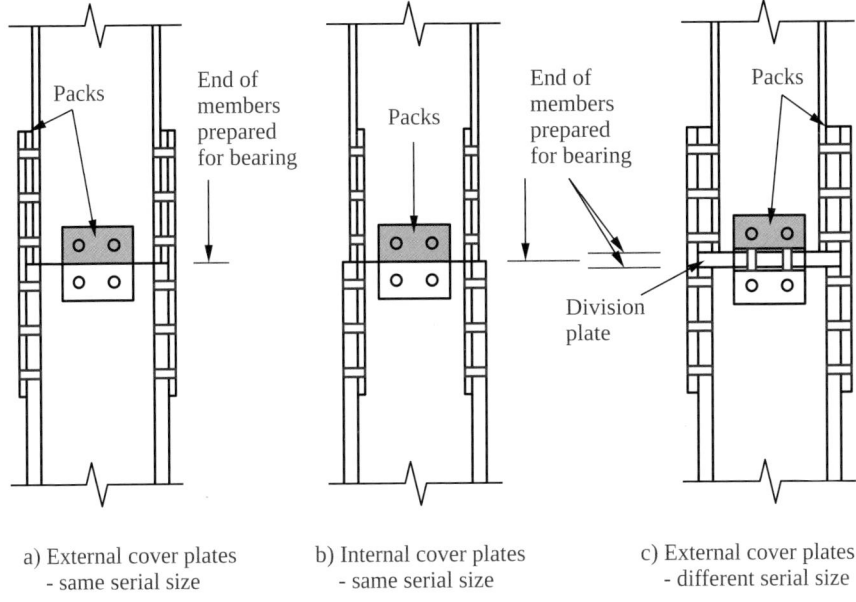

Figure 6.49 – Bearing type splices (Moreno et al, 2011)

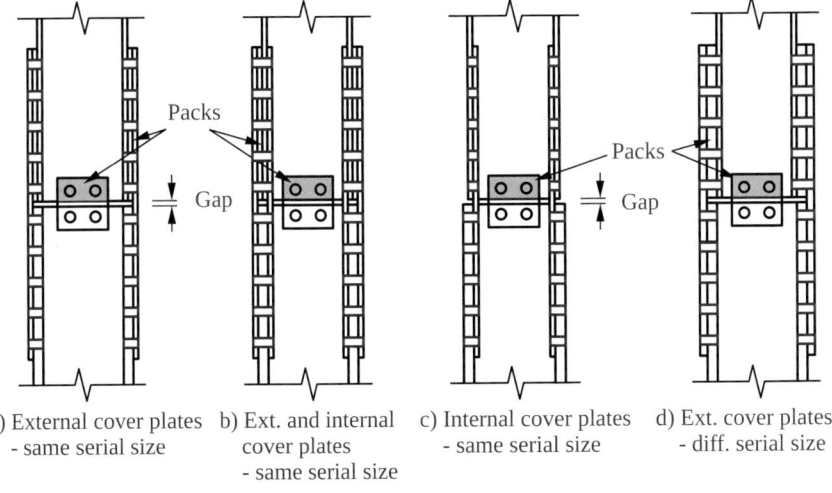

Figure 6.50 – Non-bearing type splices (Moreno et al, 2011)

6.7.2 Design considerations

A splice in a member should be designed to transmit all the moments including second order effects and forces to which the member is subjected at that point. Adequate rotational stiffness should also be provided to the splice so as to validate the design of the column based on member continuity.

Where the members are prepared for full contact in bearing, splice material should be provided to transmit 25% of the maximum compressive force in the column and friction forces between contact surfaces may not be relied upon to hold connected members in place.

Where the members are not prepared for full contact in bearing, splice material should be provided to transmit the internal forces and moments in the member at the spliced section, including the moments due to applied eccentricity, initial imperfections and second-order deformations. The internal forces and moments should be taken as not less than a moment equal to 25% of the moment capacity of the weaker section about both axes and a shear force equal to 2.5% of the normal force capacity of the weaker section in the directions of both axes.

Finally, splices with cover plates used in flexural members should comply with the following requirements:

- Compression flanges should be treated as compression members;
- Tension flanges should be treated as tension members;
- Parts subjected to shear should be designed to transmit the following effects acting together:
  - the shear force at the splice;
  - the moment resulting from the eccentricity, if any, of the centroids of the groups of fasteners on each side of the splice;
  - the proportion of moment, deformation or rotations carried by the web or part, irrespective of any shedding of stresses into adjoining parts assumed in the design of the member.

The interested reader will find a large set of interesting information about the detailing and the design of splices in (Moreno *et al*, 2011).

6.8 APPLICATION TO COLUMN BASES

6.8.1 Common column basis configurations

Typical column bases are represented in Figure 6.51 (Moreno *et al*, 2011).

6. MOMENT RESISTANT JOINTS

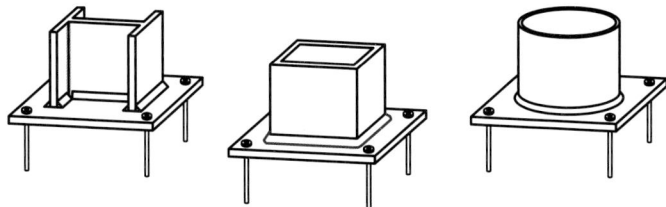

Figure 6.51 – Typical column bases (Moreno et al, 2011)

They consist of a column, a base plate and holding down bolts (anchors). In Figure 6.51, no stiffeners are added on the steel plate but stiffened base plates may be used where the connection is required to transfer high bending moments (see Figure 6.52). Alternatively the thickness of the base plate could be increased. The column base is usually supported by either a concrete slab or a sub-structure (e.g. a piled foundation) in which the holding down bolts are anchored.

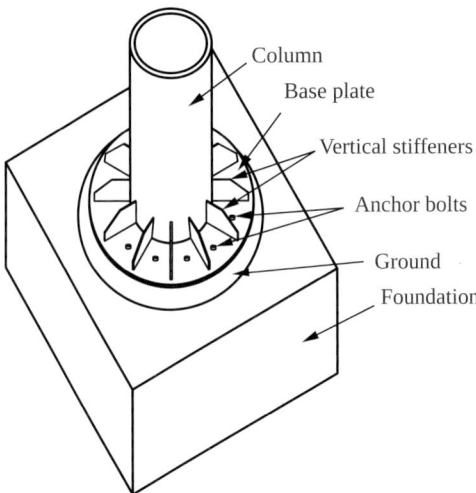

Figure 6.52 – Stiffened column base

Different types of anchors may be contemplated. Few are illustrated in Figure 5.22.

So as to facilitate a proper erection, the columns are regularly erected on steel levelling packs that are left permanently under the plate. An alternative is to use levelling nuts screwed on the holding bolts before the column is put into position. In all cases, a grout is poured into the space left free between the concrete block and the steel plate.

6.8 APPLICATION TO COLUMN BASES

Other configurations exist, as the one shown in Figure 6.53 where a so-called embedded configuration is illustrated. Info about this specific solution may be found in (Demonceau *et al*, 2012)

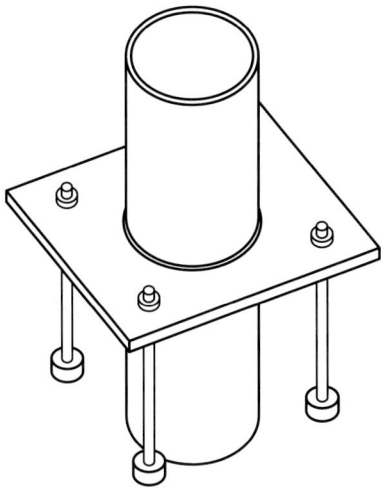

Figure 6.53 – Embedded column joint configuration

Traditionally column bases are designed as pinned or as rigid joints with a number of anchors chosen accordingly (two for pinned configurations, four for rigid configurations, see Figure 6.54)

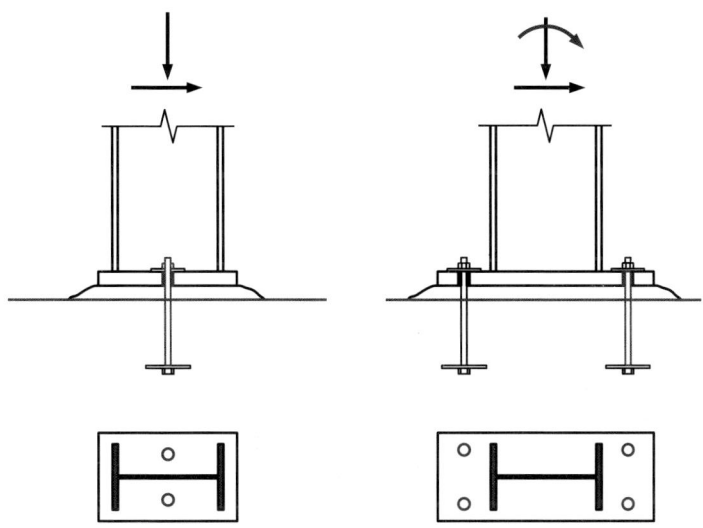

Figure 6.54 – Classical pinned and moment resisting joint configurations

6. MOMENT RESISTANT JOINTS

A pinned joint is assumed to resist to axial compression or tension forces and to shear forces while the rigid one has additionally to transfer bending moments to the foundation. In many actual cases, column bases behave in a semi-rigid manner, even if the application of axial compression forces is a quite positive factor as far as global joint stiffness is concerned.

Compression forces are transmitted to the foundation block by direct bearing while tension forces are transferred through the anchors. Shear forces may be carried out to the concrete block by friction or directly through the anchors (the rupture of the concrete in bearing and the bearing capacity of the plate should be also checked). For higher levels of shear, designers will often weld a shear stud to the underside of the base plate. Practical considerations about the detailing of column bases may be found in (Moreno *et al*, 2011).

6.8.2 Design considerations

The following paragraphs are highly inspired by a series of papers (Gresnigt *et al*, 2008; Jaspart *et al*, 2008b; Steenhuis *et al*, 2008; Wald *et al*, 2008a; Wald *et al*, 2008b) published few years ago, on which the contents of Eurocode 3 Part 1-8 is widely based as far as the design of column bases under axial force and bending moment is concerned. For the verification of column base joints under shear forces (topic which is only very briefly addressed in Part 1-8), the reader should more particularly consult paper (Gresnigt *et al*, 2008) or refer to section 5.4.4 where the problem has been already addressed.

6.8.2.1 Proportional and non-proportional loading

A characteristic of column bases is that they are loaded by a combination of the shear forces, axial forces and bending moments. The loading sequence is also relevant and, from that point of view, two usual cases are here distinguished:

- The non-proportional loading in which the axial force is applied to the base plate connection first, before the moment is increased.

- The proportional loading in which the axial force and the bending moment are applied simultaneously to the joints, with a constant ratio between the bending moment and the axial force (this is also called the eccentricity e).

In (Wald *et al*, 2008b), moment-rotation curves obtained for a typical column base with proportional and non-proportional loading are compared. It is shown, for instance, that in the case of non-proportional loading, the initial stiffness of the joint is higher than the one characterising the non-proportional loading. This is due to the presence of the axial force in the column, which keeps the base plate in contact with the concrete for low bending moments.

6.8.2.2 General procedure for the derivation of the design properties of column bases with base plates

In section 6.3.3, the general procedure to derive the design resistance of a joint subjected to axial forces and bending moments has been described. It may be applied to column bases, except that, in this case, the position of the compression force is not so obvious to estimate.

Let first consider a symmetric column base plate joint subject to pure compression (Figure 6.55). According to the T-stub component model presented in section 3.5.2, the stresses between the plate and the concrete block are assumed to be uniformly distributed on a fictitious rigid equivalent plate resulting from the sum (without overlapping) of the equivalent T-stub rigid equivalent area associated respectively to the two column flanges and to the column web (Figure 6.55).

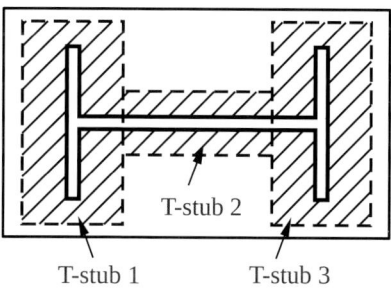

Figure 6.55 – Equivalent rigid plate under axial compression force

6. MOMENT RESISTANT JOINTS

The axial resistance N_{Rd} of the column base is simply evaluated as equal to $F_{c,Rd}$ obtained by multiplying the area of the equivalent plate by the design strength of the concrete block in compression (see Figure 6.56).

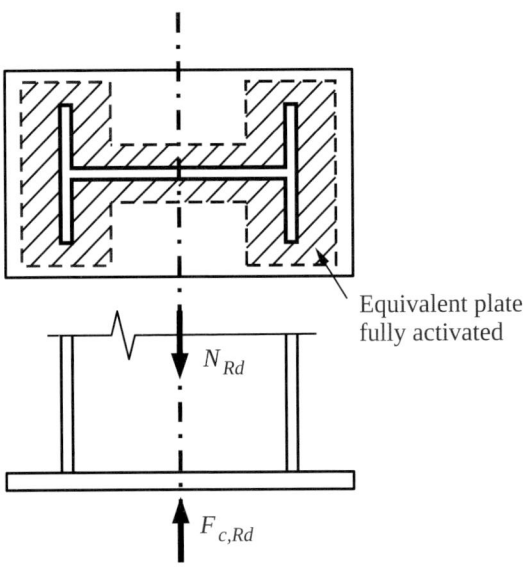

Figure 6.56 – Force equilibrium under axial force

In a second step, a constant axial force N_{Ed} is applied and a bending moment is then progressively increased until the maximum resistance M_{Rd} of the column base joint is reached (non-proportional loading). A specific situation is illustrated in Figure 6.57 where (i) only a part of the rigid equivalent plate (called the active one) is subjected to compression ($F_{c,Ed} < F_{c,Rd}$) while (ii), in the tension zone, anchors are in tension and it is assumed that they reach their design resistance ($F_{t,Rd}$). By expressing the axial and rotational equilibrium of the forces acting on the column base joints (assembly procedure), the location of the compression force (z_c) and the value of the compression force ($F_{c,Ed}$) will be obtained as well as the resistance of the joint.

The same principles could be applied to the evaluation of the rotational stiffness and axial stiffness of the column base joints.

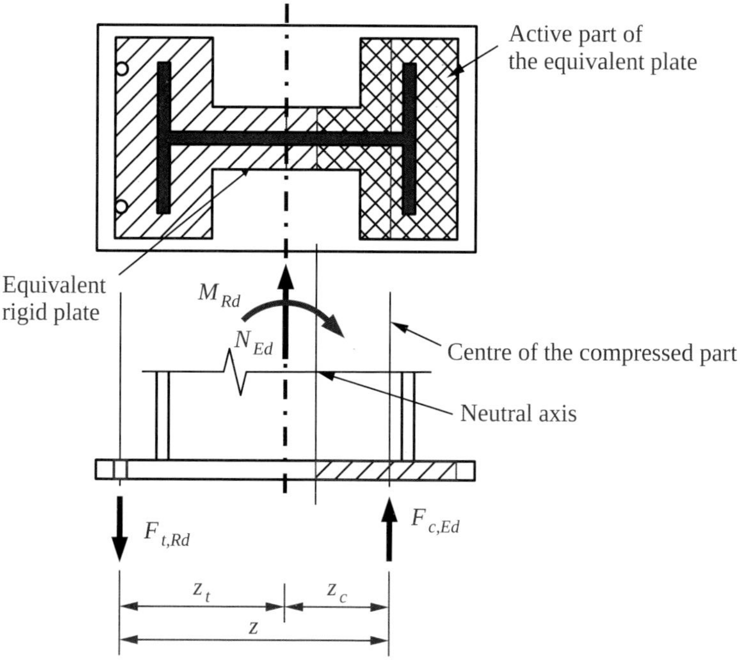

6.8 APPLICATION TO COLUMN BASES

Figure 6.57 – Force equilibrium under axial force and bending moment

6.8.2.3 Simplified procedure for the derivation of the design properties of column bases with base plates

The application of the here-above described general procedure may quickly become rather complex and hence some authors propose simplified procedures.

In (Jaspart, Vandegans, 1998), an easy-to-apply calculation procedure is suggested for the evaluation of the design resistance of the column bas joints. In Eurocode 3, another one is also presented, which is extended to the determination of the joint stiffnesses. This one, which covers proportional loading situations is described hereunder.

It is here suggested to consider that compression is transferred from the column to the concrete block under the column flanges only (Steenhuis, 1998). The effective area under the column web is so neglected, as shown in Figure 6.58. The compression forces are assumed to act at the centre of the flanges in compression, also in the cases of the limited outstand of the base, see Figure 6.58c and Figure 6.58d. The tensile force is located at the anchor bolts or midway between the bolts when there are two rows of tension bolts, as shown in Figure 6.58a.

The equilibrium of forces may be calculated by referring to Figure 6.58 where two situations are considered as examples: base plate with four anchor bolts (Figure 6.58a) and with two anchor bolts inside the column (Figure 6.58c). $F_{t,l,Rd}$ and $F_{c,r,Rd}$ represent respectively the maximum force that the left anchor bolt row(s) may carry out (equal to the weaker components in the tension zone) and the maximum force that than can be transferred to the concrete block by the right column flange in compression ($F_{c,l,Rd}$ on the left side). From Figure 6.58a and Figure 6.58c, the following formulae can be derived which express the way on how to check the resistance of the column base:

$$\frac{M_{Ed}}{z} + \frac{N_{Ed} z_{c,r}}{z} \leq F_{t,l,Rd} \tag{6.129}$$

and

$$\frac{M_{Ed}}{z} + \frac{N_{Ed} z_{t,l}}{z} \leq F_{c,r,Rd} \tag{6.130}$$

Eq. (6.129) and (6.130) can be rewritten as:

$$M_{Rd} = \min \begin{cases} \dfrac{F_{t,l,Rd} \cdot z}{\dfrac{z_{c,r}}{e} + 1} \\ \dfrac{-F_{c,r,Rd} \cdot z}{\dfrac{z_{t,l}}{e} - 1} \end{cases} \tag{6.131}$$

This expression is valid as long as $F_{t,l,Ed}$ and $F_{c,r,Ed}$ are greater (or equal) to zero, what may be expressed as follows ($M_{Ed} > 0$ is clockwise and $N_{Ed} > 0$ is tension):

$$N_{Ed} > 0 \text{ and } e = M_{Ed}/N_{Ed} > z_{t,l} \tag{6.132}$$

$$N_{Ed} \leq 0 \text{ and } e = M_{Ed}/N_{Ed} \geq -z_{c,r} \tag{6.133}$$

6.8 APPLICATION TO COLUMN BASES

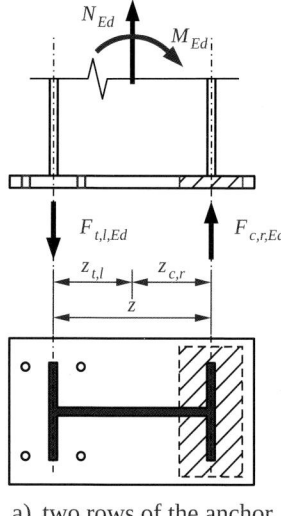

a) two rows of the anchor bolts in tension

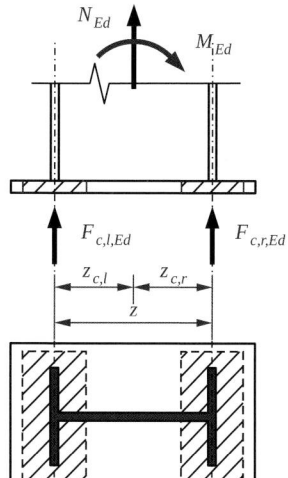

b) no tension in anchor bolts

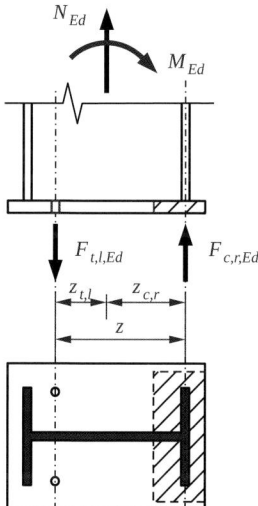

c) one row of the anchor bolts in tension and limited plate outstand

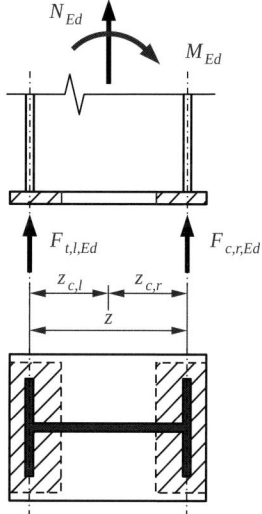

d) no tension in anchor bolts and limited plate outstand

Figure 6.58 – Equilibrium of forces on the base plate (with the effective area under the flanges only)

A similar reasoning may be achieved for the situations illustrated in Figure 6.58b and Figure 6.58d where there is no tension force in the anchor bolts but both parts of the joint are under compression. In this case, the equation may be rewritten as:

6. MOMENT RESISTANT JOINTS

$$M_{Rd} = \min \left\{ \begin{array}{c} \dfrac{-F_{c,l,Rd} \cdot z}{\dfrac{z_{c,r}}{e}+1} \\ \\ \dfrac{-F_{c,r,Rd} \cdot z}{\dfrac{z_{c,l}}{e}-1} \end{array} \right\} \quad (6.134)$$

And the field of validity expresses as:

$$N_{Ed} \leq 0 \text{ and } e = M_{Ed}/N_{Ed} \geq z_{c,l} \quad (6.135)$$

$$N_{Ed} \leq 0 \text{ and } e = M_{Ed}/N_{Ed} \leq -z_{c,r} \quad (6.136)$$

Two other cases should be contemplated:

- Left side of the column base in compression and right side in tension;
- Both sides of the column bases are in tension

The derivation of the related design moment resistances and application fields is immediate. Eurocode 3 Part 1-8 provides formulae for all cases.

Besides that, similar assumptions than those expressed before may be also considered to evaluate the bending stiffness of a column base plate joint.

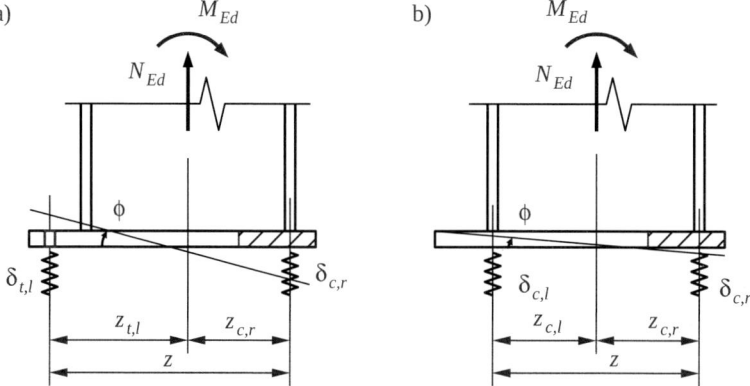

Figure 6.59 – Mechanical stiffness model for a column base plate joint

6.8 Application to column bases

The elastic deformation of the springs representing the constitutive components may be expressed as follows for the specific situation represented in the left part of Figure 6.58, as an example (Steenhuis, 1998):

$$\delta_{t,l} = \frac{\dfrac{M_{Ed}}{z} + \dfrac{N_{Ed} \cdot z_{c,r}}{z}}{E \cdot k_{t,l}} = \frac{M_{Ed} + N_{Ed} \cdot z_{c,r}}{E \cdot z \cdot k_{t,l}} \qquad (6.137)$$

$$\delta_{c,r} = \frac{\dfrac{M_{Ed}}{z} - \dfrac{N_{Ed} \cdot z_{t,l}}{z}}{E \cdot k_{c,r}} = \frac{M_{Ed} - N_{Ed} \cdot z_{t,l}}{E \cdot z \cdot k_{c,r}} \qquad (6.138)$$

So the column base rotation writes:

$$\phi = \frac{\delta_{t,l} + \delta_{c,r}}{z} = \frac{1}{E \cdot z^2} + \left(\frac{M_{Ed} + N_{Ed} \cdot z_{c,r}}{k_{t,l}} + \frac{M_{Ed} - N_{Ed} \cdot z_{t,l}}{k_{c,r}} \right) \qquad (6.139)$$

The eccentricity e_0, at which the rotation is zero, can be evaluated from Eq. (6.139) as:

$$\phi = \frac{1}{E \cdot z^2} + \left(\frac{N_{Ed} \cdot e_0 + N_{Ed} \cdot z_{c,r}}{k_{t,l}} + \frac{N_{Ed} \cdot e_0 - N_{Ed} \cdot z_{t,l}}{k_{c,r}} \right) = 0 \qquad (6.140)$$

Hence the eccentricity under zero rotation is:

$$e_0 = \frac{z_{c,r} k_{c,r} - z_{t,l} k_{t,l}}{k_{c,r} + k_{t,l}} \qquad (6.141)$$

The bending stiffness of the column base plate joint depends on the bending moment due to the change of the eccentricity of the axial force:

$$S_{j,ini} = \frac{M_{Ed}}{\phi} \qquad (6.142)$$

6. MOMENT RESISTANT JOINTS

It may so be derived based on the above formula:

$$S_{j,ini} = \frac{M_{Sd}}{M_{Sd} + N_{Sd} e_0} \frac{E z^2}{\sum \frac{1}{k}} = \frac{e}{e + e_0} \frac{E z^2}{\sum \frac{1}{k}} \quad (6.143)$$

The reader will find in Eurocode 3 Part 1-8 similar formulae to cover the whole range of application.

6.9 APPLICATION TO COMPOSITE JOINTS

6.9.1 Generalities

Several possible arrangements for composite joints are shown Figure 6.60, demonstrating the wide variety of steelwork connection that may be used. For economy, it is desirable that the steelwork connection is not significantly more complicated than that used for simple construction of steel frames. In this cases, the joints are nominally pinned at the construction phase. Later the steelwork connection combines with the slab reinforcement to form a composite joint of substantial resistance and stiffness.

A particularly straightforward arrangement arises with "boltless" steelwork connections. At the composite stage all of the tensile resistance is provided by the slab reinforcement, and no bolts act in tension. The balancing compression acts in bearing through plates or shims, inserted between the end of the lower flange of the beam and the face of the column to make up for construction tolerances. Additional means to resist vertical shear must also be provided, for example by a seating for the beam.

The benefits which result from composite joints include reduced beam depths and weight, improved service performance, including control of cracking, and greater robustness. Besides the need for a more advanced calculation method and the placement of reinforcement, the principal disadvantage is the possible need for transverse stiffeners to the column web, placed opposite the lower beam flange. These are required if the compression arising from the action of the joint exceeds the resistance of the unstiffened column web. Alternatively, stiffeners can be replaced by concrete encasement.

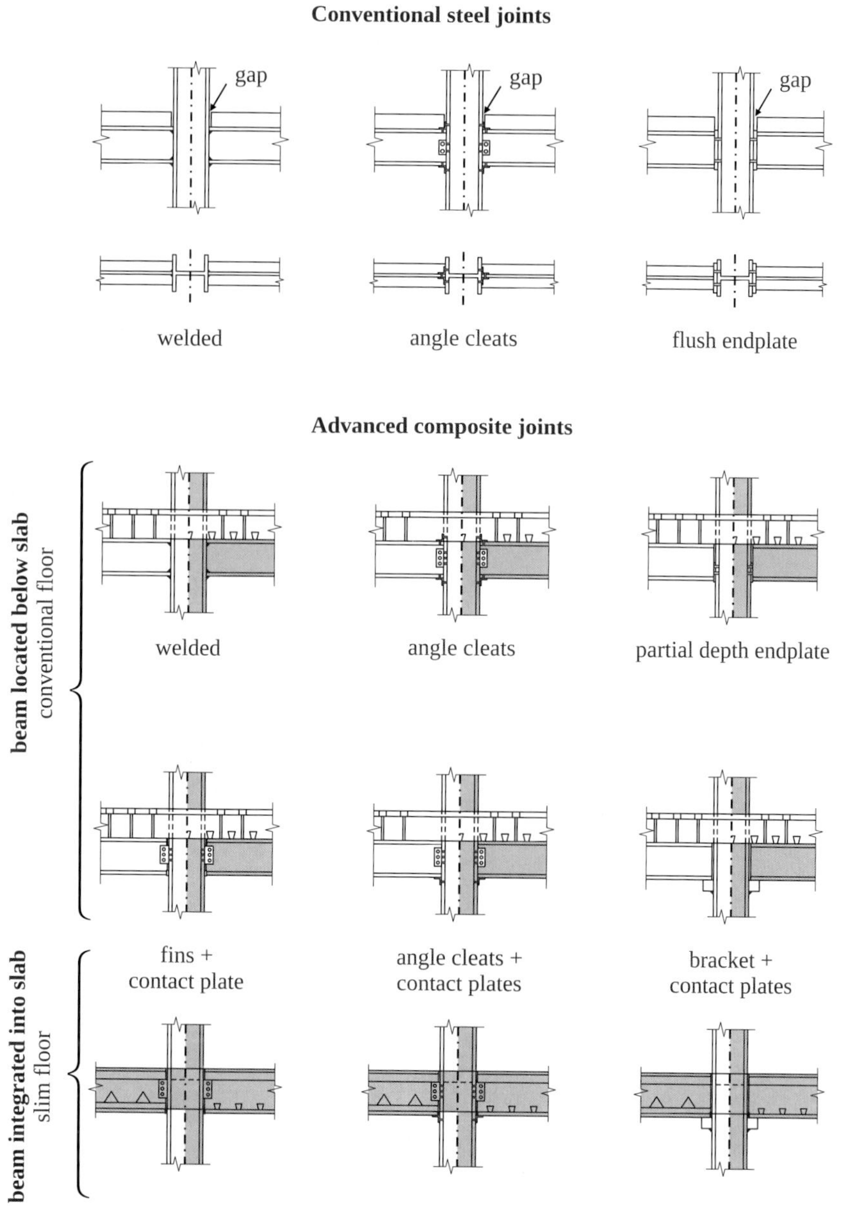

6.9 APPLICATION TO COMPOSITE JOINTS

Conventional steel joints

welded | angle cleats | flush endplate

Advanced composite joints

beam located below slab — conventional floor

welded | angle cleats | partial depth endplate

fins + contact plate | angle cleats + contact plates | bracket + contact plates

beam integrated into slab — slim floor

Figure 6.60 – Types of joints- H- shaped column (Anderson *et al*, 1999)

Composite joints offer to the designer a significant freedom to meet the particular requirements of the structure. This is characteristic of semi-continuous construction and is illustrated by the example shown in Figure 6.61 (Anderson *et al*, 1999). In the discussion which follows, it is assumed

6. MOMENT RESISTANT JOINTS

that the designer has sought to achieve beams of similar depth. It is assumed that for the building of concern, composite connections are not required purely for crack control. The numbers refer to the joint types in the lower part of the figure and are also used to identify particular beams.

All internal beams are to be designed as composite members, but as the edge beams 1-2 and 3-3 carry less load, they may be designed as non-composite. In that case it follows that connections 1, 2 and 3 are also non-composite.

Connections to perimeter columns can be composite, provided that a region of slab exists beyond the column in which to anchor the tension reinforcement or reinforcement is looped around the column. Here it is assumed that connections to the perimeter columns will be bare steel, to avoid any difficulty in achieving this.

Each of connection types 1 and 2 may be either nominally-pinned or moment-resisting, as required to achieve the aim of beams of similar depth. In the interests of overall economy, a stiffened connection should be avoided. For ease of fabrication and erection, connection 3 should be nominally-pinned. In any case, a column web has only very limited resistance under unbalanced moment. Joint 4 is also nominally pinned, because of its one-sided nature.

The remaining connections, 5-7, may be designed compositely. By adjusting the reinforcement, the designer is able to vary at will the structural properties of the joint. The moment resistance and stiffness of the steel connection can also be varied by changing its details. Indeed, joint types 5-7 can be replaced by contact-plate joints, as shown in Figure 6.61. This freedom helps the designer to achieve economically a uniform large floor grid, the beams and their connections would be repeated many times. The resulting reductions in beam sections would therefore be widespread and there would be substantial repetition in fabrication and erection.

Finally, it will be noted that composite connections are used about both axes of the internal column. With a decking say 70 mm deep and a thin overall slab depth (say 120 - 130mm), problems could arise in accommodating the two layers of reinforcement in a limited thickness of slab. In such circumstances one connection could be bare steel. In the chosen example, this should be connection 6; as it connects to the column flange, it could still be designed as moment-resistant. Alternatively, a greater thickness of concrete could be provided by using shallower decking.

6.9 APPLICATION TO COMPOSITE JOINTS

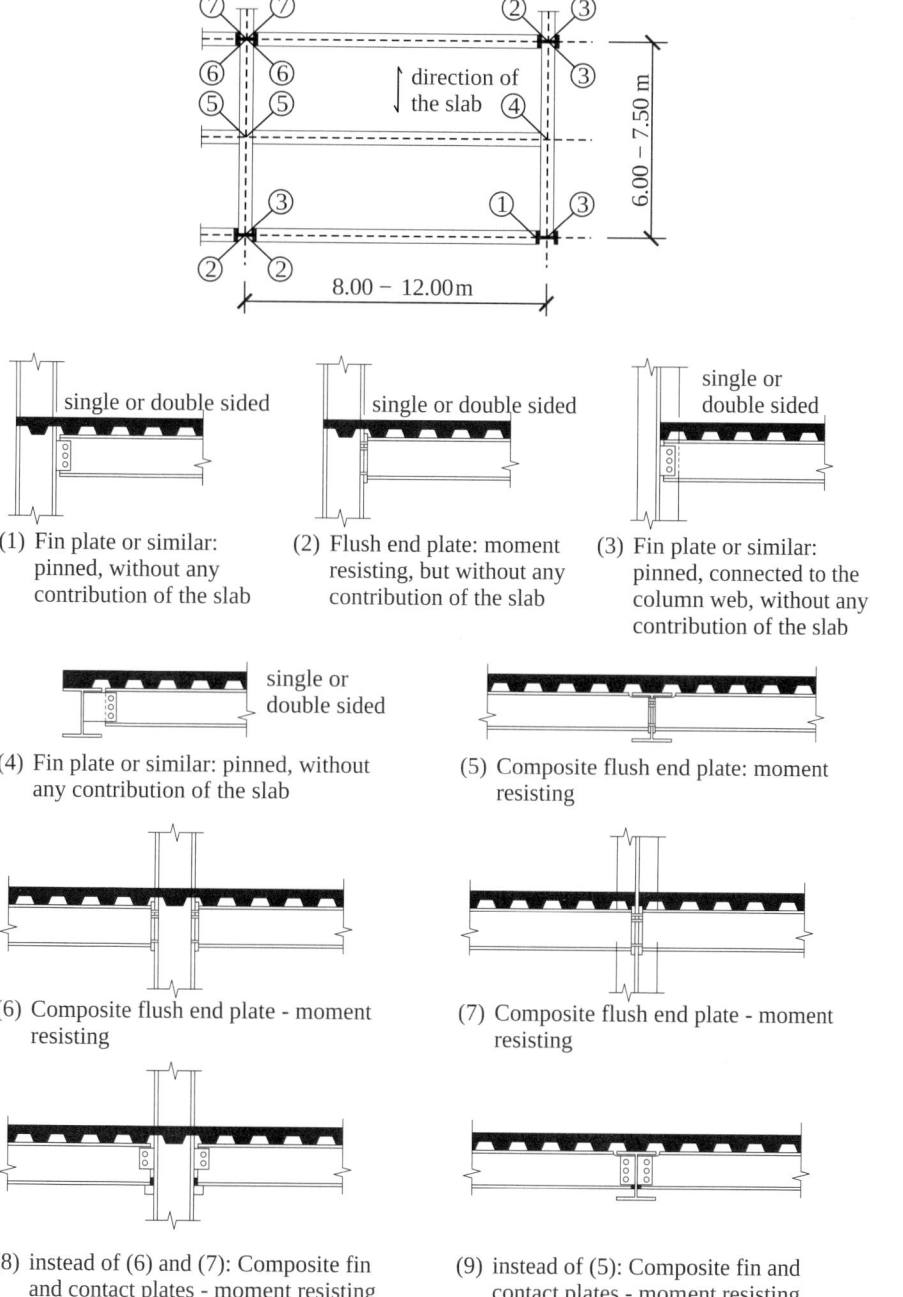

Figure 6.61 – Choice of joint configurations

6.9.2 Design properties

6.9.2.1 Resistance and rotational stiffness

As far as the evaluation of the mechanical properties of composite joints is concerned, no significant difference with the procedure applied for steel joints has to be reported. The component method is used in a quite similar way that what has been explained earlier in the present chapter for steel beam-to-column joints, beam-to-beam joints or beam splices. One of the only specificities lies in the fact that the "longitudinal steel reinforcement in tension" will be treated, in the assembly procedures, as a "bolt-row in tension" with specific resistance and stiffness tension properties and a negligible shear resistance. A second particularity, in composite constructions, is the possible development of slip between the steel beams and the concrete slaps as a result of the existence of longitudinal shear forces at the interface. Unless account is taken of the deformation of the shear connection by a more exact method, the influence of the slip on the stiffness of the joints may be determined through the definition of an additional joint component as explained in Eurocode 4 Part 1-1 section A.3.

6.9.2.2 Rotational ductility

As for other joints, rotational ductility is required from partial-strength composite joints when plastic global analysis is contemplated, what is quite usual in composite construction.

To get axial local deformation in the "steel reinforcement in tension" component, the total cross sectional area of longitudinal reinforcement A_s see 6.2.21) should not exceed the following limit:

$$A_s \leq \frac{1.1 \cdot (0.85 f_{ck} / \gamma_c) b_c d_{eff}}{\beta f_{sk} / \gamma_s} \tag{6.144}$$

where:
- for a solid slab, d_{eff} is the overall depth of the slab;
- for a composite slab, $d_{eff} = h_c$ (see Figure 6.62).

6.9 APPLICATION TO COMPOSITE JOINTS

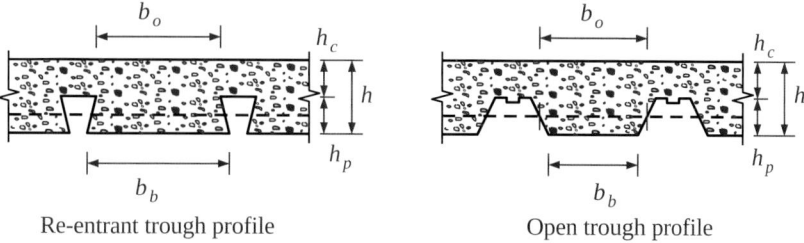

Re-entrant trough profile   Open trough profile

Figure 6.62 – Sheet and slab dimensions

Longitudinal reinforcement for a composite joint shall be positioned so that the distance e_L from the axis of the column web to the centre of gravity of the longitudinal reinforcement placed each side of the column is within the following limits:

$$0.7b_c \leq e_L \leq 2.5b_c \qquad (6.145)$$

where:

b_c is the width of the column's steel section

In addition, transverse slab reinforcement shall be provided adjacent to the column.

Transverse reinforcement shall be positioned so that the distance e_T from the face of the column's steel section to the centre of gravity of the transverse reinforcement placed each side of the column is within the following limits:

$$e_L \leq e_T \leq 1.5\, e_L \qquad (6.146)$$

Sufficient transverse reinforcement should be provided so that the design tensile resistance of the transverse reinforcement placed each side of the column fulfils the condition:

$$\frac{A_T f_{sk,T}}{\gamma_S} \geq \frac{\beta}{2\tan\delta} \cdot \frac{A_s f_{sk}}{\gamma_S} \qquad (6.147)$$

where:

A_T is the area of transverse reinforcement placed each side of the column;

$f_{sk,T}$ is the characteristic tensile strength of that reinforcement,

6. MOMENT RESISTANT JOINTS

and:

$$\tan \delta = 1.35 \left(\frac{e_T}{e_L} - 0.2 \right) \qquad (6.148)$$

All rebars shall be anchored in accordance with EN 1992-1 (CEN, 2004a) so as to develop their design tension resistance.

For a single-sided joint configuration, the longitudinal slab reinforcement in tension shall be anchored sufficiently well beyond the span of the beam to enable the design tension resistance to be developed.

6.9.3 Assembly procedure under M and N

6.9.3.1 Introduction

In the present section, the analytical procedure to predict the *M-N* interaction curve of a composite joint is detailed through a work example (Demonceau, 2008). In this section, only the part of the curve when the composite joint is subjected to tensile loads is derived; a similar procedure could be used to derive the part of the curve when the joint is subjected to a compression load.

6.9.3.2 Studied composite joint configuration

The studied composite joint configuration is given in Figure 6.63 this joint is a double-sided joint symmetrically loaded. The numbering of the rows to be considered within the analytical procedure is shown in Figure 6.64:

- The concrete slab (only activated when subjected to compression stresses) is divided in four "rows": one from the upper fibre of the concrete slab to the first layer of reinforcement (row 1), one from the first layer of reinforcement to the second layer of reinforcement (row 3), one from the second layer of reinforcement to the third layer of reinforcement (row 5) and one from the third layer of reinforcement to the lower fibre of the concrete slab (row 7).
- The slab rebars are numbered as bolt rows.

6.9 APPLICATION TO COMPOSITE JOINTS

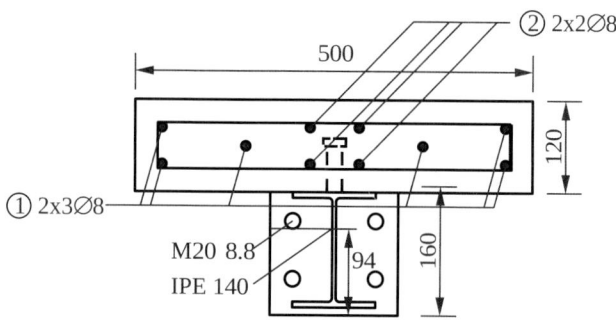

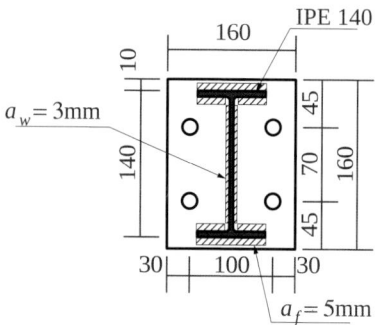

Figure 6.63 – Studied double-sided composite joint configuration (dimension in [mm])

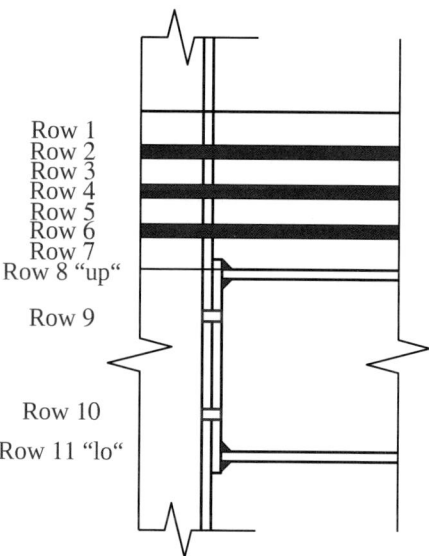

Figure 6.64 – Row numbering

6. MOMENT RESISTANT JOINTS

In order to be able to compare the analytical prediction to experimental test results, the reference point where the axial load and the bending moments are assumed to be applied is the same than the one used to compute the applied bending moment and normal force during the test, i.e. the one presented in Figure 6.65.

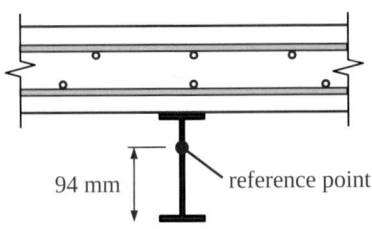

94 mm   reference point

Figure 6.65 – Considered reference point to compute the applied bending moment at the joint

So, having the position of this point, it is possible to compute the lever arm associated to each row; the associated resistance is also reported:

$h_1 \in [127\text{mm};166\text{mm}]$ $\left. \begin{array}{l} F_1^{Rd,+} = -266.4 \cdot z^+ \cdot 24.8 \\ = -6.6 \cdot z^+ \text{kN} \\ F_1^{Rd,-} = -6.6 \cdot z^- \text{kN} \end{array} \right\}$ with z^+ and z^- in mm $z^+, z^- \in [0\text{mm};39\text{mm}]$

$h_2 = 127 \text{ mm}$ $F_2^{Rd,+} = F_2^{Rd,-} = 54.4 \text{ kN}$

$h_3 \in [106\text{mm};127\text{mm}]$ $\left. \begin{array}{l} F_1^{Rd,+} = -6.6 \cdot z^+ \text{kN} \\ F_1^{Rd,-} = -6.6 \cdot z^- \text{kN} \end{array} \right\}$ with z^+ and z^- in mm $z^+, z^- \in [0\text{mm};21\text{mm}]$

$h_4 = 106 \text{ mm}$ $F_4^{Rd,+} = F_4^{Rd,-} = 54.4 \text{ kN}$

$h_5 \in [85\text{mm};106\text{mm}]$ $\left. \begin{array}{l} F_1^{Rd,+} = -6.6 \cdot z^+ \text{kN} \\ F_1^{Rd,-} = -6.6 \cdot z^- \text{kN} \end{array} \right\}$ with z^+ and z^- in mm $z^+, z^- \in [0\text{mm};21\text{mm}]$

$h_6 = 85 \text{ mm}$ $F_6^{Rd,+} = F_6^{Rd,-} = 54.4 \text{ kN}$

$h_7 \in [46\text{mm};85\text{mm}]$ $\left. \begin{array}{l} F_1^{Rd,+} = -6.6 \cdot z^+ \text{kN} \\ F_1^{Rd,-} = -6.6 \cdot z^- \text{kN} \end{array} \right\}$ with z^+ and z^- in mm $z^+, z^- \in [0\text{mm};39\text{mm}]$

$h_8 = 42.6 \text{mm}$ $F_8^{Rd,+} = F_8^{Rd,-} = -284.6 \text{kN}$

$h_9 = 11 \text{mm}$ $F_9^{Rd,+} = 169.4 \text{kN}$ and $F_9^{Rd,-} = 67.3 \text{kN}$

6.9 APPLICATION TO COMPOSITE JOINTS

$h_{10} = -59$ mm $\quad F_{10}^{Rd,+} = 67.3$ kN and $F_{10}^{Rd,-} = 169.4$ kN

$h_{11} = -90.6$ mm $\quad F_{11}^{Rd,+} = F_{11}^{Rd,-} = -284.6$ kN

where z^+ and z^- are defined in Figure 6.66. To compute the resistance of the component "concrete slab in compression", the value of $b_{eff,conn}$ is taken as equal to 266.4 mm (Demonceau, 2008). The resistance of each component has been computed with the material properties measured in laboratory in order to be able to compare to experimental test results.

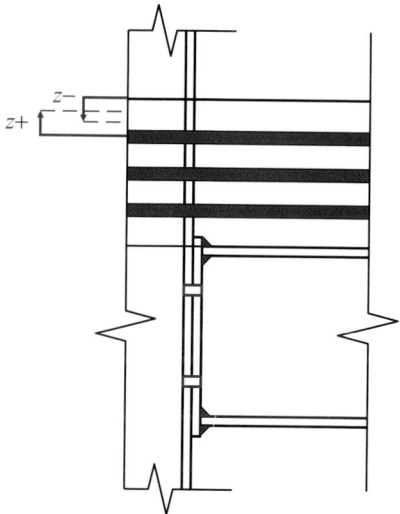

Figure 6.66 – Definition of z^+ and z^-

It is not possible for the component "concrete slab in compression" to obtain one specific value of lever arm as the latter is linked to the height of concrete which is assumed to be in compression. It is the reason why a domain and not one value is given for the lever arms associated to this component. As an example, if the height of concrete subjected to compression in row 1 is equal to 39 mm (i.e. all the row is subjected to compression), the lever arm is equal to 146.5 mm (the resultant force is assumed to be applied at mid-height of the zone subjected to compression).

For the bolt rows (i.e. row 9 and 10), the resistance is associated to a group effect. Indeed, the individual resistance of a bolt row is equal to 169.4 kN while the resistance of the group is equal to 236.7 kN (which is smaller than $2 \times 169.4 = 338.8$ kN). So, it is the reason why, when

6. MOMENT RESISTANT JOINTS

computing the values of $F^{Rd,+}$ (see Figure 6.25), the upper bolt row resistance is taken as equal to 169.4 kN ($= F_9^{Rd,+}$) and the lower bolt row resistance to 236.7 − 169.4 = 67.3 kN ($= F_{10}^{Rd,+}$); when computing the values of $F^{Rd,-}$, it is the opposite, i.e. the upper bolt row resistance is equal to 67.3 kN ($= F_9^{Rd,-}$) and the lower bolt row resistance to 169.4 kN ($= F_{10}^{Rd,-}$). The computation of these values is graphically illustrated in Figure 6.67.

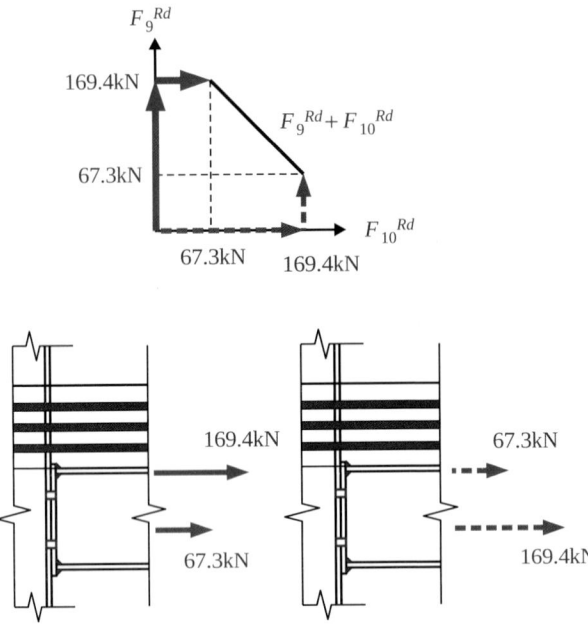

Figure 6.67 – Computation of $F^{Rd,+}$ and $F^{Rd,-}$ (dashed arrow) for the bolt rows

6.9.3.3 Computation of the M-N resistance interaction curve

6.9.3.3.1 Introduction

With the parameters characterised in the previous section, it is now possible to compute the M-N resistance interaction curve.

To compute the resistance to axial loads and the resistance to bending moments, the formulas proposed in 6.3.3.2, Eq. (6.43), are used. The different points of the M-N resistance interaction curve are computed first in 6.9.3.3.2 and 6.9.3.3.3; then, the so-obtained M-N resistance interaction curve is drawn in 6.9.3.3.4 (Figure 6.68).

6.9.3.3.2 Upper rows in tension ($F_i^{Rd,+}$)

a) Point A: all the rows are in tension

$N = \sum_i F_i^{Rd,+} = F_2^{Rd,+} + F_4^{Rd,+} + F_6^{Rd,+} + F_9^{Rd,+} + F_{10}^{Rd,+}$

$= 3 \times 54.4 + 169.4 + 67.3 = 399.9$ kN

$M = \sum_i h_i \cdot F_i^{Rd,+}$

$= (127 + 106 + 85) \cdot 54.4 + 11 \times 169.4 - 59 \times 67.3 = 15.2$ kNm

b) Point B: rows 1 to 10 in tension and row 11 in compression

$N = F_2^{Rd,+} + F_4^{Rd,+} + F_6^{Rd,+} + F_9^{Rd,+} + F_{10}^{Rd,+} + F_{11}^{Rd,+}$

$= 3 \times 54.4 + 169.4 + 67.3 - 284.6 = 115.3$ kN

$M = \sum_i h_i \cdot F_i^{Rd,+}$

$= (127 + 106 + 85) \times 54.4 + 11 \times 169.4 - 59 \times 67.3 - 90.6 \times (-284.6)$

$= 41$ kN

c) Point C: rows 1 to 9 in tension and rows 10 and 11 in compression

$N = F_2^{Rd,+} + F_4^{Rd,+} + F_6^{Rd,+} + F_9^{Rd,+} + F_{11}^{Rd,+}$

$= 3 \times 54.4 + 169.4 - 284.6 = 48$ kN

$M = \sum_i h_i \cdot F_i^{Rd,+}$

$= (127 + 106 + 85) \times 54.4 + 11 \times 169.4 - 90.6 \times (-284.6) = 45$ kNm

d) Point D: rows 1 to 8 in tension and rows 9 to 11 in compression

$N = F_2^{Rd,+} + F_4^{Rd,+} + F_6^{Rd,+} + F_{11}^{Rd,+} = 3 \times 54.4 - 284.6 = -121.4$ kN

$M = \sum_i h_i \cdot F_i^{Rd,+} = (127 + 106 + 85) \times 54.4 - 90.6 \times (-284.6) = 43.1$ kNm

6. MOMENT RESISTANT JOINTS

It can be observed that this point enters in the compression "zone" of the interaction curve. So, the computation for the case "upper rows in tension" is stopped here.

6.9.3.3.3 Lower rows in tension $\left(F_i^{Rd,-}\right)$

a) Point E: all the rows are in tension

$$N = \sum_i F_i^{Rd,+} = F_2^{Rd,-} + F_4^{Rd,-} + F_6^{Rd,-} + F_9^{Rd,-} + F_{10}^{Rd,-}$$

$$= 3 \times 54.4 + 67.3 + 169.4 = 399.9 \text{ kN}$$

$$M = \sum_i h_i \cdot F_i^{Rd,+}$$

$$= (127 + 106 + 85) \times 54.4 + 11 \times 67.3 - 59 \times 169.4 = 8 \text{ kNm}$$

b) Zone F: rows 2 to 11 in tension and row 1 in compression

$$N = F_1^{Rd,-} + F_2^{Rd,-} + F_4^{Rd,-} + F_6^{Rd,-} + F_9^{Rd,-} + F_{10}^{Rd,-}$$

$$= -6.6 \cdot z^- + 3 \times 54.4 + 67.3 + 169.4 = 399.9 - 6.6 \cdot z^- \text{ kN}$$

(with z^- in mm)

$$M = \sum_i h_i \cdot F_i^{Rd,+}$$

$$= \left(166 - \frac{z^-}{2}\right) \cdot \left(-6.6 \cdot z^-\right) + (127 + 106 + 85) \times 54.4 + 11 \cdot 67.3 - 59 \times 169.4$$

It can be observed that the M and N couples for this zone depends of the value of z^- (= height of concrete subjected to compression in row 1) which is between 0 mm and 39 mm. This zone in the M-N resistance interaction curve has been computed through an Excel sheet for different values of z^-. When z^- is equal to 0, the obtained point is point E; when z^- is equal to 39 mm, the obtained point is $N = 142.5$ kN and $M = -29.66$ kNm.

c) Zone G: rows 4 to 11 in tension and row 1 to 3 in compression

$$N = F_1^{Rd,-} + F_3^{Rd,-} + F_4^{Rd,-} + F_6^{Rd,-} + F_9^{Rd,-} + F_{10}^{Rd,-}$$

$$= -6.6 \times 39 - 6.6 \cdot z^- + 2 \times 54.4 + 67.3 + 169.4 = 142.5 - 6.6 \cdot z^- \text{ kN}$$

(with z^- in mm)

$$M = \sum_i h_i \cdot F_i^{Rd,+}$$

$$= 146.5 \times (-257.4) + \left(166 - 39 - \frac{z^-}{2}\right) \times (-6.6 \cdot z^-)$$

$$+ (127 + 106 + 85) \times 54.4 + 11 \times 67.3 - 59 \times 169.4$$

Again, as for the previous zone, it can be observed that the M and N couples for this zone depends of the value of z^- representing the height of concrete subjected to compression in row 3. This zone in the M-N resistance interaction curve has been computed through an Excel sheet for different values of z^-.

6.9.3.3.4 Obtained M-N resistance interaction curves

The computations which have been performed within this section are associated to the elastic strength of the materials. The same procedure has been followed to compute the M-N interaction curve associated to the ultimate strength of the materials. The so-obtained curves are reported in Figure 6.68 (Note: Point D has not been reported in Figure 6.68 as the latter is not in the tensile part of the graph).

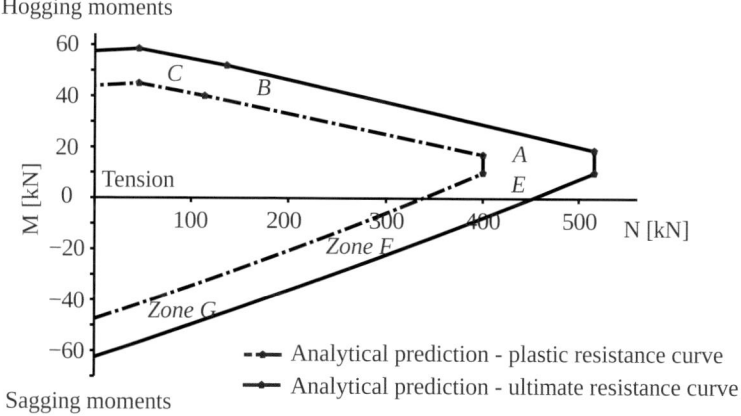

Figure 6.68 – Resistance interaction curves predicted through the proposed procedure (Demonceau, 2008)

The validity of the analytical procedure presented in the previous section was checked through comparisons to results of experimental tests performed in Stuttgart in the framework of an RFCS project (Kuhlmann *et al*, 2009). The interested reader can find more information in (Demonceau, 2008).

Chapter 7

LATTICE GIRDER JOINTS

7.1 GENERAL

EN 1993-1-8 (CEN, 2005c) provides design rules for joints between both open and hollow sections. Design rules for joints between open sections are mainly based on recommendations developed under the umbrella of ECCS (European Convention for Constructional Steelwork), whereas design rules for hollow section joints were published originally as design recommendations of CIDECT (International Committee for Research and Technical Support for Hollow Section Structures). However, these two design methods are based on different approaches: The so-called component method, see 1.6.2, recommended by ECCS is used for the design of joints between open sections and is statistically evaluated against test results, whereas design formulae for hollow section joints are based on semi-empirical investigations in which analytical models were fitted with test results.

Although in many cases the use of hollow sections offers clear advantages, engineers often decide to use open sections instead. There are various reasons for this preference. The design of hollow section joints seems to be complex and only few design tools are available, whereas these are widely available for joints of open sections. Architectural arguments seem to be less relevant at that moment.

As already mentioned in 1.2.7, design rules for joints between tubular hollow sections are based on theoretical mechanical models and they are then fitted through comparisons with results of experimental and numerical investigations. As a consequence, their field of application is often restricted to the domain for which the rules have been validated. One cannot assume that an extrapolation of the design rules outside the given field of application will lead to safe results. But the range of validity given in chapter 7 of EN 1993-1-8 must be strictly followed.

For different failure modes observed in the experimental tests, the design formulae generally give a resistance value for the joint as a whole, for specific loading cases and specific joint configurations. This restricts the field of application and hence the freedom to modify the joint detailing. It has also to be mentioned that usually no information is provided with regard to the stiffness or the ductility of the joints.

This chapter gives a short introduction to the design of hollow section joints. The interested reader will find a more comprehensive description of the background and the design rules in the CIDECT publication of Wardenier *et al* (2010). This book is freely available on the CIDECT website (http://www.cidect.com). Design tools for hollow section joints (Weynand *et al*, 2011), i.e. design tables for some standardized joints and a special edition of the computer program COP (Weynand *et al*, 2014), are also freely available.

In order to facilitate the use of hollow sections, CIDECT has launched an initiative to adapt the current rules for hollow sections to the same design approach as it is used for open sections (Jaspart, Weynand, 2015) and it is expected that the design rules for both open and hollow sections will be presented in a common approach in a next version of Eurocode 3. The more traditional "CIDECT" approach will be transferred into a separate EN standard to be used, at least for a certain period, as an alternative method.

7.2 SCOPE AND FIELD OF APPLICATION

Chapter 7 of EN 1993-1-8 gives detailed application rules to determine the static design resistances of uniplanar and multiplanar joints in lattice structures composed of circular, square or rectangular hollow sections. Types of joints covered are shown in Table 1.1. Also, uniplanar joints in lattice structures composed of combinations of hollow sections with

7.2 SCOPE AND FIELD OF APPLICATION

open sections are covered to a certain extend. Note that flattened end connections and cropped end connections are not covered.

The application rules are valid both for hot finished hollow sections to EN 10210 (CEN, 2006c, 2006d) and for cold formed hollow sections to EN 10219 (CEN, 2006a, 2006b), if the dimensions of the structural hollow sections fulfil the requirements given in chapter 7 of EN 1993-1-8. Even if the denomination of the sections are the same for hot finished hollow sections according to EN 10210 and cold formed hollow sections according to EN 10219, differences can be found with regard to cross section dimensions and the section properties. Due to this fact, a simple substitution of hot finished sections by cold formed sections cannot be done without taking these differences into account. Furthermore, EN 1993-1-8 gives in section 4.14 restricting conditions for rectangular hollow sections (RHS) regarding welding in cold-formed zones. Especially for larger brace to chord width ratios β, these restrictions have to be kept in mind, whereas for hot finished hollow sections no restrictions exist.

As already said in 7.1, the range of validity given in chapter 7 of EN 1993-1-8 must be strictly followed. The application rules may only be used if all the conditions specified in EN 1993-1-8 section 7.1.2 are satisfied:

- The compression elements of the members should satisfy the requirements for Class 1 or Class 2 given in EN 1993-1-1 for the condition of pure compression.
- The angles Θ_i between the chords and the brace members, and between adjacent brace members, should satisfy $\Theta_i \geq 30°$.
- The ends of members that meet at a joint should be prepared in such a way that their cross sectional shape is not modified.
- In gap type joints, the gap g between the brace members should satisfy $g \geq t_1 + t_2$, where t_1 and t_2 are the thicknesses of the brace members. This condition ensure that the clearance is adequate for forming satisfactory welds.
- In overlap type joints, the overlap λ_{ov} should be large enough to ensure that the interconnection of the brace members is sufficient

7. LATTICE GIRDER JOINTS

for adequate shear transfer from one brace to the other. In any case the overlap should be $\lambda_{ov} \geq 25\%$. In case the overlap exceeds a certain limit $\lambda_{ov,lim}$, the connection between the braces and the chord should be checked for shear. Detailed limits are given in EN 1993-1-8 clause 7.1.2(6).

– Where overlapping brace members have different thicknesses and/or different strength grades, the member with the lowest t_i/f_{yi} value should overlap the other member.

– Where overlapping brace members are of different widths, the narrower member should overlap the wider one;

– Application limits are given in Tables 7.1 to Table 7.24 of EN 1993-1-8.

Moments resulting from eccentricities in K, N or KT joints may be neglected in the design of tension chord members and brace members. They may also be neglected in the design of connections if the eccentricities are within the following limits:

$$-0.55 d_0 \leq e \leq 0.25 d_0 \qquad (7.1)$$

$$-0.55 h_0 \leq e \leq 0.25 h_0 \qquad (7.2)$$

where:
- e is the eccentricity;
- d_0 is the diameter of the chord;
- h_0 is the depth of the chord, in the plane of the lattice girder.

Definition of joint eccentricity e, gap g and overlap λ_{ov} are shown in Figure 1.19.

The eccentricities e depend directly on the gap g or overlap λ_{ov}. The eccentricity e of the joint can be determined from its geometry:

$$e = \left(\frac{2h_1}{\sin\Theta_1} + \frac{2h_2}{\sin\Theta_2} + g \right) \cdot \frac{\sin\Theta_1 \sin\Theta_2}{\sin(\Theta_1 + \Theta_2)} - \frac{h_0}{2} \qquad (7.3)$$

where:
- h_i is the section depth of member i ($i = 0$, 1 or 2);

Θ_i is the angle of brace members i ($i = 1$ or 2).

Note: In an overlap joint, the value g will be negative.

Hence, the given limitations concerning minimum gap on one hand and minimum overlap on the other hand may lead to the situation that moments resulting from eccentricities may not be neglected. The designer should have in mind this fact when selecting section sizes of the chord and brace members.

7.3 DESIGN MODELS

7.3.1 General

Lattice structures are often composed of combinations of circular, square or rectangular hollow sections. The design resistance of joints between the braces and chords are expressed in terms of design axial and/or moment resistances of the brace members, or, in some cases, in terms of design axial resistances of the chords.

The complex geometry of the joints, local influences of the corners of rectangular sections and residual stresses, for instance due to welding, lead to non-uniform stress distributions. Strain hardening and membrane effects are also influencing the local structural behaviour. Due to the complexity of the parameters which influence the resistance of the joints, semi-empirical approaches were used to develop design models.

Simplified analytical models which consider the most relevant parameters were developed. Based on extensive research activities, these models were calibrated with parameters based on results of experimental and numerical investigations. The design formulae are based on statistical evaluations against test results. This approach directly led to design values and not to characteristic values. Consequently, the partial safety factor γ_{M5} of EN 1993-1-8 is recommended as $\gamma_{M5} = 1.0$. The design equations are only valid within the investigated parameter ranges. Those application limits are given in EN 1993-1-8, Tables 7.1, 7.8, 7.20 and 7.23 and more specific limits given in the various design resistance tables (for example: Tables 7.3 and 7.4 give additional limits to Table 7.1).

7.3.2 Failure modes

The design resistance of a joint depends on the resistance of the weakest part of the joint. Eurocode 3 distinguishes between six different failure modes for joints in lattice structures:

- *Chord face failure* (CFF), also referred to as *chord plastification*

    Failure due to plastification of the chord face or the whole chord cross section.

- *Chord web failure* (CWF), also referred to as *chord side wall failure*

    Failure due to yielding, crushing or instability (crippling or buckling of the chord side wall or chord web) under the compression brace member.

- *Chord shear failure* (CSF)

    Failure due to shear forces in chord side wall.

- *Punching shear failure* (PSF)

    Chord failure in the flange wall.

- *Brace failure* (BF)

    Failure due to cracking in the welds or in the brace members with reduced effective width.

- *Local buckling failure*

    Failure due to instability of the chord or brace members at the joint location. However, this failure is not governing for joints within the scope of application according to tables 7.1 and 7.8 of EN 1993-1-8.

For each failure mode, Eurocode 3 provides appropriate design resistance formulae. To determine the design resistance of a joint, all relevant failure modes must be checked and the corresponding resistances should be determined. The minimum resistance is taken as the design resistance of a joint.

7.3.3 Models for CHS chords

The design formulae for CHS chord failure modes are based on the so-called *Ring Model* (Togo, 1967). This model assumes that the brace member stresses (axial stresses) into the chord section are transferred mainly at the saddle of the brace members (see point A in Figure 7.1). This loading is taken into account by a pair of concentrated forces acting vertically to the flange, each $0.5 \cdot N_1 \cdot sin\Theta_1$. In longitudinal direction, it is assumed that the loads are transferred within an effective length L_e (see Figure 7.1).

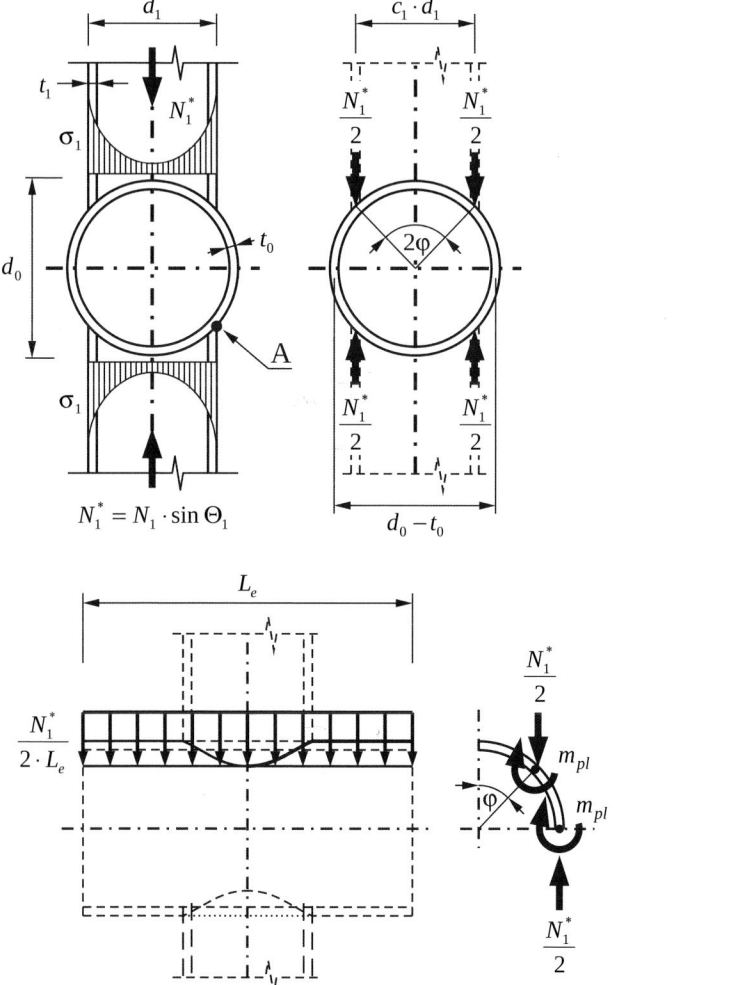

Figure 7.1 – Ring model for chord plastification under axial brace loading (Togo, 1967)

7. Lattice Girder Joints

The design formulae are derived assuming that yielding will occur in the chord section. Based on equilibrium of internal and external work, adequate equations can be derived (Togo, 1967). These basic formulae were modified and calibrated based on experimental or numerical investigations (Wardenier, 2001) to cover the various failure modes for relevant practical configurations and loading situations, see Table 7.2 and Table 7.5 of EN 1993-1-8.

7.3.4 Model for RHS chords

As mentioned in section 7.3.2 different failure modes need to be checked to determine the design resistance of a joint. For joints between braces and the flange of rectangular hollow sections, the following failure modes may be relevant: *chord face failure* (CFF), *chord web failure* (CWF), *chord shear failure* (CSF) and *punching shear failure* (PSF). The basic models for the design formulae to check those failure modes are briefly described hereafter. More details can be found in (Wardenier, 1982; Wardenier *et al*, 1991; Packer *et al*, 1992; Packer, Henderson, 1997; Wardenier, 2001).

The basic model to check the *chord face failure* (CFF) is the yield line model for T, Y and X joints under axial load, which is shown in Figure 7.2. By "equalizing" the externally applied virtual work and the internal dissipated energy, the lowest load for flange plastification can be calculated. The design values for the resistances for chord face failure given in EN 1993-1-8, Table 7.11 for T, X and Y joints are directly derived from this model. The design models for in-plane bending (IPB) and out-of-plane bending (OPB) are deduced from that model as well. As the analytical model for N and K joints result in very complex equations, a semi-empirical approach, also based on a yield line model, has been used to develop and validate the design equations in Eurocode 3 (see EN 1993-1-8, Table 7.12).

7.3 DESIGN MODELS

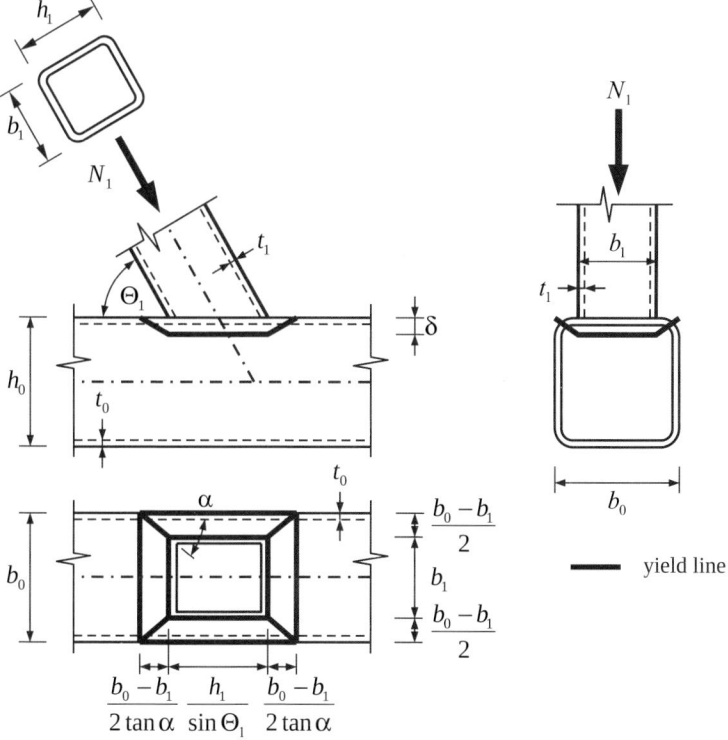

Figure 7.2 – Yield line model for T, Y and X joints

The model for *chord web failure* (CWF) of T, X and Y joints with RHS chords is based on the theory of linear elastic buckling of an isolated member (Euler case 2). The width of the isolated member is determined by an assumed load distribution into the web of 1:2.5 through the wall thickness. The buckling length is equal to the height of the web. To calculate the buckling strength f_b buckling curve *a* is used. The appropriate design equation is given in EN 1993-1-8, Table 7.11.

In N and K joints with RHS chords, the joint may fail due to high shear in the chord section, see Figure 7.3. This failure is called *chord shear failure* (CSF). In the design equations of EN 1993-1-8, Table 7.12, the influence of the gap size is considered by a factor in the determination of the shear area A_v.

7. LATTICE GIRDER JOINTS

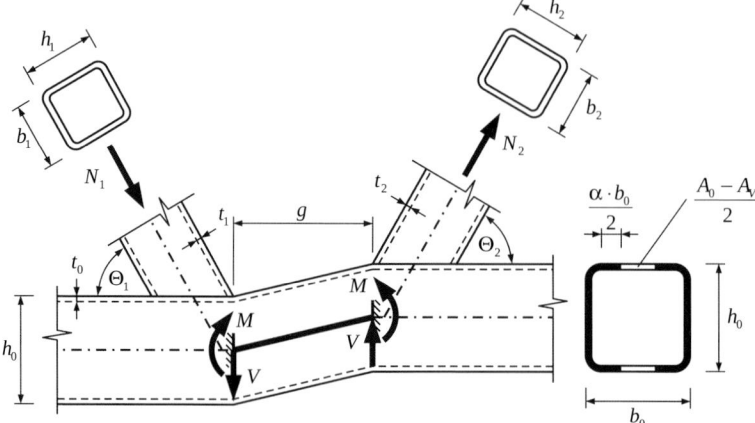

Figure 7.3 – Analytical model for chord shear failure

7.3.5 Punching shear failure

Punching shear failure (PSF) is also a chord face failure, where the flange of the chord can fail in shear due to the brace loaded in tension, compression or bending. The appropriate formulae in EN 1993-1-8 consider the normal component of the braces $N_i \cdot \sin\Theta_i$ and, if necessary, the unequal stiffness distribution along the connection between the brace and the chord flange by the effective width $b_{e,p}$, see Figure 7.4.

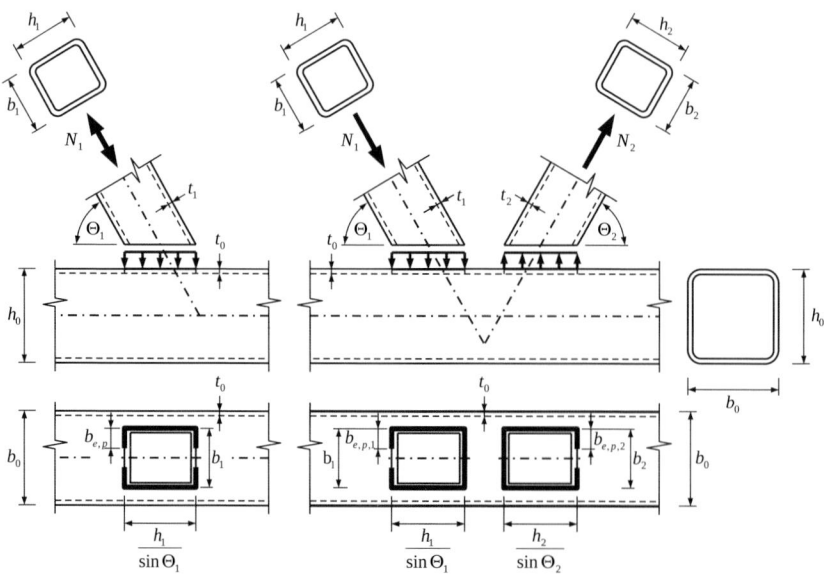

Figure 7.4 – Effective width for punching shear failure

7.3.6 Model for brace failure

Due to the unequal stiffness distribution the brace can fail by fracture close to the weld, local yielding or buckling. Like in the equations for *punching shear failure*, also for *brace failure* (BF), the effects of the unequal stiffness distribution are taken into account by an effective width b_{eff} in the appropriate formulae of EN 1993-1-8 (Tables 7.10, 7.11, 7.12 and 7.13, see Figure 7.5).

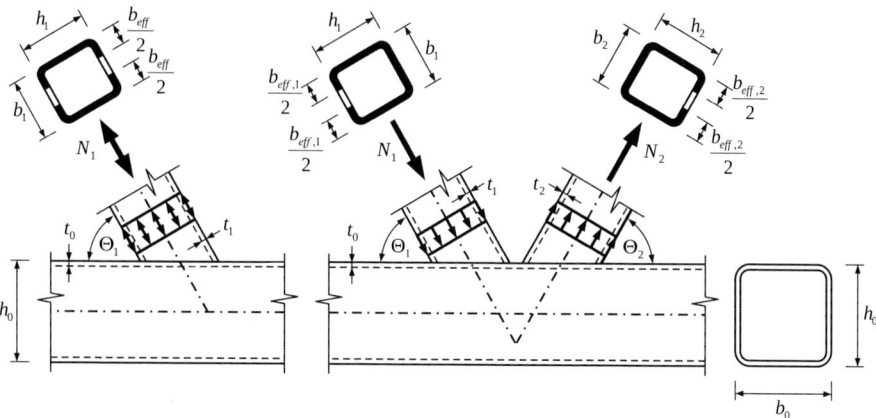

Figure 7.5 – Effective width for brace failure

7.3.7 *M-N interaction*

The design resistances provided in the aforementioned tables of EN 1993-1-8 are resistances of a joint expressed in terms of design axial resistance or design moment resistance of the brace members. When brace member connections are subjected to combined bending and axial force, the following condition should be satisfied.

For welded joints between CHS members:

$$\frac{N_{i,Ed}}{N_{i,Rd}} + \left(\frac{M_{ip,i,Ed}}{M_{ip,i,Rd}}\right)^2 + \frac{|M_{op,i,Ed}|}{M_{op,i,Rd}} \leq 1.0 \qquad (7.4)$$

For welded joints between CHS or RHS brace members and RHS chord members:

7. LATTICE GIRDER JOINTS

$$\frac{N_{i,Ed}}{N_{i,Rd}} + \frac{M_{ip,i,Ed}}{M_{ip,i,Rd}} + \frac{M_{op,i,Ed}}{M_{op,i,Rd}} \leq 1.0 \tag{7.5}$$

where:

$N_{i,Ed}$ is the design value of the internal axial force in member i ($i = 0, 1, 2$ or 3);

$N_{i,Rd}$ is the design value of the resistance of the joint, expressed in terms of the internal axial force in member i;

$M_{ip,i,Ed}$ is the design value of the in-plane internal moment in member i;

$M_{ip,i,Rd}$ is the design value of the resistance of the joint, expressed in terms of the in-plane internal moment in member i;

$M_{op,i,Ed}$ is the design value of the out-of-plane internal moment in member i;

$M_{op,i,Rd}$ is the design value of the resistance of the joint, expressed in terms of the out-of-plane internal moment in member i.

Chapter 8

JOINTS UNDER VARIOUS LOADING SITUATIONS

8.1 INTRODUCTION

In the Eurocodes (CEN, 2005c, 2004b), the component method, see section 1.6.2, is used as a reference for the design of joints in steel and composite structures. Its use enables a wide range of application as far as the individual response of the constitutive components of the studied joint are known and the so-called "assembly procedure of the components" is available. Nowadays the acquired knowledge allows covering a large set of joint configurations where the joints are subjected to bending mainly, as it is the case in the Eurocodes. In the present chapter, a review is made of recent developments making the application of the component method possible to various loading situations, including fire, earthquake, impact or explosion.

In EN 1993-1-8 (CEN, 2005c) and EN 1994-1-1 (CEN, 2004b), guidelines on how to apply the component method for the evaluation of the initial stiffness and the design moment resistance of steel and composite joints are provided. The aspects of ductility are also addressed. The combination of the components proposed in (CEN, 2005c) and (CEN, 2004b) allows one to cover a wide range of joint configurations and should be largely sufficient to satisfy the needs of practitioners (welded joints, bolted joints with end-plates or cleats, various joint stiffening including transverse column stiffeners, supplementary web plates, backing plates, column web plates and beam haunches). However, assembly design rules are provided for joints under static loading and mainly subjected to bending

8. JOINTS UNDER VARIOUS LOADING SITUATIONS

moments and shear forces, but also for joints connecting profiles with open sections. The application of the component method to joints subjected to axial forces or combinations of axial forces and bending moments has been additionally addressed in sections 6.3.3, 6.3.4 and 6.4.2.

The extension of the application field of the component method may follow separate ways. Amongst them: (i) increase the number of components for which design rules are provided to the user and (ii) derive knowledge for the characterisation of the component and the assembly of components in other loading conditions. Through (i), the field is extended to other joint configurations. In (Weynand et al, 2015) the authors have gathered relevant information for many different components for which design rules have been recently developed. In the same reference, the practical application of the component method to tubular joints is also covered, including component rules, design recommendations and works examples.

In the following sections of the paper, it is intended to investigate further the second way (ii) and to present recent works in this domain.

8.2 COMPOSITE JOINTS UNDER SAGGING MOMENT

The mechanical characterisation of composite joints subjected to hogging bending moment may be achieved by means of Eurocode 4 (CEN, 2004b). However, insufficient information is provided there to predict the behaviour of under sagging moments (Figure 8.1).

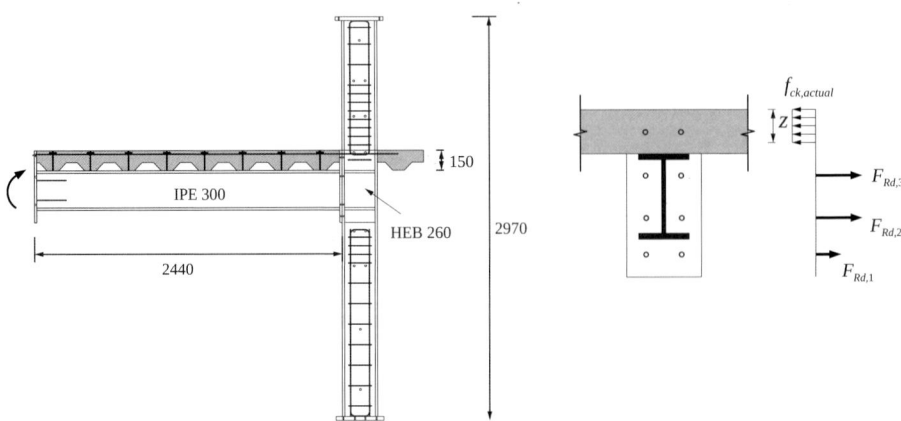

Figure 8.1 – Composite joint subjected to sagging moment

Indeed, even if most of the activated components under such a loading can be characterised using (CEN, 2005c) and (CEN, 2004b), no rule is available to characterise one of the activated components: the concrete slab in compression in the vicinity of the column, i.e. where contact forces are transferred.

In recent researches, methods to characterise this component in terms of resistance and stiffness have been proposed (Ferrario, 2004; Liew *et al*, 2004; Demonceau, 2008; Demonceau *et al*, 2008).They aim at defining a rectangular cross section of concrete participating to the joint resistance. In (Demonceau, 2008) and (Demonceau *et al*, 2008), the second author suggests to combine two methods proposed respectively by Ferrario (2004) and Liew *et al* (2004), the combination of these two methods reflecting in a more appropriate way how the concrete resists to the applied load in the vicinity of the joint.

So, through the study of one single new component (Demonceau, 2008; Demonceau *et al*, 2008), it is possible to characterise a significant number of composite joint configurations under a new specific loading (sagging moments). This demonstrates the flexibility and the adaptability of the component method.

8.3 JOINTS IN FIRE

In (Simões da Silva *et al*, 2011), Da Silva *et al* have first proposed to refer also to the component method to characterise the behaviour of steel joints at high temperature. In their study, they have demonstrated that the rotational stiffness and the bending resistance of a structural joint may be simply obtained by multiplying the corresponding properties derived at room temperature by ad-hoc reduction factors (respectively $k_{E;\theta}$ for stiffness and $k_{y;\theta}$ for resistance) evaluated according to Eurocode 3 Part 1-2 (CEN, 2005b). These factors express the decrease of the steel Young modulus E and yield strength f_y at temperature θ. However this simple approach is limited to cases where all the components are subjected to the same increase of temperature, what is not often reflecting the reality. This is why in (Demonceau *et al*, 2013) the procedure has been improved as follows: (i) in the tests used as references, temperatures θ_i have been measured in all components i, (ii) the stiffness and resistance properties of the components i have been multiplied by ad-hoc reduction factors $k_{E;\theta_i}$ and $k_{y;\theta_i}$ and (iii) the assembly of the components has been finally achieves so as to characterise the global response of the joint, duly attention being so paid to the variation of temperatures in the joint. The procedure has been validated through comparisons with test results on composite joints. So as to allow an easy application in practice, Demonceau et al (2009) have decided to go one step further by replacing the measurement of actual temperatures in the joint during laboratory tests by a thermal analysis achieved with SAFIR (Franssen, 2005). In some cases, he has even suggested analytical expressions to determine the temperature of individual components as a function of the time.

8.4 JOINTS UNDER CYCLIC LOADING

In Eurocode 8, Part 1-1 (CEN, 2004c), and in particular in chapters 6 and 7 dealing respectively with the seismic design of steel and composite structures, it is clearly stated that the use of partial strength joints is permitted but the number of requirements to be respected for this joint typology are such that it is nowadays nearly compulsory to perform experimental tests to check when these requirements are fulfilled; this fact is confirmed in Eurocode 8, Part 1-1, in the clause (6) of Section 6.5.5.

Accordingly, the use of partial strength joints in structures prone to seismic actions is very limited and, as a consequence, the practitioners have to design full strength joints taking into account the possible overstrength effects, which leads to expensive joint solutions and so limits the competiveness of steel structures compared to other structural solutions. An example of such optimised full strength joint solution is presented in Figure 8.2 (Hoang et al, 2014); this solution was developed in the framework of a recent RFCS project entitled HSS-SERF (High Strength Steel in SEismic Resistant building Frames) (Dubina, 2015).

The component method could be a solution to overcome this "full-strength" obligation. Indeed, the component method has the potential to predict the response of joints under cyclic loading but for that, it is required to know the behaviour of each component under such loading conditions. In particular, it is necessary to know the post-yielding behaviour of the components accounting for the strain-hardening effects, their ultimate resistance, their deformations capacity but also the degradation of their strength and stiffness under the applied cycles associated to phenomena of oligo-cyclic fatigue. In addition, it is needed to know the unloading behaviour of each component as such unloading may occur during the cyclic loading imposed by the seismic action.

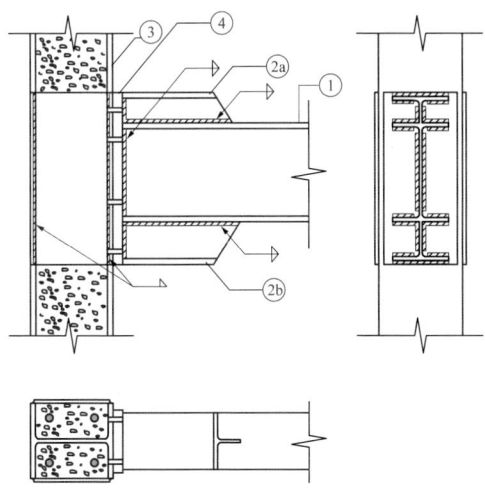

Elements		Steel materials
1	I-steel beam	Mill steel
2a, 2b	Top and bottom hammer-head	Extracted from the beam profiles
3	Partially-encased wide-flange	High strength steel may be used
4	End-plate	Mill steel
5	Bolts	High strength bolts (8.8 or 10.9)
6	Lateral plates	Same grade with the column profiles

Figure 8.2 – Full strength optimised beam-to-column joint solution

Investigations have been recently conducted at the University of Coimbra (Nogueiro *et al*, 2009) with very promising results. The proposed model consists in a numerical implementation of a hysteretic model able to simulate a generic steel or steel-composite joint behaviour. The use of a numerical implementation is here required as the evolution of the loads in each component at each step of the applied loading has to be known in order to be able to detect the strength/stiffness degradation and the moment at which the maximum deformation capacity of a component is reached. The proposed model in (Nogueiro et al, 2009) is still under development/improvement nowadays, in particular through contributions to the European RFCS project EQUALJOINTS (European pre-QUALified steel JOINTS).

8.5 JOINTS UNDER EXCEPTIONAL EVENTS

A structure should be designed to behave properly under service loads (at SLS) and to resist design factored loads (at ULS). The type and the intensity of the loads to be considered in the design process may depend on different factors such as: the intended use of the structure (type of variable loads…), the location (wind action, level of seismic risk…) and even the risk of accidental loading (explosion, impact, flood…). In practice, these individual loads are combined so as to finally derive the relevant load combination cases. In this process, the risk of an exceptional (and therefore totally unexpected) event leading to other accidental loads than those already taken into consideration in the design process in itself is not at all covered. This is a quite critical situation in which the structural integrity should be ensured, i.e. the global structure should remain globally stable even if one part of it is destroyed by the exceptional event (explosion, impact, fire as a consequence of an earthquake …). In conclusion, structural integrity is required when the structure is subjected to exceptional actions not explicitly considered in the definition of the design loads and load combination cases.

Under such exceptional actions, the structural elements and in particular the joints are generally subjected to loadings not initially foreseen through the ULS design. For instance, if the exceptional event "loss of a column" is considered, the joints will experience high tying forces after the loss of a column, as a result of the development of membrane forces in the beams located just above the damaged or destroyed column while these joints are initially designed to transfer shear forces and hogging bending moments. Moreover a reversal of moments occurs in the joints located just above the damaged column. Finally, the joints could be subjected to some dynamic effects if the column loss is for instance induced by an impact or an explosion.

In section 6.3.3, it has been shown how the $M-N$ plastic resistant curve of a joint can be predicted through the use of the component method.

Of course, the methodology presented there can be of help to predict the behaviour of the joints when subjected to exceptional events. However, it has to be pointed out that, when considering the behaviour of structures subjected to such event, the main objective is to ensure that the building will remain globally stable and so it can be accepted to go a step further in the resistance of the structural elements in comparison to what is imposed as limits for ULS. Accordingly, in addition to the prediction of the plastic resistance of the joints under $M-N$, it is also important to be able to predict the ultimate resistant curve, i.e. to be able to predict the $M-N$ combinations under which the joints fail. In (Jaspart, Demonceau, 2008a), it is explained how the $M-N$ plastic and ultimate resistance curves can be predicted for a composite joints and how the proposed model has been validated through comparisons to experimental results.

In parallel to the prediction of the resistance curves, another key issue to be considered when predicting/estimating the robustness of a structure is the prediction of the deformation capacity or the ductility of a joint. Research efforts are still required in this field. Finally, another aspect to be dealt with when considering the behaviour of joints under exceptional events is the possible dynamic effects which can be induced by these events, dynamic effects which can be associated to strain rate effects in some joint components. This aspect is presently under investigation at the University of Liège in the framework of a RFCS European project entitled ROBUSTIMPACT.

Chapter 9

DESIGN STRATEGIES

9.1 DESIGN OPPORTUNITIES FOR OPTIMISATION OF JOINTS AND FRAMES

9.1.1 Introduction

This chapter describes the various approaches which can be used to design steel frames with due attention being paid to the behaviour of the joints. In practice, this design activity is normally performed by one or two parties, according to one of the following ways:

- An engineering office (in short *engineer*) and a steel fabricator (in short *fabricator*), referred as *Case A*;
- An engineering office (*engineer*) alone, referred as *Case B1*;
- A steel fabricator (*fabricator*) alone, referred as *Case B2*.

At the end of this design phase, fabrication by the steel fabricator takes place.

The share of responsibilities for design and fabrication respectively is given in Table 9.1 for these three cases.

Table 9.1 – Parties and their roles in the design/fabrication process of a steel structure

Role	Case A	Case B1	Case B2
Design of members	Engineer	Engineer	Fabricator
Design of joints	Fabricator	Engineer	Fabricator
Fabrication	Fabricator	Fabricator	Fabricator

9. DESIGN STRATEGIES

The design process is ideally aimed at ascertaining that a given structure fulfils architectural requirements, on the one hand, and is safe, serviceable and durable for a minimum of global cost, on the other hand. The parties involved in the design activities also care about the cost of these latter, with a view to optimising their respective profits.

In *Case A*, the engineer designs the members while the steel fabricator designs the joints. It is up to the engineer to specify the mechanical requirements to be fulfilled by the joints. The fabricator has then to design the joints accordingly, keeping in mind the manufacturing aspects also. Due to the disparity in the respective involvements of both parties, the constructional solution adopted by the fabricator for the joints may reveal to be sub-optimal; indeed it is dependent on the beam and column sizing that is made previously by the engineer. The latter may for instance aim at minimum shape sizes, with the consequence that the joints then need stiffeners in order to achieve safety and serviceability requirements. If he chooses larger shapes, then joints may prove to be less elaborated and result in a better economy of the structure as a whole (Figure 9.1).

In *Case B1*, the engineer designs both the members and the joints. He is thus able to account for mechanical joint properties when designing members. He can search for global cost optimisation too. It may happen however that the engineer has only a limited knowledge of the manufacturing requisites (machinery used, available materials, bolt grades and spacing, accessibility for welding, …); then this approach may contribute to some increase in the fabrication costs.

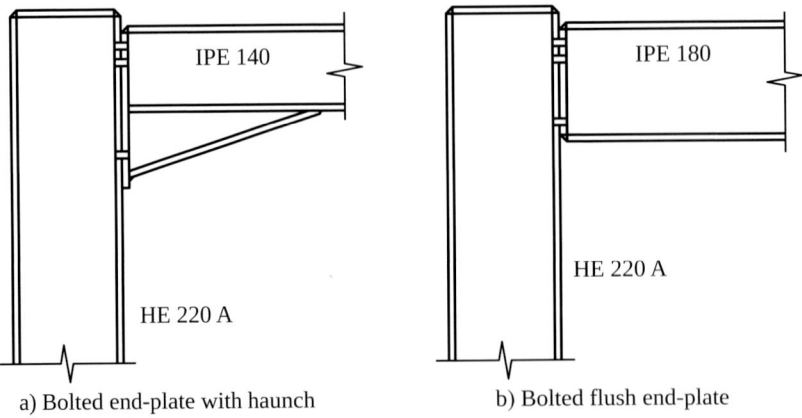

a) Bolted end-plate with haunch b) Bolted flush end-plate

Figure 9.1 – Two solutions: different economy

9.1 DESIGN OPPORTUNITIES FOR OPTIMISATION OF JOINTS AND FRAMES

Case B2 is ideal with regard to global economy. Indeed the design of both members and joints are in the hands of the fabricator who is presumably well aware of all the manufacturing aspects.

Before commenting on these various approaches, it is necessary to introduce some wording regarding joints.

A joint is termed *simple, semi-continuous* or *continuous*. This wording is general. It is concerned with resistance, with stiffness or with both. Being a novelty for most readers, some detailed explanations are given in sub-chapter 2.2 on joints. In two circumstances only, that are related to the methods of global frame analysis, this wording leads to more commonly used terms:

1. In an elastic global frame analysis, only the stiffness of the joints is involved. Then, a simple joint is a *pinned* joint, a continuous joint is a *rigid* joint while a semi-continuous joint is a *semi-rigid* joint;

2. In a rigid-plastic analysis, only the resistance of the joints is involved. Then, a simple joint is a *pinned* joint, a continuous joint is a *full-strength* joint while a semi-continuous joint is *a partial-strength* joint.

The various cases described above are commented on in the present chapter. For the sake of simplicity, it is assumed that global frame analysis is conducted based on an elastic method of analysis. This assumption is however not at all a restriction; should another kind of analysis be performed, similar conclusions would indeed be drawn.

For the design of steel frames, the designer can follow one of the following design approaches:

- *Traditional design approach:*

    The joints are presumably either simple or continuous. The members are designed first; then the joints are. Such an approach may be used in any Case A, B1 or B2; it is of common practice in almost all the European countries.

- *Consistent design approach:*

    Both member and joint properties are accounted for when starting the global frame analysis. This approach is normally used in Cases B1 and B2, and possibly in Case A.

- *Intermediate design approach:*

    Members and joints are preferably designed by a single party (Case B1 or B2).

9. DESIGN STRATEGIES

9.1.2 Traditional design approach

In the traditional design approach, any joint is assumed to be either simple or continuous. A simple joint is capable of transmitting the internal forces got from the global frame analysis but does not develop a significant moment resistance which might affect adversely the beam and/or column structural behaviour. A continuous joint exhibits only a limited relative rotation between the members connected as long as the applied bending moment does not exceed the bending resistance of the joint.

The assumption of simple and/or continuous joints results in the share of design activities into two more or less independent tasks, with limited data flow in between. The traditional design approach of any steel frame consists of eight steps (Figure 9.2).

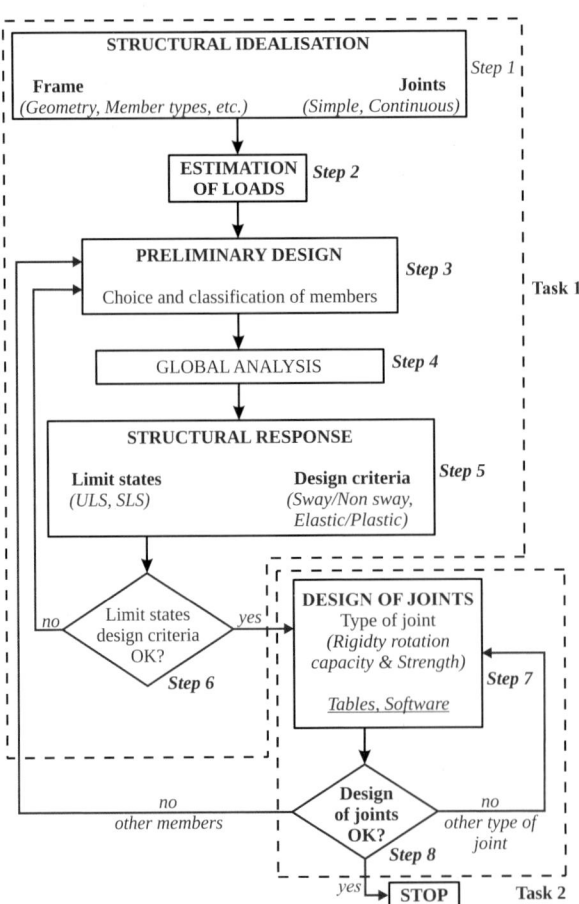

Figure 9.2 – Traditional design approach (simple/continuous joints)

9.1 DESIGN OPPORTUNITIES FOR OPTIMISATION OF JOINTS AND FRAMES

Step 1: The structural idealisation is a conversion of the real properties of the frame into the properties required for frame analysis. Beams and columns are normally modelled as bars. Dependent on the type of frame analysis which will be applied, properties need to be assigned to these bars. For example, if an elastic analysis is used, only the stiffness properties of the members are relevant. The joints are pinned or rigid and are modelled accordingly (Figure 9.3).

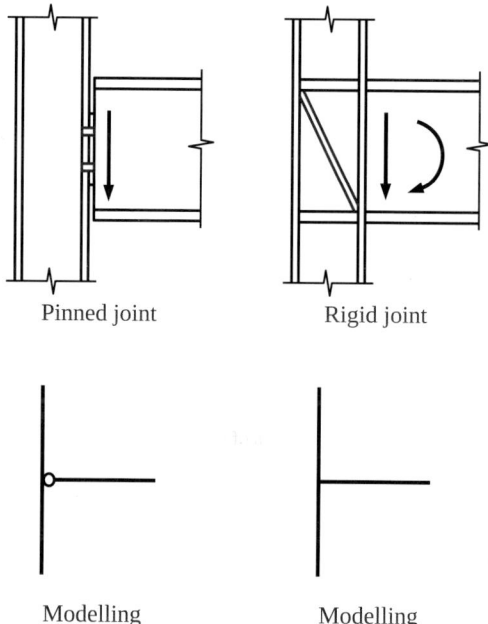

Figure 9.3 – Modelling of pinned and rigid joints (elastic global analysis)

Step 2: Loads are determined based on relevant standards.

Step 3: The designer normally performs the preliminary design - termed pre-design in short - of beams and columns by taking advantage of his own design experience from previous projects. Should he have a limited experience only, then simple design rules can help for a rough sizing of the members. In the pre-design, an assumption shall be made concerning the stress distribution within the sections (elastic, plastic), on the one hand, and,

9. Design Strategies

possibly, on the allowance for plastic redistribution between sections, on the other hand. Therefore classes need to be assumed for the structural shapes composing the frame; the validity of this assumption shall be verified later in Step 5.

Step 4: The input for global frame analysis is dependent on the type of analysis. In an elastic analysis, the input is the geometry of the frame, the loads and the flexural stiffness of the members. In a rigid-plastic analysis, the input is the geometry of the frame, the loads and the resistance of the members. An elastic-plastic analysis requires both resistance and flexural stiffness of the members. Whatever the type of global frame analysis, the distribution and the magnitude of the internal forces and displacements are the output (however rigid-plastic analysis does not allow for any information regarding displacements).

Step 5: Limit state verifications consist normally in checking the displacements of the frame and of the members under service loading conditions (Serviceability Limit States, in short *SLS*), the resistance of the member sections (Ultimate Limit States, in short *ULS*), as well as the frame and member stability (*ULS*). The assumptions made regarding the section classes (see Step 3) are checked also.

Step 6: The adjustment of member sizing is to be carried out when the limit state verifications fail, or when undue under-loading occurs in a part of the structure. Member sizing is adjusted by choosing larger shapes in the first case, smaller ones in the second case. Normally the designer's experience and know-how form the basis for the decisions made in this respect.

Steps 7/8: The member sizes and the magnitude of the internal forces that are experienced by the joints are the starting point for the design of joints. The purpose of any joint design task is to find a conception which allows for a safe and sound transmission of the internal forces between the connected members. Additionally, when a simple joint is adopted, the fabricator shall verify that no significant bending moment develops in the joint. For a continuous joint, the fabricator shall check whether the joint satisfies the assumptions made in Step 1 (for instance, whether the joint stiffness is sufficiently large when an elastic global frame analysis is performed). In addition, the rotation capacity shall be appropriate when necessary.

The determination of the mechanical properties of a joint is called *joint characterisation* (see sub-chapter 1.6). To check whether a joint may be considered as simple or continuous, reference will be made to *joint classification* (see sub-chapter 2.4).

Rules for joint characterisation in compliance with Eurocode 3 are available (see chapter 3 to chapter 6), but design tools can be very helpful during the design process because they enlighten drastically the design tasks. In many cases the designer may select the joints out of tables which provide the strength and stiffness properties of the relevant joints, as well as, when necessary, the rotation capacity. Dedicated software may be an alternative to the latter. They require the whole layout of the joint as input and provide strength, stiffness and rotational capacity as output. Computer based design proceeds interactively by trial and error. For instance, the designer first tries a simple solution and improves it by adjusting the joint layout until the strength and stiffness criteria are fulfilled.

The mechanical properties of a joint shall be consistent with those required by the modelling of this joint in view of global frame analysis. Either the design of the joint and/or that of the members may need to be adjusted. In any case, some steps of the design approach need to be repeated.

9.1.3 Consistent design approach

In the consistent design approach, the global analysis is carried out in full consistency with the presumed real joint response (Figure 9.4).

It is therefore different from the traditional design approach described in section 9.1.2 in several respects:

– *Structural conception:*

In the structural conception phase, the real mechanical behaviour of the joints is modelled;

- *Preliminary design:*

    In the pre-design phase, joints are selected by the practitioner based on his experience. Proportions for the joint components are determined: end-plate or cleat dimensions, location of bolts, number and diameter of bolts, sizes of column and beam flanges, thickness and depth of column web, etc.;

- *Determination of the mechanical properties:*

    In Step 4, the structural response of both the selected members and joints is determined. First the joints are characterised (see sub-chapter 1.6) with the possible consequence of having a non-linear behaviour. This characterisation is followed by an idealisation, for instance according to a linear or bi-linear joint response curve (see sub-chapter 2.3), which becomes a part of the input for global frame analysis;

- *Global frame analysis*:

    For the purpose of global frame analysis, any joint structural response is assigned to a relevant spring in the frame model. This activity is called *modelling* (see sub-chapter 2.2).

Of course the consistent design approach is only possible when both members and joints are designed by a single party, because the mechanical properties of the joints must be accounted for when starting the global frame analysis. In other words, this approach suits both Case B1 and Case B2. In sub-chapter 9.2, information is given on intermediate forms.

As it accounts for any kind of behaviour, the consistent design approach is especially applicable to frames with so-called semi-continuous joints. The modelling of semi-continuous joints is presented in sub-chapter 2.2 in relation with the global frame analysis. It may also be applied when designing frames with simple or continuous joints.

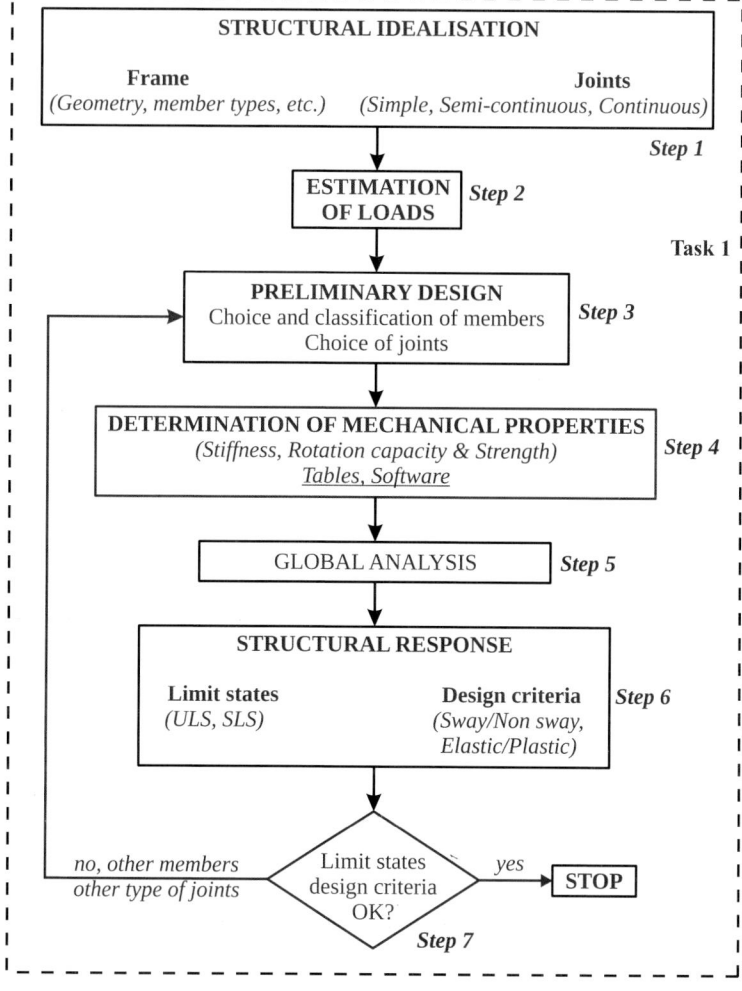

Figure 9.4 – Consistent design approach

9.1.4 Intermediate design approaches

The two design approaches described in section 9.1.2 (for frames with simple or continuous joints) and in section 9.1.3 (for frames with semi-continuous joints) correspond to extreme situations. Intermediate approaches

9. Design Strategies

can be used. For example, the procedure given in Figure 9.1 can also be applied for semi-continuous joints. In that case, during the first pass through the design process, the joints are assumed to behave as simple or continuous joints. Joints are then chosen, the real properties of which are then accounted for in a second pass of the global analysis (i.e. after Step 8). The design process is then pursued similarly to the one described in Figure 9.4. But for sure such a way to proceed is not recommended as it involves iterations and so extra calculation costs.

That is why more "clever" applications of intermediate design approaches are commented on in sub-chapter 9.2.

9.1.5 Economic considerations

9.1.5.1 Savings of fabrication and erection costs

a) Optimal detailing of rigid joints

A first very efficient strategy can be summarized as follows: *"Optimise the joint detailing such that the joint stiffness comes close to the 'rigid' classification boundary but remains higher"*. This is illustrated in Figure 9.5.

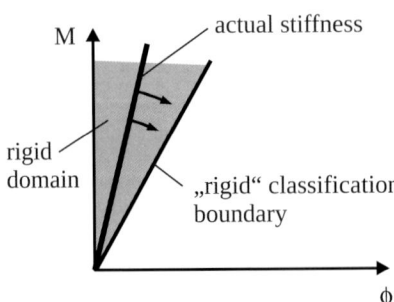

Figure 9.5 – Optimization of rigid joints

The actual stiffness of a joint as well the classification boundary for rigid joints can be calculated according to Eurocode 3. The classification boundary is the minimum stiffness required to model a joint as rigid. If the actual joint stiffness is significantly higher, e.g. due to stiffeners, it should be checked whether it is not possible to omit some of the stiffeners while still

9.1 DESIGN OPPORTUNITIES FOR OPTIMISATION OF JOINTS AND FRAMES

fulfilling the criterion for rigid joints. This will not change the overall design at all, but it will directly reduce the fabrication costs of the joints (e.g. less welding). This procedure is used in the example described below:

The joints of a typical portal frame (see Figure 9.6) were designed using traditional design practice.

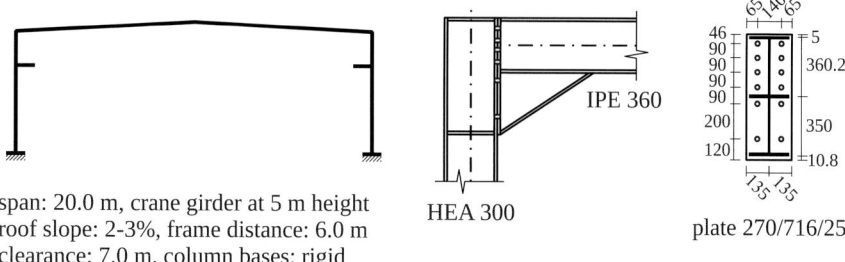

span: 20.0 m, crane girder at 5 m height
roof slope: 2-3%, frame distance: 6.0 m
clearance: 7.0 m, column bases: rigid

HEA 300

IPE 360

plate 270/716/25

Figure 9.6 – Example for the optimization of rigid joints

To classify the joints as rigid, EC 3 requires for the given unbraced frame that:

$$S_{j,ini} \geq 25\frac{EI_b}{L_b} = 85\,628 \text{ kNm/rad}$$

According to the EN 1993-1-8 the joint characteristics in Figure 9.6 are calculated as follows:

- design moment resistance $M_{j,Rd} = 281.6$ kNm
- initial stiffness $S_{j,ini} = 144\,971$ kNm/rad

Hence the joint is classified as rigid. But in order to optimise the joint, the detailing of the joint is modified step by step as follows:

1. omit the stiffeners at compression side,

2. in addition, omit the stiffeners at tension side,

3. in addition, omit the lowest bolt row in tension.

For all variations it is requested that the joint may still be classified as rigid, i.e. the initial stiffness of the joint should be higher than $85\,628$ kNm/rad. Table 1 gives the resistance and stiffness for all variations.

9. DESIGN STRATEGIES

It is seen that the design moment resistance is less than the applied moment and the joint behaves as a rigid one. The different joint detailings will have no influence on the design of the member as the joint remains rigid. Therefore, differences in the fabrication costs of the joint give direct indication of economic benefits. The fabrication costs (material and labour) are given for all investigated solutions in Table 9.2 as a percentage of the fabrication costs of the original joint layout. Beside the reduction of fabrication costs, the modified joints are seen to be more ductile. Another possibility to optimize the stiffness is the use of thinner plates. This can also lead to a more ductile behaviour. Other advantages are: smaller weld, the preparation of hole drilling is easier or hole punching becomes possible.

Table 9.2 – Variations in joint detailing and relative savings in fabrication costs

Variations	$M_{j,Rd}$ [kNm]	$S_{j,ini}$ [kNm]	Stiffness classification	Relative fabrication costs *)	Saving
stiffener, IPE 360, HEA 300	255.0	92 706	rigid	87 %	13 %
no stiffener, IPE 360, HEA 300	250.6	89 022	rigid	73 %	27 %
IPE 360, bolt row ommited, HEA 300	247.8	87 919	rigid	72 %	28 %

*) fabrication costs of joints relative to those of the configuration in Figure 9.6

b) Economic benefits from semi-rigid joints

A second strategy to profit from the extended possibilities in design can be expressed as follows: *"Use semi-rigid joints in order to have any freedom to optimise the global frame and joint design"*.

9.1 DESIGN OPPORTUNITIES FOR OPTIMISATION OF JOINTS AND FRAMES

The ideal assumption that the joints are rigid can lead often to situations where it is not possible to work without stiffeners. In consequence the joints are very expensive due to high fabrication costs (e.g. welding). In this case more economical solutions can be found by 'crossing the rigid classification boundary' to semi-rigid joints (see Figure 9.7).

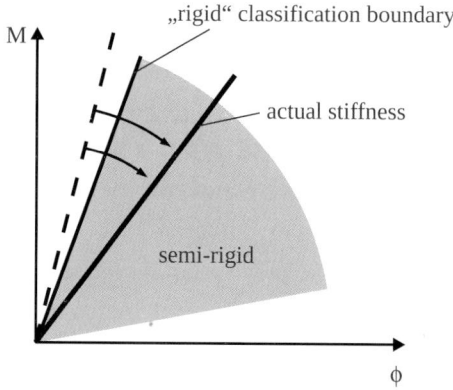

Figure 9.7 – Optimisation with semi-rigid joints

The layout of the joints is then chosen for economy. Usually, this leads to more flexible, i.e. semi-rigid joints. The joint behaviour has to be taken into account in the frame analysis. This is possible by modelling the joints with rotational springs at beam ends. Their behaviour influences the global response of the frame, i.e. moment redistributions and displacements. As a consequence the size of the members may increase in comparison with a design with rigid joints. The decrease of the fabrication cost for the joints on one side has to be compared with the increase of the weight of the structure due to larger profiles on the other side. The optimum solution can only be found when a detailed calculation of the costs is carried out. Beside the positive effect in view of economy the use of joint without stiffeners provides more advantages: it becomes easier to connect secondary beams, service pipes may be installed between the column flanges, easier coating and less problems with corrosion. The steel construction can appear more aesthetic and lighter and again, joint without stiffeners are usually more ductile. With respect to ultimate limit state design this is an important aspect as well, for instance in the case of seismic or robust requirements.

9. Design Strategies

9.1.5.2 Savings of material costs

In the previous section, strategies where discussed when a frame is designed with so-called moment connections. This is usually required for unbraced frames. However for braced frames simple joints are normally more economic. However the moment diagram for the (simple supported) beams leads to beam sizes which are optimized for the mid-span moment, but 'oversized' at their ends. Further - due to the fact that the joints do not transfer any moment (this is the assumption for the frame analysis) - it becomes sometimes necessary to install additional temporary bracings during erection. However in many cases, joints which are assumed to behave as nominally pinned (e.g. flush end-plates) can be treated as semi-rigid joint. The strategy therefore is: *"Simple joints may have some inherent stiffness and may transfer moments - take profit from that actual behaviour"*

This can improve the distribution of moments and reinforce the frame, i.e. lighter members (also for the bracing systems) without any change in joint detailing. And - as a consequent step - a further strategy is as follows: *"Check if small reinforcements of simple joints may strongly reinforce the frame"*

Sometimes it will be possible to reinforce simple joints without a large increase of fabrication cost (e.g. flush end-plates instead of a short end-plates). As those joints can transfer significant moments, the bending diagram is more balanced and the dimensions of the members can be lighter. Here again, increase of fabrication costs for the joints and decrease of weight of the structure are two competing aspects and a check of balance is necessary. However recent investigations show that the most economical solutions can be found if the contributions of the stiffness and resistance of the joints are considered, i.e. semi-rigid joints are used.

9.1.5.3 Summary and conclusions

With respect to joint design basically two different strategies can be identified when minimum costs of steel structures are of interest:

- Simplification of the joint detailing, i.e. reduction of fabrication costs. Typically this is relevant for unbraced frames when the joints transfer significant moments (traditionally rigid joints).
- Reduction of profile dimensions, i.e. reduction of material costs. Typically this is relevant for braced frames with simple joints.

9.1 DESIGN OPPORTUNITIES FOR OPTIMISATION OF JOINTS AND FRAMES

In general both strategies would lead to the use of semi-rigid joints. In case of rigid joints an economic solution may already be found if the stiffness of the joint is close to the classification boundary.

Economy studies in various countries have shown possible benefits from the use of the concept of semi-rigid joints. More significant savings can be achieved when moment connections are optimized in view of economy. is remarkable that all studies came to similar values when the saving due to the use of the new concept is compared to traditional design solutions. However it should be understood that the savings depend on the preferences of the steel fabricators to design the joints and how the cost are calculated. From the different studies it can be concluded that the possible savings due to semi-rigid design can be 20 - 25 % in case of unbraced frames and 5 - 9 % in case of braced frames. With the assumption that the costs of the pure steel frames are about 10% of the total costs for office buildings and about 20 % for industrial buildings, the reduction of the total building costs could be estimated to 4-5% for unbraced frames. For braced systems savings of 1-2% are possible.

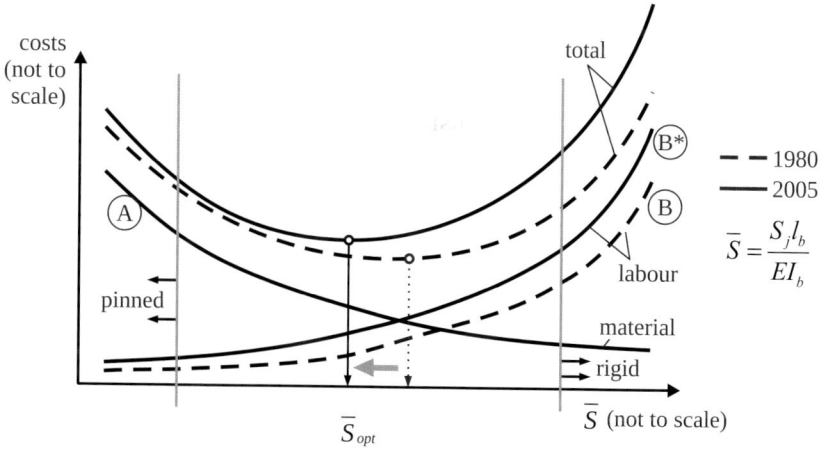

Figure 9.8 – Costs of steel structures depending on the relative joint stiffness

Of course the results of the investigations presented in this paper cannot be compared directly, particularly because different types of frames are used. However, the following conclusion can be drawn as shown in Figure 9.8: the costs for material and fabrication (labour) are dependent on the relative stiffness of the joints. While the material costs decrease (curve A), the labour costs increase (curve B) with an increasing joint stiffness. For the total costs

9. DESIGN STRATEGIES

which are the sum of these both curves, a minimum can be found and from this, an "optimum joint stiffness". In many cases the value (which leads to an optimized design of the structure with respect to minimum total costs) is neither pinned nor rigid. Following the tendency of the last decades, it is obvious that the labour costs are increasing in comparison to the material costs (see dashed curve B*). From Figure 9.8 it becomes clear that as a consequence there is a progressive evolution of the "optimum joints stiffness" towards more flexible joints. Hence to find economical solutions for steel structures the use of semi-rigid joints will become more and more interesting.

9.2 APPLICATION PROCEDURES

9.2.1 Guidelines for design methodology

This section treats the strategy for an appropriate application of *the design methodology*, i.e. of the various *design strategies* introduced in sub-chapter 9.1. The design strategy followed by the designer is of vital importance for an efficient design process. A good strategy leads normally to economical solutions for both members and joints.

In sub-chapter 9.1, several approaches for the design process have been given. The traditional design approach reflects current practice where joints are modelled as either simple or continuous. In the consistent design approach, global frame analysis is started with due account taken of the joint response.

When the joints are semi-continuous, the traditional design approach may be adopted too but in an iterative way (see section 9.1.4). Then it is of course far better that only a single party be in charge of the whole design process; indeed the actual joint properties, that depend on the joint sizing and detailing - i.e. on the fabricator's task - need to be included in the global frame analysis - i.e. in the engineer's task. Such a use of the traditional design approach is neither very efficient nor suits well for a rather common practice where the respective designs of members and joints are carried out by different parties. For these reasons this section gives design strategies which focus on frames with semi-continuous joints. The purpose of these three strategies is to allow for a two task process. The design of the frame is separated from the design of the joints, as in the traditional design approach, with the aim of efficiency and allowance for semi-continuity.

These strategies are:

- Use of a *good guess* for joint stiffness with a view to elastic global frame analysis;
- Use of the fixity factor in the traditional design approach;
- Design of braced frames with rigid-plastic analysis.

The first two strategies are especially applicable when elastic frame design, possibly when elasto-plastic one is adopted. They address mainly unbraced frames but also braced frames. The third strategy focuses on one type of plastic frame design; its use may be recommended for braced frames only.

9.2.2 Use of a *good guess* for joint stiffness

This strategy refers to the traditional design approach (see Figure 9.2). However, two changes are needed to make it suitable for semi-continuous design (Figure 9.9):

1. *Account for joint stiffness in the elastic global frame analysis*:

    In Step 3, use is made of a *good guess* of the initial joint stiffness. The way on how to evaluate it for joints not intended to be considered as rigid is provided in section 6.4.1.2. This stiffness, named as $S_{j,app}$, is the one to be used for the elastic global frame analysis (Step 4 in Figure 9.9) when the designer intends to perform elastic design checks; it shall be divided by an appropriate η factor (see sub-chapter 2.3) when plastic design checks are carried out.

2. *Verification of stiffness in the design of joints*:

    In Step 8 of Figure 9.9, it shall be verified whether the actual stiffness of any of the joints is in reasonable agreement with the relevant approximate stiffness that is accounted for in the elastic global frame analysis. This replaces the verifications for rigid and/or pinned joints in the traditional design. Section 9.2.3 gives some rules aimed at enabling this verification; their philosophy is quite similar to the classification diagrams of Eurocode 3. These rules can be applied easily in combination with a global analysis software.

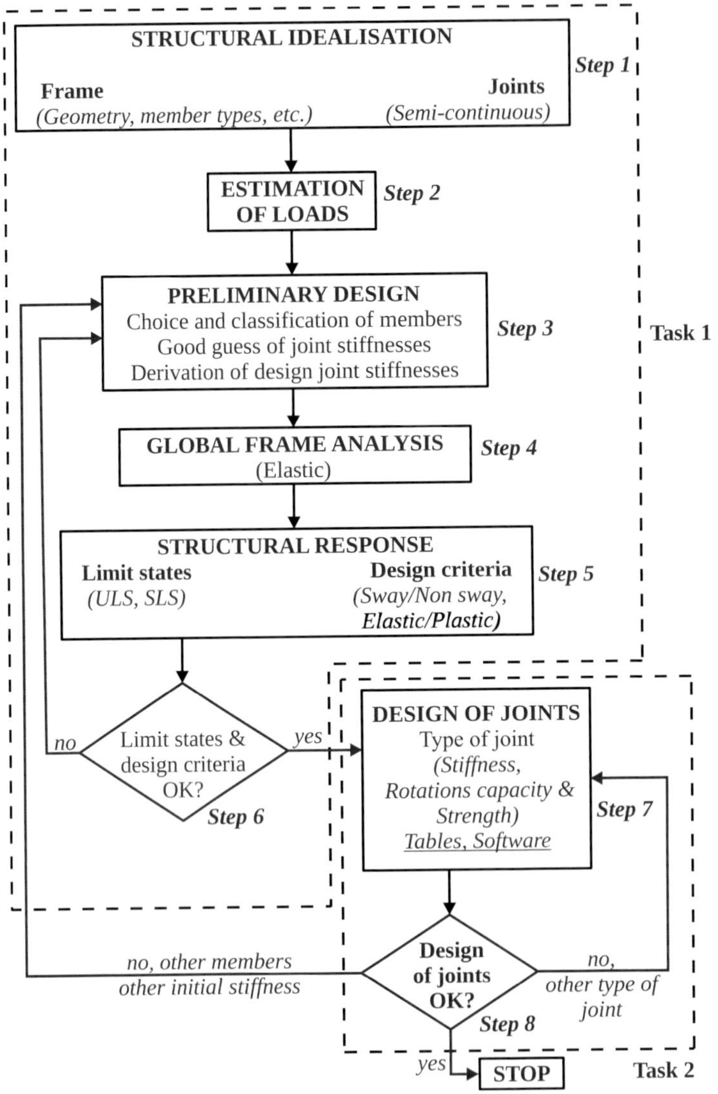

9. DESIGN STRATEGIES

Figure 9.9 – Design strategy when semi-continuous joints (elastic global analysis)

9.2.3 Required joint stiffness

Eurocode 3 provides the designer with two diagrams which enable the classification of joints according to their stiffness (pinned, semi-rigid, rigid): one for braced and one for unbraced frames.

Accordingly, for braced frames, a joint may be regarded as rigid if the actual initial joint stiffness fulfils:

9.2 APPLICATION PROCEDURES

$$S_{j,ini} \geq \frac{8EI_b}{L_b} \tag{9.1}$$

That condition ensures that the real flexibility of the joints does not result in a more than 5% drop on the critical Euler load, what corresponds to a maximum difference of 2% on the actual bearing capacity of the frame. The acceptance of this drop magnitude permits not to restart the global frame analysis with a finite value of the joint stiffness. Here, the wording *actual initial stiffness* is the best value a designer can obtain for the initial stiffness of a particular joint. This is, for example, the value obtained from experiments, from numerical simulations or computed based on EN 1993 Part 1-8.

To check whether a joint is rigid needs a three steps procedure (see Figure 9.10):

- *Step a:* frame analysis conducted with the assumption of rigid joints (Step 3 of Figure 9.9).
- *Step b:* range in which the actual initial stiffness should be (in Step 8 of Figure 9.9).
- *Step c:* check whether the actual initial stiffness is in this range (in Step 8 of Figure 9.9).

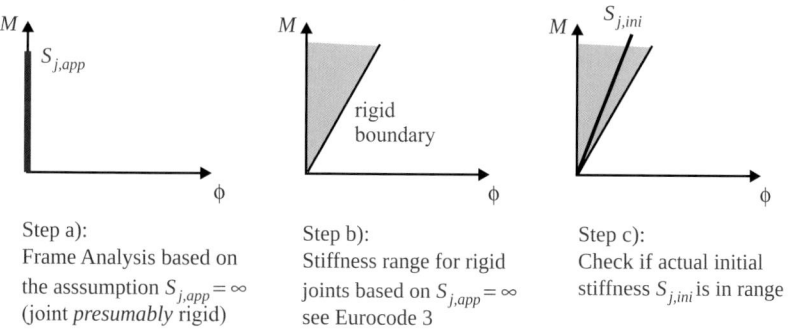

Step a):
Frame Analysis based on the asssumption $S_{j,app} = \infty$ (joint *presumably* rigid)

Step b):
Stiffness range for rigid joints based on $S_{j,app} = \infty$ see Eurocode 3

Step c):
Check if actual initial stiffness $S_{j,ini}$ is in range

Figure 9.10 – Check of the stiffness requirement for a rigid joint

This concept can be generalised to check whether a difference between the approximate joint stiffness and the actual joint stiffness of semi-rigid joints has a significant influence on the frame behaviour (Figure 9.11). The corresponding formulae for the variance between the approximate joint stiffness and the actual initial one are given in Table 9.3. These criteria may

9. DESIGN STRATEGIES

be used to check whether a difference between these stiffnesses has a no more than 5% effect on the frame elastic Euler load (i.e. no more than 2% of the frame bearing capacity).

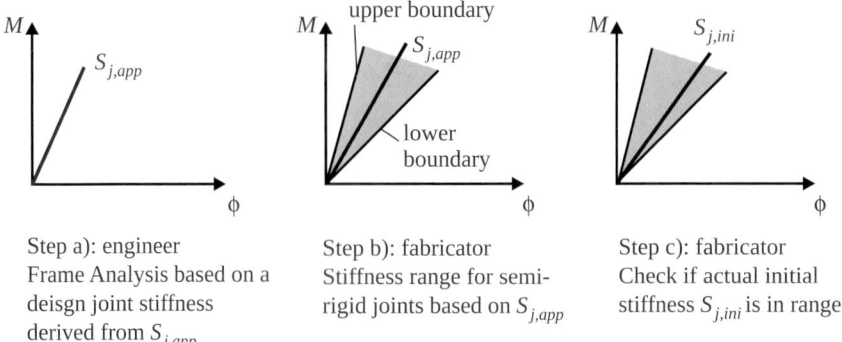

Step a): engineer Frame Analysis based on a design joint stiffness derived from $S_{j,app}$

Step b): fabricator Stiffness range for semi-rigid joints based on $S_{j,app}$

Step c): fabricator Check if actual initial stiffness $S_{j,ini}$ is in range

Figure 9.11 – Check of stiffness requirement of a semi-rigid joint

Table 9.3 – Boundaries for variance between *actual* and *approximate* initial stiffnesses

Frame	Lower boundary	Upper boundary
Braced	$S_{j,ini} \geq \dfrac{8 S_{j,app} EI_b}{10 EI_b + S_{j,app} L_b}$	If $S_{j,ini} \leq \dfrac{8 EI_b}{L_b}$ then $S_{j,ini} \leq \dfrac{10 S_{j,app} EI_b}{8 EI_b - S_{j,app} L_b}$ else $S_{j,ini} \leq \infty$
Unbraced	$S_{j,ini} \geq \dfrac{24 S_{j,app} EI_b}{30 EI_b - S_{j,app} L_b}$	If $S_{j,ini} \leq \dfrac{24 EI_b}{L_b}$ then*) $S_{j,ini} \leq \dfrac{30 S_{j,app} EI_b}{24 EI_b - S_{j,app} L_b}$ else $S_{j,ini} \leq \infty$
in which: $S_{j,app}$ = approximate joint stiffness (estimate of the initial one) $S_{j,ini}$ = actual initial joint stiffness E = Young modulus L_b = beam length I_b = second moment of area of the beam cross section *) For sake of simplicity, the stiffness boundary of Eurocode 3 is modified from $S_j \geq \dfrac{25 EI_b}{L_b}$ to $S_j \geq \dfrac{24 EI_b}{L_b}$ for unbraced frames.		

9.2.4 Use of the fixity factor concept (traditional design approach)

Another strategy for a preliminary design is the use of the so-called *fixity factor* f. The fixity factor f is defined as the rotation ϕ_b of the beam end, due to a unit end moment applied at the same end, divided by the corresponding rotation ϕ_t of the beam plus the joint.

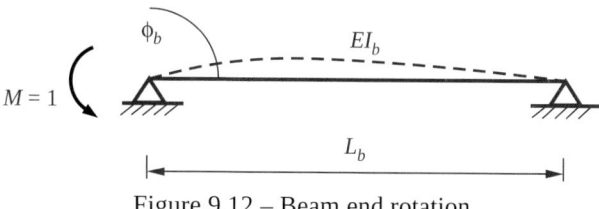

Figure 9.12 – Beam end rotation

The end rotation of the beam for a unit moment is (Figure 9.12):

$$\phi_b = \frac{L_b}{3EI_b} \tag{9.2}$$

The rotation of the beam plus the joint for the same moment is:

$$\phi_t = \frac{L_b}{3EI_b} + \frac{1}{S_j} \tag{9.3}$$

The fixity factor is then defined as:

$$f = \frac{\phi_b}{\phi_t} = \frac{1}{1+1.5\alpha} \tag{9.4}$$

where the symbols E, L_b, I_b and S_j are defined in Table 9.3 and:

$$\alpha = \frac{2EI_b}{L_b S_j} \tag{9.5}$$

For a *truly pinned* joint, f is equal to 0 while for a *truly rigid* joint, f amounts 1.

To speed up the process of converting to a solution, the designer can decide to adopt a fixity factor between 0 and 1 and start the global frame analysis accordingly. Recommended values are $0.1 \leq f \leq 0.6$ for braced frames and $0.7 \leq f \leq 0.9$ for unbraced frames. Let us assume $f = 0.5$ for braced frames; then a joint stiffness of $3EI_b/L_b$ should be adopted in the global frame analysis. If $f = 0.8$ is used for unbraced frames, the corresponding value of the joint stiffness is $12EI_b/L_b$.

These respective values can be considered as a good guess in the design procedure of Figure 9.9. It shall be verified that this good guess is in reasonable agreement with the actual initial stiffness of the joints. For this verification, Table 9.3 can be used, but it merges to Table 9.4 if a fixity factor $f = 0.5$ for braced frames or $f = 0.8$ for unbraced frames is adopted.

Table 9.4 – Boundaries for actual initial stiffness (given fixity factors)

Frame	Lower boundary	Upper boundary
Braced ($f = 0.5$) $S_{j,app} = \dfrac{3EI_b}{L_b}$	$S_{j,ini} \geq \dfrac{24EI_b}{13L_b}$	$S_{j,ini} \leq \dfrac{6EI_b}{L_b}$
Unbraced ($f = 0.8$) $S_{j,app} = \dfrac{12EI_b}{L_b}$	$S_{j,ini} \geq \dfrac{48EI_b}{7L_b}$	$S_{j,ini} \leq \dfrac{30EI_b}{L_b}$

in which:
$S_{j,app}$ = approximate joint stiffness used for the global frame analysis
$S_{j,ini}$ = actual initial joint stiffness
E = Young modulus
L_b = beam length
I_b = second moment of area of the beam cross section

9.2.5 Design of non-sway frames with rigid-plastic global frame analysis

The strategy described above focuses on the joint stiffness only and therefore suits especially for elastic global frame analysis. When plastic design is used - the latter is especially appropriate to braced non-sway frames -, the resistance of the joints is governing the global frame analysis too.

For this reason, an alternative strategy can be contemplated; it is illustrated in Figure 9.13. In a first step, the frame is designed assuming simple joints. In a second step, the beam section depth is reduced by one size. So any joint has to transfer some bending moment. If this moment is small enough, simple partial-strength joints are sufficient. This strategy focuses very much on the economy of the frame: it is presumed that savings in material make more than compensate the additional cost due to the use of

9.2 APPLICATION PROCEDURES

partial-strength joints. Should these partial-strength joints need to be stiffened, then increase the beam size reveals often a better solution.

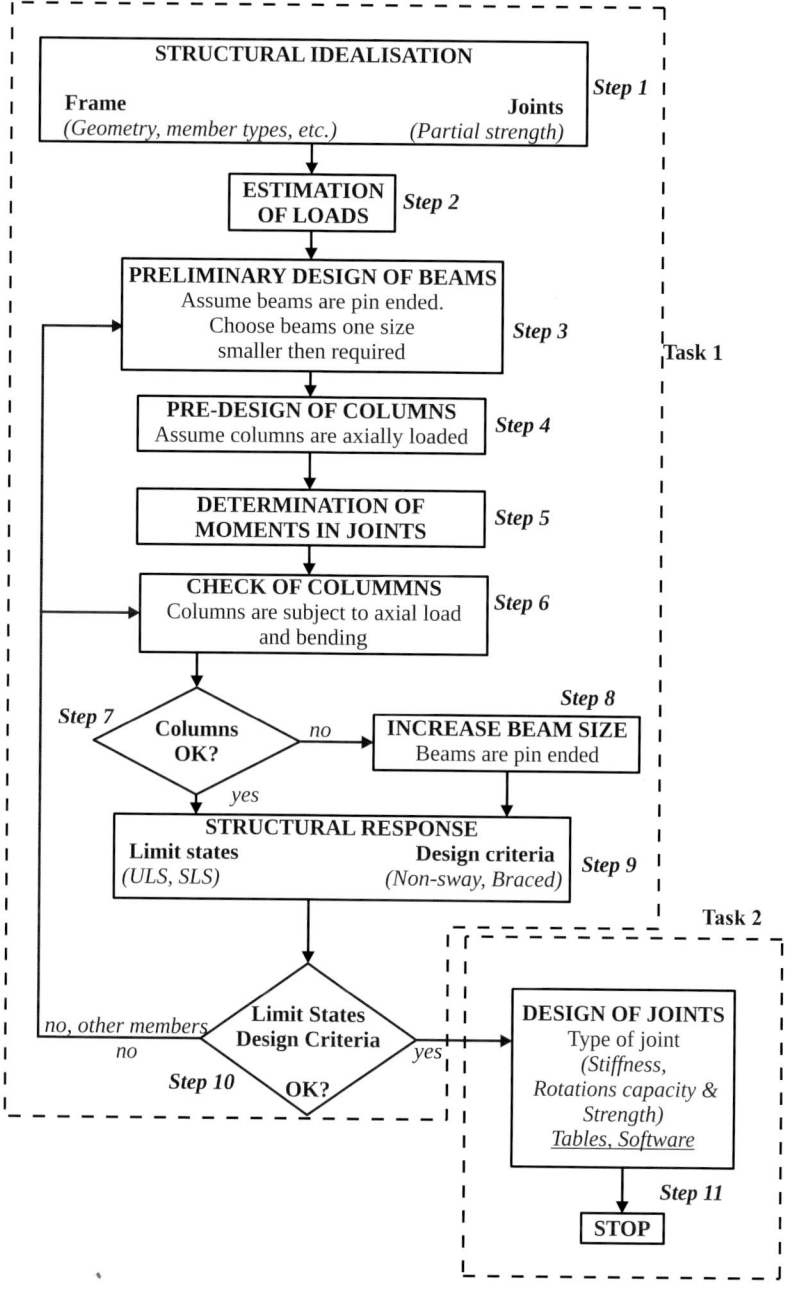

Figure 9.13 – Design strategy for partial-strength joints in non-sway frames

9. Design Strategies

The strategy is conducted as follows:

Steps 1/2: Quite similar to those of the strategy depicted in Figure 9.9.

Step 3: First design of the beams as fitted with presumably simple joints at the ends; for instance a beam subject to a uniformly distributed load is first sized with regard to the field moment only. The beam size is however chosen one size smaller than the one resisting just the maximum field moment (As a consequence, the joints will have to transfer some bending moment).

Step 4: Design of columns as if they were pinned connected to the beams.

Step 5: Determination of the bending moments to be transferred through the joints.

Step 6: Columns are checked for coincident axial force and moments (section resistance and stability).

Steps 7/8: If any check of step 6 fails, economy commands probably to increase the beam size and to have simple joints rather than increase the column size and adopt partial-strength joints.

Steps 9/10: The checks concerned with the limit states are carried out. If the beams do not fulfil the serviceability limit states, they can be chambered. The stiffness of the joints can also be taken into account for this purpose. An estimate can be found using Table 6.7. Alternatively the beam sizes can be increased. These options are a matter of economy.

Step 11: If the checks of the ultimate limit states is successful, the joints are designed in such a way that they are able to resist the relevant bending moments. If a design joint stiffness derived from an approximate one is used for the check of the serviceability limit states, it shall be checked whether the actual joint stiffness is in compliance with this estimate (see also section 9.2.3).

In the conceptual stage of the design process of Figure 9.13, it is useful for the designer to have a rapid indication of the type of joint he will end-up with the last step (Step 11). It is important to have this indication as early as possible, preferably before the check of beams and columns. If it appears that the joint needs stiffeners, it may be more economical to adopt simple joints combined with an increased beam size.

To obtain such a rapid indication, Table 9.5 can be used.

Table 9.5 – Strength recommendations for joints during preliminary design

Joint detailing	Single-sided	Double-sided
	$M_{j,Rd}$	$M_{j,Rd}$
Simple	0	0
Intermediate	$\leq 5 f_y z t_{fc}^2 / \gamma_{M0}$	$\leq 7 f_y z t_{fc}^2 / \gamma_{M0}$
Complex	$> 5 f_y z t_{fc}^2 2/ \gamma_{M0}$	$> 7 f_y z t_{fc}^2 / \gamma_{M0}$

where:
- z distance between centres of compression and tension;
- f_y yield strength of the column flange;
- t_{fc} column flange thickness;
- γ_{M0} partial safety factor for resistance of members.

Simply detailed joints are those joints which traditionally are considered as nominally pinned. Joints with complex detailing are able to transfer bending moments but require stiffening. Joints which are able to transfer bending moments but do not require stiffening are described as intermediate. In general, stiffening is labour expensive and thus "complex" joints may not lead to economical solutions. Partial-strength joints in general fall in the "intermediate" or "complex" categories.

With help of this table a designer can check whether he may expect to end up with "moderate" joints during the design process of Figure 9.13. This can be done directly after Step 5. In this step, the moments M_{Ed} which shall be transmitted from beam to column have been determined. In general, the design moment resistance M_{Rd} of the joint shall be greater than or equal to this M_{Ed}. In other words, for example in the case of a single-sided joint, whenever the design moment at the joint M_{Ed} is such that $M_{Ed} < 5 f_y z t_{fc}^2 / \gamma_{M0}$, it may be expected that the final solution for the joint can be without stiffeners. Otherwise, stiffening is likely to be required and it may be better to increase the beam depth and to choose simple connections in that case.

9. DESIGN STRATEGIES

Table 9.5 is especially useful for the design of non-sway frames by rigid-plastic global frame analysis. However, it may also be helpful in the design of braced or unbraced (sway) frames using either elastic or plastic global frame analysis.

BIBLIOGRAPHIC REFERENCES

Anderson D, Aribert J-M, Bode H, Huber G, Jaspart J-P, Kronenberger H-J, Tschemmernegg F. (1999). *Design of composite joints for buildings*, edited by ECCS TC 11, Publ. No. 109, ECCS - European convention for constructional steelwork, Brussels, Belgium.

Astaneh A, Bergsma G, Shen JR. (1992). *Behavior and Design of Base Plates for Gravity, Wind and Seismic Loads*. In *National Steel Construction Conference*, edited by AISC. Las Vegas.

BCSA/SCI Connections Group. (1992). *Joints in Steel Construction: Volume 2: Practical Applications*, 2nd Ed., in: SCI Publ. P206, The Steel Construction Institute and The British Constructional Steelwork Association.

BCSA/SCI Connections Group. (1993). *Joints in Steel Construction: Volume 1: Design Methods*, 2nd Ed., in: SCI Publ. P205, The Steel Construction Institute and The British Constructional Steelwork Association.

BCSA/SCI Connections Group, Hole E, Malik A, Way A. (2002). *Joints in Steel Construction: Simple Connections*, SCI Publ. P212, The Steel Construction Institute and The British Constructional Steelwork Association.

Bouwman LP, Gresnigt N, Romeijn A. (1989). *Onderzoek naar de bevestiging van stalen voetplaten aan funderingen van beton. (Research into the connection of steel base plates to concrete foundations)*, Report 25.6.89.05/c6, TU-Delft Stevin Laboratory, Delft.

Brettle M. (2008). *Steel Building Design: Worked Examples - Hollow Sections*, SCI Publ. P374, The Steel Construction Institute.

Brettle M. (2009). *Steel Building Design: Worked Examples - Open Sections*, SCI Publ. P364, The Steel Construction Institute.

Brown D, Iles D, Brettle M, Malik A, BCSA/SCI Connections Group. (2013). *Joints in Steel Construction: Moment-Resisting Joints to Eurocode 3*, SCI Publ. P398, The Steel Construction Institute and The British Constructional Steelwork Association.

BSI (2000). BS 5950-1:2000 – Structural use of steelwork in building - Part 1: Code of practice for design. Rolled and welded sections, *British Standards Institution*, United Kingdom.

CEB. (1994). *Fastenings to Concrete and Masonry Structures - State of the Art Report*, edited by CEB, (Comité Euro-International du Béton), Thomas Telford Services Ltd, London.

CEB. (1996). *Design of fastenings in concrete - Design Guide*, edited by CEB, (Comité Euro-International du Béton), Thomas Telford Services Ltd, London.

BIBLIOGRAPHIC REFERENCES

CEN (2002). EN 1990:2002 – Eurocode - Basis of structural design, *European Committee for Standardization*, Brussels, Belgium.

CEN (2004a). EN 1992-1-1:2004 – Eurocode 2: Design of concrete structures - Part 1-1: General rules and rules for buildings, *European Committee for Standardization*, Brussels, Belgium.

CEN (2004b). EN 1994-1-1:2004 – Eurocode 4: Design of composite steel and concrete structures - Part 1-1: General rules and rules for buildings, *European Committee for Standardization*, Brussels, Belgium.

CEN (2004c). EN 1998-1-1:2004 – Eurocode 8: Design of structures for earthquake resistance - Part 1: General rules, seismic actions and rules for buildings, *European Committee for Standardization*, Brussels, Belgium.

CEN (2004d). EN 10025-1:2004 – Hot rolled products of structural steels - Part 1: General technical delivery conditions, *European Committee for Standardization*, Brussels, Belgium.

CEN (2004e). EN 10164:2004 – Steel products with improved deformation properties perpendicular to the surface of the product - Technical delivery conditions, *European Committee for Standardization*, Brussels, Belgium.

CEN (2005a). EN 1993-1-1:2005 – Eurocode 3: Design of steel structures - Part 1-1: General rules and rules for buildings, *European Committee for Standardization*, Brussels, Belgium.

CEN (2005b). EN 1993-1-2:2005 – Eurocode 3: Design of steel structures - Part 1-2: General rules – Structural fire design, *European Committee for Standardization*, Brussels, Belgium.

CEN (2005c). EN 1993-1-8:2005 – Eurocode 3: Design of steel structures - Part 1-8: Design of joints, *European Committee for Standardization*, Brussels, Belgium.

CEN (2005d). EN 1993-1-9:2005 – Eurocode 3: Design of steel structures - Part 1-9: Fatigue, *European Committee for Standardization*, Brussels, Belgium.

CEN (2005e). EN 1993-1-10:2005 – Eurocode 3: Design of steel structures - Part 1-10: Material toughness and through-thickness properties, *European Committee for Standardization*, Brussels, Belgium.

CEN (2006a). EN 10219-1:2006 – Cold formed welded hollow sections of non-alloy and fine grain steels - Part 1: Technical delivery conditions, *European Committee for Standardization*, Brussels, Belgium.

CEN (2006b). EN 10219-2:2006 – Cold formed welded hollow sections of non-alloy and fine grain steels - Part 2: Tolerances dimensions and sectional properties, *European Committee for Standardization*, Brussels, Belgium.

CEN (2006c). EN 10210-1:2006 – Hot finished structural hollow sections of non-alloy and fine grain steels – Part 1: Technical delivery conditions, *European Committee for Standardization*, Brussels, Belgium.

CEN (2006d). EN 10210-2:2006 – Hot finished structural hollow sections of non-alloy and fine grain steels – Part 2: Tolerances dimensions and sectional properties, *European Committee for Standardization*, Brussels, Belgium.

CEN (2006e). EN 1993-2:2006 – Eurocode 3 - Design of steel structures - Part 2: Steel Bridges), *European Committee for Standardization*, Brussels, Belgium.

CEN (2007a). EN 1993-1-12:2007 – Eurocode 3: Design of steel structures - Part 1-12: Additional rules for the extension of EN 1993 up to steel grades S 700, *European Committee for Standardization*, Brussels, Belgium.

CEN (2007b). EN 15048-1:2007 – Non-preloaded structural bolting assemblies - Part 1: General requirements, *European Committee for Standardization*, Brussels, Belgium.

CEN (2009a). EN 14399-9:2009 – High-strength structural bolting assemblies for preloading - Part 9: System HR or HV - Direct tension indicators for bolt and nut assemblies

CEN (2009b). EN 14399-10:2009 – High-strength structural bolting assemblies for preloading - Part 10: System HRC - Bolt and nut assemblies with calibrated preload, *European Committee for Standardization*, Brussels, Belgium.

CEN (2011). EN 1090-2:2008+A1:2011 – Execution of steel structures and aluminium structures - Part 2: Technical requirements for steel structures, *European Committee for Standardization*, Brussels, Belgium.

CEN (2013). EN ISO 898-1:2013 – Mechanical properties of fasteners made of carbon steel and alloy steel - Part 1: Bolts, screws and studs with specified property classes - Coarse thread and fine pitch thread (ISO 898-1:2013), *European Committee for Standardization*, Brussels, Belgium.

CEN (2014). EN ISO 5817:2014 – Welding - Fusion-welded joints in steel, nickel, titanium and their alloys (beam welding excluded) - Quality levels for imperfections (ISO 5817:2014), *European Committee for Standardization*, Brussels, Belgium.

CEN (2015). EN 14399-1:2015 – High-strength structural bolting assemblies for preloading - Part 1: General requirements, *European Committee for Standardization*, Brussels, Belgium.

Cerfontaine F (2003). *Study of the interaction between bending moment and axial force in bolted joints (in French)*, Ph.D. Thesis, Liège University.

CRIF, Université de Liege - Département M.S.M, CTICM Sait Rémy-les-Chevreuse, ENSAIS Strasbourg, Universita degli studi di Trento, LABEIN Bilbao. *Steel Moment Connections according to Eurocode 3, Simple design aids for rigid and semi-rigid joints*, Report SPRINT Contract RA 351.

Demonceau J-F (2008). *Steel and composite building frames: sway response under conventional loading and development of membrane effects in beams further to an exceptional action*, Ph.D. Thesis, Liège University.

BIBLIOGRAPHIC REFERENCES

Demonceau J-F, Hanus F, Jaspart J-P, Franssen JM. (2009). *Behaviour of single-sided composite joints at room temperature and in case of fire after an earthquake*, International Journal of Steel Structures, Vol. 9(4), pp. 329-342.

Demonceau J-F, Haremza C, Jaspart J-P, Santiago A, Simões da Silva L (2013). *Composite joints under M-N at elevated temperatures*, In: Proceedings of the Composite Construction VII Conference (Composite Construction VII), Palm Cove, Australia.

Demonceau J-F, Hoang V-L, Jaspart J-P, Sommariva M, Maio D, Zilli G, Demofonti G, Ferino J, Pournara A-E, Varelis GE, Papatheocharis C, Perdikaris P, Pappa P, Karamanos SA, Ferrario F, Bursi OS, Zanon G (2012). *Performance-based approaches for high strength tubular columns and connections under earthquake and fire loadings (ATTEL)*, Report, RFCS Reserch Found for coal and steel, Brussels, http://bookshop.europa.eu/is-bin/INTERSHOP.enfinity/WFS/EU-Bookshop-Site/en_GB/-/EUR/ViewPublication-Start?PublicationKey=KINA25867.

Demonceau J-F, Jaspart J-P, Klinkhammer R, Weynand K, Labory F, Cajot LG (2008). *Recent developments on composite connections*, Steel Construction – Design and Research Journal, pp. 71-76.

Demonceau J-F, Jaspart J-P, Weynand K, Oerder R, Müller C (2011). *Connections with four bolts per horizontal row* (Eurosteel 2011), Budapest, Hungary.

DeWolf JT (1978). *Axially Loaded Column Base Plates*, Journal of the Structural Division ASCE, Vol. 104, pp. 781-794.

DeWolf JT, Ricker DT (1990). *Column Base Plates*, Steel Design Guide (Series 1), edited by AISC, Chicago.

Disque, Robert O (1984). *Engineering for Steel Construction*, American Institute of Steel Construction, Chicago.

Dubina D (2015). *High strength steel in seismic resistant building frames (HSS-SERF)*, (RFCS project - Final report), European Commission.

Dutch Commissie SG/TC-10a (Verbindingen) (1998). *Verbindingen: Aanbevelingen voor normaalkrachtverbindingen en dwarskrachtverbindingen*, Report not published.

ECCS (1994). *European Recommendations for Bolted Connections with Injection Bolts*, Publ. No. 79, ECCS - European Convention for Constructional Steelwork, Brussels, Belgium.

ESDEP. *Lecture 11.2.1*.

Ferrario F (2004). *Analysis and modelling of the seismic behaviour of high ductility steel-concrete compo-site structures*, Ph.D. Thesis, Trento University.

Franssen JM (2005). *SAFIR A thermal/structural program modelling structures under fire*, Engineering Journal AISC, pp. 143-158.

Gaboriau M (1995). *Recherche d'une méthode simple de prédimensionnement des ossatures contreventées à assemblages semi-rigides dans l'optique de l'approche élastique de dimensionnement (in French)*, Diploma work, University of Liège.

Gibbons C, Nethercot D, Kirby P, Wang Y (1993). *An appraisal of partially restrained column behaviour in non-sway steel frames*, Proc. Instn Civ. Engrs Structs & Bldgs, 99, pp. 15-28.

Gresnigt N, Romeijn A, Wald F, Steenhuis M (2008). *Column bases in shear and normal force*, HERON, Vol. 53, pp. 145-166.

Gresnigt N, Sedlacek G, Paschen M (2000). *Injection Bolts to Repair Old Bridges*, In: *Proceeding of the Fourth International Workshop on Connections in Steel Structures* (Connection in Steel Structures IV), edited by R Leon and W S Easterling, American Institute of Steel Construction (AISC), Roanoke, Virginia, USA, https://www.aisc.org/content.aspx?id=3626.

Gross LG (1990). *Experimental study of gusseted connections*, Engineering Journal AISC, pp. 89-97.

Guillaume M-L (2000). *Development of an European procedure for the design of simple joints (in French)*, Diploma work, University of Liège / CUST Clermont-Ferrand.

Hawkins MN (1968a). *The bearing strength of concrete loaded through flexible plates*, Magazine of Concrete Research, Vol. 20, pp. 95-102.

Hawkins MN (1968b). *The bearing strength of concrete loaded through rigid plates*, Magazine of Concrete Research, Vol. 20, pp. 31-40.

Hoang V-L, Jaspart J-P, Demonceau J-F (2013). *Behaviour of bolted flange joints in tubular structures under monotonic, repeated and fatigue loadings I: Experimental tests*, Journal of Constructional Steel Research, Vol 85, pp. 1–11.

Hoang V-L, Jaspart J-P, Demonceau J-F (2014). *Hammer head beam solution for beam-to-column joints in seismic resistant building frames*, Journal of Constructional Steel Research, Vol. 103, pp. 49-60.

Jaspart J-P (1991). *Etude de la semi-rigidité des noeuds poutre-colonne et son influence sur la résistance et la stabilité des ossatures en acier*, Ph.D. Thesis, Department MSM, University of Liège.

Jaspart J-P (1997). *Recent advances in the field of steel joints. Column bases and further configurations for beam-to-column joints and beam splices*, Habilitation, Department MSM, University of Liège.

Jaspart J-P, Demonceau J-F (2008a). *Composite joints in robust building frames*, In: *Proceedings of the sixth International Composite Construction in Steel and Concrete Conference* (Composite Construction in Steel and Concrete VI), Colorado, USA.

Jaspart J-P, Demonceau J-F, Renkin S, Guillaume ML (2009). *European recommendations for the design of simple joints in steel structures*, edited by ECCS TC 10, Publ. No. 126, ECCS - European convention for constructional steelwork, Brussels, Belgium.

Jaspart J-P, Pietrpertosa C, Weynand K, Busse E, Klinkhammer R (2005). *Development of a Full Consistent Design Apprach for Bolted and Welded Joints - Application of the Component Method*, Report 5BP-4/05, CIDECT - Comité International pour le Développement et l'Etude de la Construction Tubulaire.

Jaspart J-P, Vandegans D (1998). *Application of the component method to column bases*, Journal of Constructional Steel Research, Vol. 48, pp. 89-106.

Jaspart J-P, Wald F, Weynand K, Gresnigt N (2008b). *Steel column base classification*, HERON, Vol. 53, pp. 127-144.

Jaspart J-P, Weynand K (2015). *Design of Hollow Section Joints using the Component Method*, In: *Proceedings of the 15th International Symposium on Tubular Structures, Rio de Janeiro, 27-29 May 2015* (ISTS 15), edited by E Batista, P Vellasco and L Lima, Rio de Janeiro, Brazil.

Jordão S (2008). *Behaviour of internal node beam-to-column welded joint with beams of unequal height and high strength steel*, Ph.D. Thesis, University of Coimbra.

Ju S-H, Fan C-Y, Wu GH (2004). *Three-dimensional finite elements of steel bolted connections*, Engineering Structures 26, pp. 403-413.

Kuhlmann U, Rölle L, Jaspart J-P, Demonceau J-F, Vassart O, Weynand K, Ziller C, Busse E, Lendering M, Zandonini R, Baldassino N (2009). *Robust structures by joint ductility*, Report EUR 23611 EN, RFCS Research, European Commission, Brussels, http://bookshop.europa.eu/is-bin/INTERSHOP.enfinity/WFS/EU-Bookshop-Site/en_GB/-/EUR/ViewPublication-Start?PublicationKey=KINA23611.

Kurobane Y, Packer JA, Wardenier J, Yeomans N (2004). *DG 9: Design guide for structural hollow section column connections, Construction with hollow sections*, edited by CIDECT - Comité International pour le Développement et l'Etude de la Construction Tubulaire, TÜV-Verlag GmbH, Köln, Germany.

Liew RJY, Teo TH, Shanmugam NE (2004). *Composite joints subject to reversal of loading - Part 2: analytical assessments*, Journal of Constructional Steel Research, pp. 247-268.

Maquoi R, Chabrolin B (1998). *Frame Design Including Joint Behaviour*, EU 18563 EN, Commision of The European Communities, Brussels, Belgium, http://bookshop.europa.eu/en/frame-design-including-joint-behaviour-pbCGNA18563/.

Moreno EN, Tarbé C, Brown D, Malik A, BCSA/SCI Connections Group (2011). *Joints in Steel Construction: Simple Joints to Eurocode 3*, SCI Publ. P358, The Steel Construction Institute and The British Constructional Steelwork Association.

Murray TM (1983). *Design of Lightly Loaded Steel Column Base Plates*, Engineering Journal AISC, Vol. 20, pp. 143-152.

NEN 6770 (1997). *TGB 1990 - Staalconstructies - Basiseisen en basisrekenregels voor overwegend statisch belaste constructies*. Published by NEN (Nederlandse Normalisatie-instituut),

Nogueiro P, Simões da Silva L, Bento R, Simões R (2009). *Calibration of model parameters for the cyclic response of end-plate beam-to-column Steel-concrete composite joints*, Steel and Composite Structures Journal, Vol. 9, pp. 39-58.

Owens GW, Cheal BD (1989). *Structural Steelwork Connections*, Butterworth & Co. Ltd.

Packer JA (1996). *Nailed tubular connections under axial loading*, Journal of Structural Engineering, Vol. 122, pp. 458-467.

Packer JA, Henderson JE (1997). *Hollow structural section connections and trusses: A design guide*, 2nd Ed., in, Canadian Institute of Steel Constructions, Willowdale, Ontario.

Packer JA, Wardenier J, Kurobane Y, Dutta D, Yeomans N (1992). *DG 3: Design guide for rectangular hollow section (RHS) joints under predominantly static loading, Construction with hollow sections*, edited by CIDECT - Comité International pour le Développement et l'Etude de la Construction Tubulaire, Verlag TÜV Rheinland GmbH, Köln, Germany.

Packer JA, Wardenier J, Zhao X-L, van der Vegte A, Kurobane Y (2009). *DG 3: Design guide for rectangular hollow section (RHS) joints under predominantly static loading*, 2nd Ed., in: *Construction with hollow sections*, edited by CIDECT - Comité International pour le Développement et l'Etude de la Construction Tubulaire, www.cidect.com.

Penserini P, Colson A (1989). *Ultimate Limit Strength of Column-Base Connection*, Journal of Constructional Steel Research, Vol. 14, pp. 301-320.

Renkin S (2003). *Development of an European process for the design of simple structural joint in steel frames (in French)*, Diploma work, University of Liège.

Sedlacek G, Weynand K, Oerder S (2000). *Typisierte Anschlüsse im Stahlhochbau, Band 1: Gelenkige I-Trägeranschlüsse*, Band 1, Stahlbau Verlagsgesellschaft mbH, Düsseldorf, Germany.

Shelson W (1957). *Bearing Capacity of Concrete*, Journal of the American Concrete Institute, Vol. 29, pp. 405-414.

Simões da Silva L, Santiago A, Villa Real P (2011). *A component model for the behaviour of steel joints at elevated temperatures*, Journal of Constructional Steel Research, Vol. 57, pp. 1169-1195.

Simões da Silva L, Simões R, Gervásio H (2010). *Design of Steel Structures, ECCS Eurocode Design Manuals*, edited by ECCS - Europen Convention for Constructural Steelwork, Wilhelm Ernst & Sohn Verlag GmbH & Co KG, Berlin.

BIBLIOGRAPHIC REFERENCES

Stark JWB (1995). *Criteria for the use of preloaded bolts in structural joints*, In: *Proceeding of 3rd International Workshop on Connections* (Connection in Steel Structures III), edited by R Bjorhovde, A Colson and R Zandonini, Pergamon, Trento, Italy.

Steenhuis CM (1998). *Assembly Procedure for Base Plates*, Report 98-R-0477, TNO Building and Construction, Delft

Steenhuis M, Bijlaard FSK (1999). *Tests On Column Bases in Compression*, In: *Commemorative Publication for Prof. Dr. F. Tschemmemegg* (Timber and Mixed Building Technology), edited by G Huber.

Steenhuis M, Wald F, Sokol Z, Stark JWB (2008). *Concrete in compression and base plate in bending*, HERON, Vol. 53, pp. 109-126.

Stockwell FW, Jr (1975). *Preliminary Base Plate Selection*, Engineering Journal AISC, Vol. 21, pp. 92-99.

Togo T (1967). *Experimental study on mechanical behaviour of tubular joints*, Ph.D. Thesis, Osaka University.

Veljkovic M (2015). *Eurocodes Workshop: Design of Steel Buildings with Worked Examples*.

Wald F (1993). *Column-Base Connections, A Comprehensive State of the Art Review*, Report, CTU, Praha.

Wald F, Obata M, Matsuura S, Goto Y (1993). *Flexible Baseplate Behaviour using FE Strip Analysis*, Acta Polytechnic a CTU, Prague, Vol. 33, pp. 83-98.

Wald F, Sokol Z, Jaspart J-P (2008a). *Base plate in bending and anchor bolts in tension*, HERON, Vol. 53, pp. 79-108.

Wald F, Sokol Z, Steenhuis CM, Jaspart J-P (2008b). *Component method for steel column bases*, HERON, Vol. 53, pp. 61-78.

Wardenier J. (1982). *Hollow section joints*, Delft University Press, Delft, The Netherlands.

Wardenier J (2001). *Hollow sections in structural applications*, edited by CIDECT - Comité International pour le Développement et l'Etude de la Construction Tubulaire, www.cidect.com.

Wardenier J, Kurobane Y, Packer JA, Dutta D, Yeomans N (1991). *DG 1: Design guide for circular hollow section (CHS) joints under predominantly static loading, Construction with hollow sections*, edited by CIDECT - Comité International pour le Développement et l'Etude de la Construction Tubulaire, Verlag TÜV Rheinland, Köln, Germany.

Wardenier J, Kurobane Y, Packer JA, van der Vegte A, Zhao X-L (2008). *DG 1: Design guide for circular hollow section (CHS) joints under predominantly static loading*, 2nd Ed., in: *Construction with hollow sections*, edited by CIDECT - Comité International pour le Développement et l'Etude de la Construction Tubulaire, www.cidect.com.

Wardenier J, Packer JA, Zhao X-L, van der Vegte A (2010). *Hollow sections in structural applications*, edited by CIDECT - Comité International pour le Développement et l'Etude de la Construction Tubulaire, Bouwen met Staal, Zoetermeer, The Netherlands.

Weynand K, Jaspart J-P, Demonceau J-F, Zhang L (2015). *Component method for tubular joints*, Report 16F - 3/15, CIDECT - Comité International pour le Développement et l'Etude de la Construction Tubulaire.

Weynand K, Kuck J, Oerder R, Herion S, Fleischer O, Rode M (2011). *Design Tools for Hollow Section Joints*, 2nd Ed., in, V&M Deutschland GmbH, Düsseldorf, Germany, www.vmtubes.com.

Weynand K, Oerder R (2013). *Standardised Joints in Steel Structures to DIN EN 1993-1-8*, Complete Edition 2013 Ed., in, Stahlbau Verlags- und Service GmbH, Düsseldorf, Germany.

Weynand K, Oerder R, Demonceau J-F (2012). *Resistance tables for standardized joints in accordance with EN 1993-1-8*, In: Proceeding of 7th International Workshop on Connections (Connection in Steel Structures VII), Timosoara, Rumania.

Weynand K, Oerder R, Jaspart J-P (2014). *COP - The Connection Program*, Program to design steel and composite joints in accordance with EC3 and EC4, Version 1.7.2, Feldmann + Weynand GmbH, Germany, http://cop.fw-ing.com.

Whitmore RE (1952). *Experimental investigation of stresses in gusset plates*, Report Bulletin 16, Engineering Experiment Station, University of Tennessee.

Zoetemeijer P (1974). *A design method for the tension side of statically loaded bolted beam-to-column connections*, Vol. 20, Heron, Delft, The Netherlands.

Annex A

Practical Values for Required Rotation Capacity

The following Table gives some practical values for required rotation capacity $\phi_{required}$.

System of loading	M_{max}	$\phi_{required}$
Beam A-B with moment M at B	M	$\phi_A = \dfrac{\gamma M L}{6 E I}$ $\phi_B = -\dfrac{\gamma M L}{3 E I}$
Beam with point load P at mid-span	$\dfrac{P L}{4}$	$\pm \dfrac{\gamma P L^2}{16 E I}$
Beam with uniformly distributed load P	$\dfrac{p L^2}{8}$	$\pm \dfrac{\gamma p L^3}{24 E I}$
Total loading P (triangular), span L	$\dfrac{2 P L}{9 \sqrt{3}}$	$\phi_A = \dfrac{7 \gamma P L^2}{180 E I}$ $\phi_B = -\dfrac{8 \gamma P L^2}{180 E I}$

Where

- E is the elastic modulus of the material from which the beam is formed;
- I is the second moment area of a beam;
- L is the span of a beam (centre-to-centre of columns);
- γ is the loading factor at ULS.

Annex B

Values for Lateral Torsional Buckling Strength of a Fin Plate

Lateral torsional buckling strength of a fin plate f_{pLT} (BS 5950-1 Table 17 (BSI, 2000)).

λ_{LT}	Steel grade and design strength f_y (N/mm²)													
	S 275					S 355					S 460			
	235	245	255	265	275	315	325	335	345	355	400	410	430	440	460
25	235	245	255	265	275	315	325	335	345	355	400	410	430	440	460
30	235	245	255	265	275	315	325	335	345	355	390	397	412	419	434
35	235	245	255	265	272	300	307	314	321	328	358	365	378	385	398
40	224	231	237	244	250	276	282	288	295	301	328	334	346	352	364
45	206	212	218	224	230	253	259	265	270	276	300	306	316	321	332
50	190	196	201	207	212	233	238	243	248	253	275	279	288	293	302
55	175	180	185	190	195	214	219	223	227	232	251	255	263	269	281
60	162	167	171	176	180	197	201	205	209	212	237	242	253	258	269
65	150	154	158	162	166	183	188	194	199	204	227	232	242	247	256
70	139	142	146	150	155	177	182	187	192	196	217	222	230	234	242
75	130	135	140	145	151	170	175	179	184	188	207	210	218	221	228
80	126	131	136	141	146	163	168	172	176	179	196	199	205	208	214
85	122	127	131	136	140	156	160	164	167	171	185	187	190	192	195
90	118	123	127	131	135	149	152	156	159	162	170	172	175	176	179
95	114	118	122	125	129	142	144	146	148	150	157	158	161	162	164
100	110	113	117	120	123	132	134	136	137	139	145	146	148	149	151
105	106	109	112	115	117	123	125	126	128	129	134	135	137	138	140
110	101	104	106	107	109	115	116	117	119	120	124	125	127	128	129
115	96	97	99	101	102	107	108	109	110	111	115	116	118	118	120
120	90	91	93	94	96	100	101	102	103	104	107	108	109	110	111

B. Values for Lateral Torsional Buckling Strength of a Fin Plate

λ_{LT}	S 275					S 355					S 460				
	235	245	255	265	275	315	325	335	345	355	400	410	430	440	460
125	85	86	87	89	90	94	95	96	96	97	100	101	102	103	104
130	80	81	82	83	84	88	89	90	90	91	94	94	95	96	97
135	75	76	77	78	79	83	83	84	85	85	88	88	89	90	90
140	71	72	73	74	75	78	78	79	80	80	82	83	84	84	85
145	67	68	69	70	71	73	74	74	75	75	77	78	79	79	80
150	64	64	65	66	67	69	70	70	71	71	73	73	74	74	75
155	60	61	62	62	63	65	66	66	67	67	69	69	70	70	71
160	57	58	59	59	60	62	62	63	63	63	65	65	66	66	67
165	54	55	56	56	57	59	59	59	60	60	61	62	62	62	63
170	52	52	53	53	54	56	56	56	57	57	58	58	59	59	60
175	49	50	50	51	51	53	53	53	54	54	55	55	56	56	56
180	47	47	48	48	49	50	51	51	51	51	52	53	53	53	54
185	45	45	46	46	46	48	48	48	49	49	50	50	50	51	51
190	43	43	44	44	44	46	46	46	46	47	48	48	48	48	48
195	41	41	42	42	42	43	44	44	44	44	45	45	46	46	46
200	39	39	40	40	40	42	42	42	42	42	43	43	44	44	44
210	36	36	37	37	37	38	38	38	39	39	39	40	40	40	40
220	33	33	34	34	34	35	35	35	34	36	36	36	37	37	37
230	31	31	31	31	31	32	32	33	33	33	33	33	34	34	34
240	28	29	29	29	29	30	30	30	30	30	31	31	31	31	31
250	26	27	27	27	27	28	28	28	28	28	29	29	29	29	29
λ_{L0}	37.1	36.3	35.6	35.0	34.3	32.1	31.6	31.1	30.6	30.2	28.4	28.1	27.4	27.1	26.5

where $\lambda_{LT} = 2.8 \left(\dfrac{z_p h_p}{1.5 t_p^2} \right)^{1/2}$